VOLUME SIXTY ONE

ADVANCES IN
GEOPHYSICS
Machine Learning in Geosciences

VOLUME SIXTY ONE

ADVANCES IN
GEOPHYSICS
Machine Learning in Geosciences

Series Editor

BEN MOSELEY
*Department of Computer Science,
University of Oxford,
Oxford, United Kingdom*

LION KRISCHER
*ETH Zurich,
Department of Earth Sciences,
Zurich, Switzerland*

Academic Press is an imprint of Elsevier
50 Hampshire Street, 5th Floor, Cambridge, MA 02139, United States
525 B Street, Suite 1650, San Diego, CA 92101, United States
The Boulevard, Langford Lane, Kidlington, Oxford OX5 1GB, United Kingdom
125 London Wall, London, EC2Y 5AS, United Kingdom

First edition 2020

Copyright © 2020 Elsevier Inc. All rights reserved.

No part of this publication may be reproduced or transmitted in any form or by any means, electronic or mechanical, including photocopying, recording, or any information storage and retrieval system, without permission in writing from the publisher. Details on how to seek permission, further information about the Publisher's permissions policies and our arrangements with organizations such as the Copyright Clearance Center and the Copyright Licensing Agency, can be found at our website: www.elsevier.com/permissions.

This book and the individual contributions contained in it are protected under copyright by the Publisher (other than as may be noted herein).

Notices
Knowledge and best practice in this field are constantly changing. As new research and experience broaden our understanding, changes in research methods, professional practices, or medical treatment may become necessary.

Practitioners and researchers must always rely on their own experience and knowledge in evaluating and using any information, methods, compounds, or experiments described herein. In using such information or methods they should be mindful of their own safety and the safety of others, including parties for whom they have a professional responsibility.

To the fullest extent of the law, neither the Publisher nor the authors, contributors, or editors, assume any liability for any injury and/or damage to persons or property as a matter of products liability, negligence or otherwise, or from any use or operation of any methods, products, instructions, or ideas contained in the material herein.

ISBN: 978-0-12-821669-9
ISSN: 0065-2687

For information on all Academic Press publications
visit our website at https://www.elsevier.com/books-and-journals

Publisher: Zoe Kruze
Acquisitions Editor: Jason Mitchell
Editorial Project Manager: Chris Hockaday
Production Project Manager: Denny Mansingh
Cover Designer: Victoria Pearson

Typeset by SPi Global, India

Contents

Contributors	*vii*
Preface	*ix*

1. 70 years of machine learning in geoscience in review   1
Jesper Sören Dramsch

1. Historic machine learning in geoscience	3
2. Contemporary machine learning in geoscience	10
References	43

2. Machine learning and fault rupture: A review   57
Christopher X. Ren, Claudia Hulbert, Paul A. Johnson,
and Bertrand Rouet-Leduc

1. Introduction	57
2. Machine learning: A shallow dive	61
3. Laboratory studies	66
4. Field studies	75
5. Conclusion	95
Acknowledgments	97
References	97

3. Machine learning techniques for fractured media   109
Shriram Srinivasan, Jeffrey D. Hyman, Daniel O'Malley, Satish Karra,
Hari S. Viswanathan, and Gowri Srinivasan

1. Introduction	109
2. Preliminaries	113
3. Graph as a DFN reduced-order model	122
4. Pruned DFN as a reduced-order model	122
5. Machine learning methods for backbone identification	124
6. Further scope for ML in fractured media	144
References	144
Further reading	150

4. Seismic signal augmentation to improve generalization of deep neural networks — 151

Weiqiang Zhu, S. Mostafa Mousavi, and Gregory C. Beroza

1. Introduction	151
2. Benchmark data and training procedure	154
3. Augmentations	156
4. Discussion	170
5. Conclusions	173
Acknowledgments	174
References	174

5. Deep generator priors for Bayesian seismic inversion — 179

Zhilong Fang, Hongjian Fang, and Laurent Demanet

1. Introduction	179
2. Methodology	183
3. Seismic inversion applications	187
4. Numerical examples	188
5. Conclusions and discussion	212
Acknowledgments	214
References	214

6. An introduction to the two-scale homogenization method for seismology — 217

Yann Capdeville, Paul Cupillard, and Sneha Singh

1. Introduction	218
2. Mathematical notions and notations	222
3. A numerical introduction to the subject	226
4. Two-scale homogenization: the 1-D periodic case	237
5. Two-scale homogenization: The 1-D nonperiodic case	252
6. Two-scale homogenization: Higher dimensions	266
7. What we skipped	273
8. Examples of applications	276
9. Discussion and conclusions	299
Acknowledgments	302
References	302

Contributors

Gregory C. Beroza
Department of Geophysics, Stanford University, Stanford, CA, United States

Yann Capdeville
LPG, CNRS, Université de Nantes, Nantes, France

Paul Cupillard
GeoRessources, Université de Lorraine, Nancy, France

Laurent Demanet
Department of Mathematics; Department of Earth, Atmospheric, and Planetary Sciences; Earth Resource Laboratory, Massachusetts Institute of Technology, Cambridge, MA, United States

Jesper Sören Dramsch
Technical University of Denmark, Copenhagen, Denmark

Hongjian Fang
Department of Earth, Atmospheric, and Planetary Sciences; Earth Resource Laboratory, Massachusetts Institute of Technology, Cambridge, MA, United States

Zhilong Fang
Department of Mathematics; Earth Resource Laboratory, Massachusetts Institute of Technology, Cambridge, MA, United States

Claudia Hulbert
Laboratoire de Géologie, Département de Géosciences, École Normale Supérieure, PSL Université, Paris, France

Jeffrey D. Hyman
Computational Earth Science (EES-16), Earth and Environmental Sciences Division, Los Alamos National Laboratory, Los Alamos, NM, United States

Paul A. Johnson
Geophysics Group, Los Alamos National Laboratory, Los Alamos, NM, United States

Satish Karra
Computational Earth Science (EES-16), Earth and Environmental Sciences Division, Los Alamos National Laboratory, Los Alamos, NM, United States

S. Mostafa Mousavi
Department of Geophysics, Stanford University, Stanford, CA, United States

Daniel O'Malley
Computational Earth Science (EES-16), Earth and Environmental Sciences Division, Los Alamos National Laboratory, Los Alamos, NM, United States

Christopher X. Ren
Intelligence and Space Research Division, Los Alamos National Laboratory, Los Alamos, NM, United States

Bertrand Rouet-Leduc
Geophysics Group, Los Alamos National Laboratory, Los Alamos, NM, United States

Sneha Singh
LPG, CNRS, Université de Nantes, Nantes, France

Gowri Srinivasan
Computational Physics Division (XCP-8), Los Alamos National Laboratory, Los Alamos, NM, United States

Shriram Srinivasan
Center for Nonlinear Studies (CNLS) & Computational Earth Science (EES-16), Earth and Environmental Sciences Division, Los Alamos National Laboratory, Los Alamos, NM, United States

Hari S. Viswanathan
Computational Earth Science (EES-16), Earth and Environmental Sciences Division, Los Alamos National Laboratory, Los Alamos, NM, United States

Weiqiang Zhu
Department of Geophysics, Stanford University, Stanford, CA, United States

Preface

Volume 61 of *Advances in Geophysics* includes a collection of chapters around the topic of machine learning in geosciences as well as one review chapter focusing on homogenization techniques in seismology.

Historically, geosciences and in particular geophysics have been data-driven fields. The vast amounts of geoscientific data now available, combined with recent developments in the field of machine learning as well as access to increasingly powerful computers, have led to a renewed and significantly growing interest in machine learning for addressing the outstanding challenges in these fields. This special volume is meant to provide a review on the evolution and application of current machine learning techniques in various subdisciplines of geosciences, with a focus on geophysics.

Dramsch (2020) provides an extensive review of the developments of machine learning in geosciences over the last 70 years. They begin with the "roots" of machine learning in applied statistics and its earliest geoscientific applications such as Kriging, before providing a wide-ranging analysis covering mathematical foundations, algorithmic concepts all the way to present-day applications of modern machine-learning techniques.

Machine learning has successfully been employed to study seismic signals originating from fault slip, rupture, and failure. Ren, Hulbert, Johnson, and Rouet-Leduc (2020) summarize machine learning studies ranging from the laboratory-scale to large field studies in geodesy and seismology, and discuss aspects such as data exploration, signal analysis, and denoising, as well as applications in earthquake early warning and monitoring induced seismicity.

Studying the transport of a solute through a fracture network is a challenging task because of questions arising on how to represent the heterogeneous medium and how to describe the interaction between the solute and the porous medium. Srinivasan, Hyman, O'Malley, Karra, Viswanathan, and Srinivasan (2020) discusses the development of reduced-order models of fracture networks with machine learning techniques.

The generalization of deep learning models beyond their training data is key for the effective use of deep neural networks, for example, when monitoring earthquakes. Zhu, Mousavi, and Beroza (2020) present a series of augmentation methods designed for seismic waveform data to reduce the bias and increase the performance of deep neural networks.

Fang, Fang, and Demanet (2020) present machine learning-based prior knowledge generation methods for Bayesian inference to overcome the limitations of user-defined prior knowledge when inverting geophysical data in subsurface imaging. They demonstrate the benefits of deep generator priors on seismic traveltime tomography and full waveform inversion.

The last chapter of this volume covers a topic outside the theme of machine learning, namely the computation of an effective medium for both forward modeling and inverse problems. In their review on the homogenization method for seismology, Capdeville, Cupillard, and Singh (2020) discuss a theoretical framework to compute effective media for deterministic, multiscale media. Such effective media allow, for example, to compute a seismic wavefield that accurately represents the true wavefield up to a desired accuracy for all wave types.

Enjoy reading this volume!

BEN MOSELEY, Special Volume Editor

LION KRISCHER, Special Volume Editor

CEDRIC SCHMELZBACH, Serial Editor

References

Capdeville, Y., Cupillard, P., & Singh, S. (2020). An introduction to the two-scale homogenization method for seismology. *Advances in Geophysics, 61*, 217–306.

Dramsch, J. S. (2020). 70 Years of machine learning in geoscience in review. *Advances in Geophysics, 61*, 1–55.

Fang, Z., Fang, H., & Demanet, L. (2020). Deep generator priors for Bayesian seismic inversion. *Advances in Geophysics, 61*, 179–216.

Ren, C. X., Hulbert, C., Johnson, P. A., & Rouet-Leduc, B. (2020). Machine learning and fault rupture: A review. *Advances in Geophysics, 61*, 57–107.

Srinivasan, S., Hyman, J. D., O'Malley, D., Karra, S., Viswanathan, H. S., & Srinivasan, G. (2020). Machine learning techniques for fractured media. *Advances in Geophysics, 61*, 109–150.

Zhu, W., Mousavi, S. M., & Berosa, G. C. (2020). Seismic signal augmentation to improve generalization of deep neural networks. *Advances in Geophysics, 61*, 151–177.

CHAPTER ONE

70 years of machine learning in geoscience in review

Jesper Sören Dramsch*

Technical University of Denmark, Copenhagen, Denmark
*Corresponding author: e-mail address: jesper@dramsch.net

Contents

1. Historic machine learning in geoscience	3
1.1 Expert systems to knowledge-driven AI	4
1.2 Neural networks	6
1.3 Kriging and Gaussian processes	8
2. Contemporary machine learning in geoscience	10
2.1 Modern machine learning tools	11
2.2 Support-vector machines	12
2.3 Random forests	17
2.4 Modern deep learning	20
2.5 Neural network architectures	28
2.6 Convolutional neural network architectures	28
2.7 Generative adversarial networks	33
2.8 Recurrent neural network architectures	35
2.9 The state of ML on geoscience	37
References	43

In recent years machine learning has become an increasingly important interdisciplinary tool that has advanced several fields of science, such as biology (Ching et al., 2018), chemistry (Schütt, Arbabzadah, Chmiela, Müller, & Tkatchenko, 2017), medicine (Shen, Wu, & Suk, 2017), and pharmacology (Kadurin, Nikolenko, Khrabrov, Aliper, & Zhavoronkov, 2017). Specifically, the method of deep neural networks has found wide application. While geoscience was slower in the adoption, bibliometrics show the adoption of deep learning in all aspects of geoscience. Most subdisciplines of geoscience have been treated to a review of machine learning. Remote sensing has been an early adopter (Lary, Alavi, Gandomi, & Walker, 2016), with geomorphology (Valentine & Kalnins, 2016), solid Earth geoscience (Bergen, Johnson, Maarten, & Beroza, 2019), hydrogeo physics (Shen, 2018), seismology (Kong et al., 2019), seismic interpretation

(Wang, Di, Shafiq, Alaudah, & AlRegib, 2018), and geochemistry (Zuo, Xiong, Wang, & Carranza, 2019) following suite. Climate change, in particular, has received a thorough treatment of the potential impact of varying machine learning methods for modeling, engineering, and mitigation to address the problem (Rolnick et al., 2019). This review addresses the development of applied statistics and machine learning in the wider discipline of geoscience in the past 70 years and aims to provide context for the recent increase in interest and successes in machine learning and its challenges.[a]

Machine learning (ML) is deeply rooted in applied statistics, building computational models that use inference and pattern recognition instead of explicit sets of rules. Machine learning is generally regarded as a subfield of artificial intelligence (AI), with the notion of AI first being introduced by Turing (1950). Samuel (1959) coined the term machine learning itself, with Mitchell et al. (1997) providing a commonly quoted definition:

A computer program is said to learn from experience E with respect to some class of tasks T and performance measure P if its performance at tasks in T, as measured by P, improves with experience E.

Mitchell et al. (1997)

This means that a machine learning model is defined by a combination of requirements. A task such as, classification, regression, or clustering is improved by conditioning of the model on a training data set. The performance of the model is measured with regard to a loss, also called metric, which quantifies the performance of a machine learning model on the provided data. In regression, this would be measuring the misfit of the data from the expected values. Commonly, the model improves with exposure to additional samples of data. Eventually, a good model generalizes to unseen data, which was not part of the training set, on the same task the model was trained to perform.

Accordingly, many mathematical and statistical methods and concepts, including Bayes' rule (Bayes, 1763), least-squares (Legendre, 1805), and Markov models (Markov, 1906, 1971), are applied in machine learning. Gaussian processes stand out as they originate in time series applications (Kolmogorov, 1939) and geostatistics (Krige, 1951), which roots this machine learning application in geoscience (Rasmussen, 2003). "Kriging" originally applied two-dimensional Gaussian processes to the prediction

---

[a] The author of this chapter has a background in geophysics, exploration geoscience, and active source 4D seismic. While this skews the expertise, they attempt to give a full overview over developments in all of geoscience with the minimum amount of bias possible.

of gold mine valuation and has since found wide application in geostatistics. Generally, Matheron (1963) is credited with formalizing the mathematics of kriging and developing it further in the following decades.

Between 1950 and 2020 much has changed. Computational resources are now widely available both as hardware and software, with high-performance compute being affordable to anyone from cloud computing vendors. High-quality software for machine learning is widely available through the free- and open-source software movement, with major companies (Google, Facebook, Microsoft) competing for the usage of their open-source machine learning frameworks (Tensorflow, Pytorch, CNTK[b]) and independent developments reaching wide applications such as scikit-learn (Pedregosa et al., 2011) and xgboost (Chen & Guestrin, 2016).

Nevertheless, investigations of machine learning in geoscience are not a novel development. The research into machine learning follows interest in artificial intelligence closely. Since its inception, artificial intelligence has experienced two periods of a decline in interest and trust, which has impacted negatively upon its funding. Developments in geoscience follow this wide-spread cycle of enthusiasm and loss of interest with a time lag of a few years. This may be the result of a variety of factors, including research funding availability and a change in willingness to publish results.

1. Historic machine learning in geoscience

The 1950s and 1960s were decades of machine learning optimism, with machines learning to play simple games and perform tasks like route mapping. Intuitive methods such as k-means, Markov models, and decision trees have been used as early as the 1960s in geoscience. K-means was used to describe the cyclicity of sediment deposits (Preston & Henderson, 1964). Krumbein and Dacey (1969) give a thorough treatment of the mathematical foundations of Markov chains and embedded Markov chains in a geological context through application to sedimentological processes, which also provides a comprehensive bibliography of Markov processes in geology. Some selected examples of early applications of Markov chains are found in sedimentology (Schwarzacher, 1972), well-log analysis (Agterberg, 1966), hydrology (Matalas, 1967), and volcanology (Wickman, 1968). Decision tree-based methods found early applications in economic geology and prospectivity mapping (Newendorp, 1976; Reddy & Bonham-Carter, 1991).

---

[b] Deprecated 2019.

The 1970s were left with few developments in both the methods of machine learning, as well as, applications and adoption in geoscience (cf. Fig. 1), due to the "first AI winter" after initial expectations were not met. Nevertheless, as kriging was not considered an AI technology, it was unaffected by this cultural shift and found applications in mining (Huijbregts & Matheron, 1970), oceanography (Chiles & Chauvet, 1975), and hydrology (Delhomme, 1978). This was in part due to superior results over other interpolation techniques, but also the provision of uncertainty measures.

1.1 Expert systems to knowledge-driven AI

The 1980s marked uptake in interest in machine learning and artificial intelligence through so-called expert systems and corresponding specialized hardware. While neural networks were introduced in 1950, the tools of automatic differentiation and backpropagation for error-correcting machine learning were necessary to spark their adoption in geophysics in the late 1980s. Zhao and Mendel (1988) performed seismic deconvolution with a recurrent neural network (Hopfield network). Dowla, Taylor, and Anderson (1990) discriminated between natural earthquakes and underground nuclear explosions using feed-forward neural networks. An ensemble of networks was able

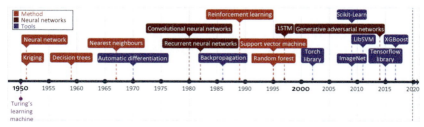

Fig. 1 Machine learning timeline from Dramsch (2019). Neural networks: Russell and Norvig (2010); Kriging: Krige (1951); decision trees: Belson (1959); nearest neighbors: Cover and Hart (1967); automatic differentiation: Linnainmaa (1970); convolutional neural networks: Fukushima (1980); LeCun, Bengio, and Hinton (2015); recurrent neural networks: Hopfield (1982); Backpropagation: Kelley (1960), Bryson (1961), Dreyfus (1962), and Rumelhart, Hinton, Williams, et al. (1988); reinforcement learning: Watkins (1989); support-vector machines: Cortes and Vapnik (1995); random forests: Ho (1995); LSTM: Hochreiter and Schmidhuber (1997); torch library: Collobert, Bengio, and Mariéthoz (2002); ImageNet: Deng et al. (2009); scikit-learn: Pedregosa et al. (2011); LibSVM: (Chang & Lin, 2011); generative adversarial networks: Goodfellow et al. (2014); Tensorflow: Abadi et al. (2015); XGBoost: Chen and Guestrin (2016).

to achieve 97% accuracy for nuclear monitoring. Moreover, the researchers inspected the network to gain the insight that the ratio of particular input spectra was beneficial to the discrimination of seismological events to the network. However, in practice the neural networks underperformed on uncurated data, which is often the case in comparison to published results. Huang, Chang, and Yen (1990) presented work on self-organizing maps (also Kohonen networks), a special type of unsupervised neural network applied to pick seismic horizons. The field of geostatistics saw a formalization of theory and an uptake in interest with Matheron et al. (1981) formalizing the relationship of spline-interpolation and kriging and Dubrule (1984) further develop the theory and apply it to well data. At this point, kriging is well-established in the mining industry as well as other disciplines that rely on spatial data, including the successful analysis and construction of the channel tunnel (Chilès & Desassis, 2018). The late 1980s then marked the second AI winter, where expensive machines tuned to run "expert systems" were outperformed by desktop hardware from nonspecialist vendors, causing the collapse of a half-billion-dollar hardware industry. Moreover, government agencies cut funding in AI specifically.

The 1990s are generally regarded as the shift from a knowledge-driven to a data-driven approach in machine learning. The term AI and especially expert systems were almost exclusively used in computer gaming and regarded with cynicism and as a failure in the scientific world. In the background, however, with research into applied statistics and machine learning, this decade marked the inception of support-vector machines (SVM) (Cortes & Vapnik, 1995), the tree-based method random forests (RF) (Ho, 1995), and a specific type of recurrent neural network (RNN) long short-term memories (LSTM) (Hochreiter & Schmidhuber, 1997). SVMs were utilized for land usage classification in remote sensing early on (Hermes, Frieauff, Puzicha, & Buhmann, 1999). Geophysics applied SVMs a few years later to approximate the Zoeppritz equations for AVO inversion, outperforming linearized inversion (Kuzma, 2003). Random forests, however, were delayed in broader adoption, due to the term "random forests" only being coined in 2001 (Breiman, 2001) and the statistical basis initially being less rigorous and implementation being more complicated. LSTMs necessitate large amounts of data for training and can be expensive to train, after further development in 2011 (Ciresan, Meier, Masci, Gambardella, & Schmidhuber, 2011) it gained popularity in commercial time series applications particularly speech and audio analysis.

1.2 Neural networks

McCormack (1991) marks the first review of the emerging tool of neural networks in geophysics. The paper goes into the mathematical details and explores pattern recognition. The author summarizes neural network applications over the 30 years prior to the review and presents worked examples in automated well-log analysis and seismic trace editing. The review comes to the conclusion that neural networks are, in fact, good function approximators, taking over tasks that were previously reserved for human work. He criticizes slow training, the cost of retraining networks upon new knowledge, imprecision of outputs, nonoptimal training results, and the black box property of neural networks. The main conclusion sees the implementation of neural networks in conventional computation and expert systems to leverage the pattern recognition of networks with the advantages of conventional computer systems.

Neural networks are the primary subject of the modern day machine learning interest, however, significant developments leading up to these successes were made prior to the 1990s. The first neural network machine was constructed by Minsky [described in Russell and Norvig (2010)] and soon followed by the "Perceptron," a binary decision boundary learner (Rosenblatt, 1958). This decision was calculated as follows:

$$
\begin{aligned}
o_j &= \sigma \left(\Sigma_j \, w_{ij} x_i + b \right) \\
&= \sigma \left(a_j \right) \\
&= \begin{cases} 1 & a_j > 0 \\ 0 & \text{otherwise} \end{cases}
\end{aligned}
\tag{1}
$$

It describes a linear system with the output o, the linear activation a of the input data x, the index of the source i and target node j, the trainable weights w, the trainable bias b and a binary activation function σ (Fig. 2). The activation function σ in particular has received ample attention since its inception. During this period, a binary σ became uncommon and was replaced by nonlinear mathematical functions. Neural networks are commonly trained by gradient descent, therefore, differentiable functions like sigmoid or tanh, allowing for the activation o of each neuron in a neural network to be continuous (Fig. 4).

Deep learning (Dechter, 1986) expands on this concept. It is the combination of multiple layers of neurons in a neural network (Fig. 3).

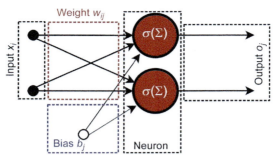

Fig. 2 Single layer neural network as described in Eq. (1). Two inputs x_i are multiplied by the weights w_{ij} and summed with the biases b_j. Subsequently an activation function σ is applied to obtain out outputs o_j.

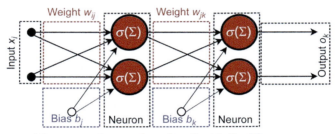

Fig. 3 Deep multilayer neural network as described in Eq. (2).

These deep networks learn representations with multiple levels of abstraction and can be expressed using Eq. (1) as input neurons to the next layer

$$\begin{aligned} o_k &= \sigma \left(\sum_k w_{jk} \cdot o_j + b \right) \\ &= \sigma \left(\sum_k w_{jk} \cdot \sigma \left(\sum_j w_{ij} x_i + b \right) + b \right) \end{aligned} \quad (2)$$

Röth and Tarantola (1994) apply these building blocks of multilayered neural networks with sigmoid activation to perform seismic inversion. They successfully invert low-noise and noise-free data on small training data. The authors note that the approach is susceptible to errors at low signal-to-noise ratios and coherent noise sources. Further applications include electromagnetic subsurface localization (Poulton, Sternberg, & Glass, 1992), magnetotelluric inversion via Hopfield neural networks (Zhang & Paulson, 1997), and geomechanical microfractures modeling in triaxial compression tests (Feng & Seto, 1998) (Fig. 4).

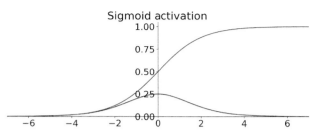

Fig. 4 Sigmoid activation function (*red*) and derivative (*blue*) to train multilayer neural network described in Eq. (2).

1.3 Kriging and Gaussian processes

Cressie (1990) reviews the history of kriging, prompted by the uptake of interest in geostatistics. The author defines kriging as best linear unbiased prediction and reviews the historical codevelopment of disciplines. Similar concepts were developed with mining, meteorology, physics, plant and animal breeding, and geodesy that relied on optimal spatial prediction. Later, Williams (1998) provide a thorough treatment of Gaussian processes, in the light of recent successes of neural networks.

> *An alternative method of putting a prior over functions is to use a Gaussian process (GP) prior over functions. This idea has been used for a long time in the spatial statistics community under the name of "kriging," although it seems to have been largely ignored as a general-purpose regression method.*
>
> **Williams (1998)**

Overall, Gaussian processes benefit from the fact that a Gaussian distribution will stay Gaussian under conditioning. That means that we can use Gaussian distributions in this machine learning process and they will produce a smooth Gaussian result after conditioning on the training data. To become a universal machine learning model, Gaussian processes have to be able to describe infinitely many dimensions. Instead of storing infinite values to describe this random process, Gaussian processes go the path of describing a distribution over functions that can produce each value when required.

$$p(x) \approx \mathcal{GP}(\mu(x), k(x, x')), \qquad (3)$$

The multivariate distribution over functions $p(x)$ is described by the Gaussian process depends on mean a function $\mu(x)$ and a covariance function $k(x, x')$. It follows that choosing an appropriate mean and covariance function, also known as kernel, is essential. Very commonly, the mean function is chosen to be zero, as this simplifies some of the math. Therefore, data with a nonzero mean are commonly centered to comply with this assumption

(Görtler, Kehlbeck, & Deussen, 2019). Choosing an appropriate kernel for the machine learning task is one of the benefits of the Gaussian process. The kernel is where expert knowledge can be incorporated into data, e.g., seasonality meteorological data can be described by a periodic covariance function.

Fig. 5 presents a 2D slice of 3D data with two classes. This binary problem can be approached by applying a Gaussian process to it. In the second panel, a linear kernel is shown, which predicts the data relatively poorly with an accuracy of 71%. A radial basis function (RBF) kernel, shown in the third panel generalizes to unseen test data with an accuracy of 90%. This figure shows how a trained Gaussian process would predict any new data point presented to the model. The linear kernel would predict any data in the top part to be blue (Class 0) and any data in the bottom part to be red (Class 1). The RBF kernel, which we explore further in the section introducing support-vector machines, separates the prediction into four uneven quadrants. The choice of kernel is very important in Gaussian processes and research into extracting specific kernels is ongoing (Duvenaud, 2014).

In a more practical sense, Gaussian processes are computationally expensive, as an $n \times n$ matrix must be inverted, with n being the number of samples. This results in a space complexity of $\mathcal{O}(n^2)$ and a time complexity $\mathcal{O}(n^3)$ (Williams & Rasmussen, 2006). This makes Gaussian processes most feasible for smaller data problems, which is one explanation for their rapid uptake in geoscience. An approximate computation of the inverted matrix is possible using the conjugate gradient (CG) optimization method, which can be stopped early with a maximum time cost of $\mathcal{O}(n^3)$ (Williams & Rasmussen, 2006). For problems with larger data sets, neural networks

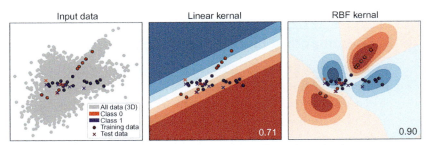

Fig. 5 Gaussian process separating two classes with different kernels. This image presents a 2D slice out of a 3D decision space. The decision boundary learnt from the data is visible, as well as the prediction in every location of the 2D slice. The two kernels presented are a linear kernel and a radial basis function (RBF) kernel, which show a significant discrepancy in performance. The bottom right number shows the accuracy on unseen test data. The linear kernel achieves 71% accuracy, while the RBF kernel achieves 90%.

become feasible due to being computationally cheaper than Gaussian processes, regularization on large data sets being viable, as well as, their flexibility to model a wide variety of functions and objectives. Regularization being essential as neural networks tend to not "overfit" and simply memorize the training data, instead of learning a generalizable relationship of the data. Interestingly, Hornik, Stinchcombe, and White (1989) showed that neural networks are a universal function approximator as the number of weights tend to infinity, and Neal (1996) was able to show that the infinitely wide stochastic neural network converges to a Gaussian process. Oftentimes Gaussian processes are trained on a subset of a large data set to avoid the computational cost. Gaussian processes have seen successful application on a wide variety of problems and domains that benefit from expert knowledge.

The 2000s were opened with a review by van der Baan and Jutten (2000) recapitulating the most recent geophysical applications in neural networks. They went into much detail on the neural networks theory and the difficulties in building and training these models. The authors identify the following subsurface geoscience applications through history: First-break picking, electromagnetics, magnetotellurics, seismic inversion, shear-wave splitting, well-log analysis, trace editing, seismic deconvolution, and event classification. They reveal a strong focus on exploration geophysics. The authors evaluated the application of neural networks as subpar to physics-based approaches and concluded that neural networks are too expensive and complex to be of real value in geoscience. This sentiment is consistent with the broader perception of artificial intelligence during this decade. Artificial intelligence and expert systems over-promised human-like performance, causing a shift in focus on research into specialized subfields, e.g., machine learning, fuzzy logic, and cognitive systems.

2. Contemporary machine learning in geoscience

Mjolsness and DeCoste (2001) review machine learning in a broader context outside of exploration geoscience. The authors discuss recent successes in applications of remote sensing and robotic geology using machine learning models. They review graphical models (hidden), Markov models, and SVMs and go on to disseminate the limitations of applications to vector data and poor performance when applied to rich data, such as graphs and text data. Moreover, the authors from NASA JPL go into detail on pattern recognition in automated rovers to identify geological prospects on Mars. They state:

> *The scientific need for geological feature catalogs has led to multiyear human surveys of Mars orbital imagery yielding tens of thousands of cataloged, characterized features including impact craters, faults, and ridges.*
>
> **Mjolsness and DeCoste (2001)**

The review points out the profound impact SVMs have on identifying geomorphological features without modeling the underlying processes.

2.1 Modern machine learning tools

This decade of the 2000s introduces a shift in tooling, which is a direct contributor to the recent increase in adoption and research of both shallow and deep machine learning research.

Machine learning software has been primarily comprised of proprietary software like MATLAB® with the Neural Networks Toolbox and Wolfram Mathematica or independent university projects like the Stuttgart Neural Network Simulator (SNNS). These tools were generally closed source and hard or impossible to extend and could be difficult to operate due to limited accompanying documentation. Early open-source projects include WEKA (Witten, Frank, & Hall, 2005), a graphical user interface to build machine learning and data mining projects. Shortly after that, LibSVM was released as free open-source software (FOSS) (Chang & Lin, 2011), which implements support-vector machines efficiently. It is still used in many other libraries to this day, including WEKA (Chang & Lin, 2011). Torch was then released in 2002, which is a machine learning library with a focus on neural networks. While it has been discontinued in its original implementation in the programming language Lua (Collobert et al., 2002), PyTorch, the reimplementation in the programming language Python, is one of the leading deep learning frameworks at the time of writing (Paszke et al., 2017). In 2007, the libraries Theano and scikit-learn were released openly licensed in Python (Pedregosa et al., 2011; Theano Development Team, 2016). Theano is a neural network library that was a tool developed at the Montreal Institute for Learning Algorithms (MILA) and ceased development in 2017 after strong industrial developers had released openly licensed deep learning frameworks. Scikit-learn implements many different machine learning algorithms, including SVMs, Random forests and single-layer neural networks, as well as utility functions including cross-validation, stratification, metrics and train-test splitting, necessary for robust machine learning model building and evaluation.

2.2 Support-vector machines

The impact of scikit-learn has shaped the current machine learning software package by implementing a unified application programming interface (API) (Buitinck et al., 2013). This API is explored by example in the following code snippets, the code can be obtained at Dramsch (2020b). First, we generate a classification dataset using a utility function. The `make_classification` function takes different arguments to adjust the desired arguments, we are generating 5000 samples (`n_samples`) for two classes, with five features (`n_features`), of which three features are actually relevant to the classification (`n_informative`). The data are stored in X, whereas the labels are contained in y.

```
# Generate random classification dataset for example
from sklearn.datasets import make_classification
X, y = make_classification(n_samples=5000, n_features=5,
                           n_informative=3, n_redundant=0,
                           random_state=0, shuffle=False)
```

It is good practice to divide the available labeled data into a training data set and a validation or test data set. This split ensures that models can be evaluated on unseen data to test the generalization to unseen samples. The utility function `train_test_split` takes an arbitrary amount of input arrays and separates them according to specified arguments. In this case 25% of the data are kept for the hold-out validation set and not used in training. The `random_state` is fixed to make these examples reproducible.

```
# Split data into train and validation set
from sklearn.model_selection import train_test_split
X_train, X_test, y_train, y_test = train_test_split(X, y,
                                                    test_size=.25,
                                                    random_state=0)
```

Then we need to define a machine learning model, considering the previous discussion of high impact machine learning models, the first example is an SVM classifier. This example uses the default values for hyperparameters of the SVM classifier, for best results on real-world problems these have to be adjusted. The machine learning training is always done by calling `classifier.fit(X, y)` on the classifier object, which in this case is the SVM object. In more detail, the `.fit()` method implements an optimization loop that will condition the model to the training data by minimizing the defined loss function. In the case of the SVM classification the parameters are adjusted to optimize a hinge loss, outlined in Eq. (5). The trained model scikit-learn model contains information about all its hyperparameters in addition to the trained model, shown below. The exact meaning of all these hyperparameters is laid out in the scikit-learn documentation (Buitinck et al., 2013).

```
# Define and train a Support Vector Machine Classifier
from sklearn.svm import SVC
svm = SVC(random_state=0)
svm.fit(X_train, y_train)

>>> SVC(C=1.0, break_ties=False, cache_size=200,
        class_weight=None, coef0=0.0, degree=3,
        decision_function_shape='ovr', gamma='scale',
        kernel='rbf', max_iter=-1, probability=False,
        random_state=0, shrinking=True, tol=0.001,
        verbose=False)
```

The trained SVM can be used to predict on new data, by calling `classifier.predict(data)` on the trained classifier object. The new data has to contain four features like the training data did. Generally, machine learning models always need to be trained on the same set of input features as the data available for prediction. The `.predict()` method outputs the most likely estimate on the new data to generate predictions. In the following code snippet, three predictions on three input vectors are performed on the previously trained model.

```
# Predict on new data with trained SVM
print(svm.predict([[0, 0, 0, 0, 0],
                   [-1, -1, -1, -1, -1],
                   [1, 1, 1, 1, 1]]))
>>> [1 0 1]
```

The blackbox model should be evaluated with the `classifier.score()` function. Evaluating the performance on the training data set gives an indication how well the model is performing, but this is generally not enough to gauge the performance of machine learning models. In addition, the trained model has to be evaluated on the hold-out set, a dataset the model has not been exposed to during training. This avoids that the model only performs well on the training data by "memorization" instead of extracting meaningful generalizable relationships, an effect called overfitting. In this example the hyperparameters are left to the default values, in real-life applications hyperparameters are usually adjusted to build better models. This can lead to an addition meta-level of overfitting on the hold-out set, which necessitates an additional third hold-out set to test the generalizability of the trained model with optimized hyperparameters. The default score uses the class accuracy, which suggests our model is approximately 90% correct. Similar train and test scores indicate that the model learned a generalizable model, enabling prediction on unseen data without a performance loss. Large differences between the training score and test score indicate either overfitting, in the case of a better training score. A higher test score than training score can be an indication of a deeper problem with the data split, scoring, class imbalances, and needs to be investigated by means of external cross-validation, building standard "dummy" models, independence tests, and further manual investigations.

```
# Score SVM on train and test data
print(svm.score(X_train, y_train))
print(svm.score(X_test, y_test))
>>> 0.9098666666666667
>>> 0.9032
```

Support-vector machines can be employed for each class of machine learning problem, i.e., classification, regression, and clustering. In a two-class problem, the algorithm considers the n-dimensional input and attempts to find a $(n - 1)$-dimensional hyperplane that separates these input data points. The problem is trivial if the two classes are linearly separable, also called a hard margin. The plane can pass the two classes of data without ambiguity. For data with an overlap, which is usually the case, the problem becomes an optimization problem to fit the ideal hyperplane. The hinge loss provides the ideal loss function for this problem, yielding 0 if none of the data overlap, but a linear residual for overlapping points that can be minimized:

$$\max\left(0, \left(1 - y_i(\vec{w} \cdot \vec{x}_i - b)\right)\right), \tag{4}$$

with y_i being the current target label and $\vec{w} \cdot \vec{x}_i - b$ being the hyperplane under consideration. The hyperplane consists of w the normal vector and point x, with the offset b. This leads the algorithm to optimize

$$\left[\frac{1}{n}\sum_{i=1}^{n}\max\left(0, 1 - y_i(w \cdot x_i - b)\right)\right] + \lambda \|w\|^2, \tag{5}$$

with λ being a scaling factor. For small λ the loss becomes the hard margin classifier for linearly separable problems. The nature of the algorithm dictates that only values for $\vec{x}$ close to the hyperplane define the hyperplane itself; these values are called the support vectors.

The SVM algorithm would not be as successful if it were simply a linear classifier. Some data can become linearly separable in higher dimensions. This, however, poses the question of how many dimensions should be searched, because of the exponential cost in computation that follows due to the increase of dimensionality (also known as the curse of dimensionality). Instead, the "kernel trick" was proposed (Aizerman, 1964), which defines a set of values that are applied to the input data simply via the dot product. A common kernel is the radial basis function (RBF), which is also the kernel we applied in the example. The kernel is defined as:

$$k(\vec{x}_i, \vec{x}_j) \rightarrow \exp\left(-\gamma \parallel \vec{x}_i - \vec{x}_j \parallel^2\right) \tag{6}$$

This specifically defines the Gaussian radial basis function of every input data point with regard to a central point. This transformation can be performed with other functions (or kernels), such as, polynomials or the sigmoid function. The RBF will transform the data according to the distance between x_i and X_j, this can be seen in Fig. 6. This results in the decision surface in Fig. 7 consisting of various Gaussian areas. The RBF is generally regarded as a good default, in part, due to being translation invariant (i.e., stationary) and smoothly varying.

An important topic in machine learning is explainability, which inspects the influence of input variables on the prediction. We can employ the utility function `permutation_importance` to inspect any model and how they perform with regard to their input features (Breiman, 2001). The permutation importance evaluates how well the blackbox model performs, when a feature is not available. Practically, a feature is replaced with random noise. Subsequently, the score is calculated, which provides a representation how

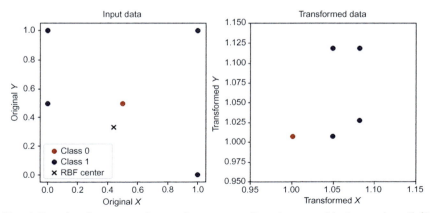

Fig. 6 Samples from two classes that are not linearly separable input data (*left*). Applying a Gaussian radial basis function centered around (0.4, 0.33) with $\lambda = 0.5$ results in the two classes being linearly separable.

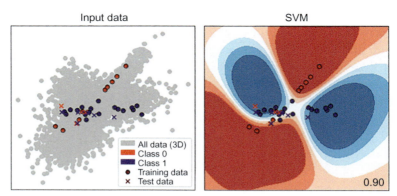

Fig. 7 Example of support-vector machine separating two classes, showing the decision boundary learnt from the data. The data contain three informative features, the decision boundary is therefore three dimensional, shown is a central slice of data points in 2D. (A video is available in Dramsch (2020a).)

informative a feature is compared to noise. The data we generated in the first example contain three informative features and two random data columns. The mean values of the calculated importances show that three features are estimated to be three magnitudes more important, with the second feature containing the maximum amount of information to predict the labels.

```
# Calculate permutation importance of SVM model
from sklearn.inspection import permutation_importance
importances = permutation_importance(svm, X_train, y_train,
                                     n_repeats=10, random_state=0)

# Show mean value of importances and the ranking
print(importances.importances_mean)
print(importances.importances_mean.argsort())
>>> [ 2.1787e-01  2.8712e-01  1.2293e-01 -1.8667e-04  7.7333e-04]
>>> [3 4 2 0 1]
```

Support-vector machines were applied to seismic data analysis (Li & Castagna, 2004) and the automatic seismic interpretation (Di, Shafiq, & AlRegib, 2017b; Liu et al., 2015; Mardan, Javaherian, et al., 2017). Compared to convolutional neural networks, these approaches usually do not perform as well, when the CNN can gain information from adjacent samples. Seismological volcanic tremor classification (Masotti et al., 2008; Masotti, Falsaperla, Langer, Spampinato, & Campanini, 2006) and analysis of ground-penetrating radar (Pasolli, Melgani, & Donelli, 2009; Xie, Qin, Yu, & Liu, 2013) were other notable applications of SVM in Geoscience. The 2016 Society of Exploration Geophysicists (SEG) machine learning challenge was held using a SVM baseline (Hall, 2016). Several other authors investigated well-log analysis (Anifowose, Ayadiuno, & Rashedian, 2017; Caté, Schetselaar, Mercier-Langevin, & Ross, 2018; Gupta, Rai, Sondergeld, & Devegowda, 2018; Saporetti, da Fonseca, Pereira, & de Oliveira, 2018), as well as seismology for event classification (Malfante et al., 2018) and magnitude determination (Ochoa, Niño, & Vargas, 2018). These rely on SVMs being capable of regression on time series data. Generally, many applications in geoscience have been enabled by the strong mathematical foundation of SVMs, such as microseismic event classification (Zhao & Gross, 2017), seismic well ties (Chaki, Routray, & Mohanty, 2018), landslide susceptibility (Ballabio & Sterlacchini, 2012; Marjanović, Kovačević, Bajat, & Voženílek, 2011), digital rock models (Ma, Jiang, Tian, & Couples, 2012), and lithology mapping (Cracknell & Reading, 2013).

2.3 Random forests

The following example shows the application of random forests, to illustrate the similarity of the API for different machine learning algorithms in the scikit-learn library. The random forest classifier is instantiated with a maximum depth of seven, and the random state is fixed to zero again. Limiting the depth of the forest forces the random forest to conform to a simpler model. Random forests have the capability to become highly complex models that are very powerful predictive models (Fig. 8). This is not conducive to this small example dataset, but easy to modify for the inclined reader. The classifier is then trained using the same API of all classifiers in scikit-learn. The example shows a very high number of hyperparameters, however, random forests work well without further optimization of these.

```
# Define and train a Random Forest Classifier
from sklearn.ensemble import RandomForestClassifier
rf = RandomForestClassifier(max_depth=7, random_state=0)
rf.fit(X_train, y_train)

>>> RandomForestClassifier(bootstrap=True, ccp_alpha=0.0,
              class_weight=None, criterion='gini', max_depth=7,
              max_features='auto', max_leaf_nodes=None,
              max_samples=None, min_impurity_decrease=0.0,
              min_impurity_split=None, min_samples_leaf=1,
              min_samples_split=2, min_weight_fraction_leaf=0.0,
              n_estimators=100, n_jobs=None, oob_score=False,
              random_state=0, verbose=0, warm_start=False)
```

The prediction of the random forest is performed in the same API call again, also consistent with all classifiers available. The values are slightly different from the prediction of the SVM.

```
# Predict on new data with trained Random Forest
print(rf.predict([[0, 0, 0, 0, 0],

                  [-1, -1, -1, -1, -1],
                  [1, 1, 1, 1, 1]]))
>>> [1 0 1]
```

The training score of the random forest model is 2.5% better than the SVM in this instance, this score however not informative. Comparing the test scores shows only a 0.88% difference, which is the relevant value to evaluate, as it shows the performance of a model on data it has not seen during the training stage. The random forest performed slightly better on the training set than the test data set. This slight discrepancy is usually not an indicator of an overfit model. Overfit models "memorize" the training data and do not generalize well, which results in poor performance on unseen data. Generally, overfitting is to be avoided in real application, but can be seen in competitions, on benchmarks, and show-cases of new algorithms and architectures to oversell the improvement over state-of-the-art methods (Recht, Roelofs, Schmidt, & Shankar, 2019).

```
# Score Random Forest on train and test data
print(rf.score(X_train, y_train))
print(rf.score(X_test, y_test))
>>> 0.9306
>>> 0.912
```

Random forests have specialized methods available for introspection, which can be used to calculate feature importance. These are based on the decision process the random forest used to build the machine learning model. The feature importance in random forests uses the same method as permutation importance, which is dropping out features to estimate their importance on the model performance. Random forests use a measure to determine the split between classes at each node of the trees called Gini impurity. While the permutation importance uses the accuracy score of the prediction, in random forests this Gini impurity can be used to measure how informative a feature is in a model. It is important to note that this impurity-based process can be susceptible to noise and overestimate high number of classes in features. Using the permutation importance instead is a valid choice. In this instance as opposed to the permutation importance, the random forest estimates the two noninformative features to be one magnitude less useful than the informative features, instead of two magnitudes.

```
# Inspect random forest for feature importance
print(rf.feature_importances_)
print(rf.feature_importances_.argsort())
>>> [0.2324 0.4877 0.2527 0.0141 0.0129]
>>> [4 3 0 2 1]
```

Random forests and other tree-based methods, including gradient boosting, a specialized version of random forests, have generally found wider application with the implementation into scikit-learn and packages for the statistical languages R and SPSS. Similar to neural networks, this method is applied to ASI (Guillen, Larrazabal, González, Boumber, & Vilalta, 2015) with limited success, which is due to the independent treatment of samples, like SVMs. Random forests have the ability to approximate regression problems and time series, which made them suitable for seismological applications including localization (Dodge & Harris, 2016), event classification in volcanic tremors (Maggi et al., 2017), and slow slip analysis (Hulbert et al., 2018). They have also been applied to geomechanical applications in fracture modeling (Valera et al., 2017) and fault failure prediction (Rouet-Leduc, Hulbert, & Bolton, 2018; Rouet-Leduc, Hulbert, & Lubbers, 2017), as well as, detection of reservoir property changes from 4D seismic data (Cao & Roy, 2017). Gradient boosted trees were the winning models in the 2016 SEG machine learning challenge (Hall & Hall, 2017) for well-log analysis, propelling a variety of publications in facies prediction (Bestagini, Lipari, & Tubaro, 2017; Blouin, Caté, Perozzi, & Gloaguen, 2017; Caté et al., 2018; Saporetti et al., 2018).

Furthermore, various methods that have been introduced into scikit-learn have been applied to a multitude of geoscience problems. Hidden Markov models were used on seismological event classification (Beyreuther & Wassermann, 2008; Bicego, Acosta-Muñoz, & Orozco-Alzate, 2013; Ohrnberger, 2001), well-log classification (Jeong, Park, Han, & Kim, 2014; Wang, Wellmann, Li, Wang, & Liang, 2017), and landslide detection from seismic monitoring (Dammeier, Moore, Hammer, Haslinger, & Loew, 2016). These hidden Markov models are highly performant on time series and spatially coherent problems. The "hidden" part of Markov models enables the model to assume influences on the predictions that are not directly represented in the input data. The K-nearest neighbors method has been used for well-log analysis (Caté, Perozzi, Gloaguen, & Blouin, 2017; Saporetti et al., 2018), seismic well ties (Wang, Lomask, & Segovia, 2017) combined with dynamic

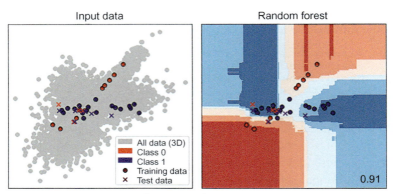

Fig. 8 Binary decision boundary for random forest in 2D. This is the same central slice of the 3D decision volume used in Fig. 7.

time warping and fault extraction in seismic interpretation (Hale, 2013), which is highly dependent on choosing the right hyperparameter k. The unsupervised k-NN equivalent, k-means has been applied to seismic interpretation (Di, Shafiq, & AlRegib, 2017 a), ground motion model validation (Khoshnevis & Taborda, 2018), and seismic velocity picking (Wei, Yonglin, Qingcai, Jiaqiang, et al., 2018). These are very simple machine learning models that are useful for baseline models. Graphical modeling in the form of Bayesian networks has been applied to seismology in modeling earthquake parameters (Kuehn & Riggelsen, 2011), basin modeling (Martinelli, Eidsvik, Sinding-Larsen, et al., 2013), seismic interpretation (Ferreira, Brazil, Silva, et al., 2018), and flow modeling in discrete fracture networks (Karra, O'Malley, Hyman, Viswanathan, & Srinivasan, 2018). These graphical models are effective in causal modeling and gained popularity in modern applications of machine learning explainability, interpretability, and generalization in combination with do-calculus (Pearl, 2012).

2.4 Modern deep learning

The 2010s marked a renaissance of deep learning and particularly convolutional neural networks. The convolutional neural network (CNN) architecture AlexNet (Krizhevsky, Sutskever, & Hinton, 2012) was the first CNN to enter the ImageNet challenge (Deng et al., 2009). The ImageNet challenge is considered a benchmark competition and database of natural images established in the field of computer vision. This improved the classification error rate from 25.8% to 16.4% (top-5 accuracy). This has propelled research in CNNs, resulting in error rates on ImageNet of 2.25% on top-5 accuracy in 2017 (Russakovsky et al., 2015). The Tensorflow library (Abadi et al., 2015) was introduced for open-source deep learning models, with some different software design compared to the Theano and Torch libraries.

The following example shows an application of deep learning to the data presented in the previous examples. The classification data set we use has independent samples, which leads to the use of simple densely connected feed-forward networks. Image data or spatially correlated datasets would ideally be fed to a convolutional neural network (CNN), whereas time series are often best approached with recurrent neural networks (RNN). This example is written using the Tensorflow library. PyTorch would be an equally good library to use.

All modern deep learning libraries take a modular approach to building deep neural networks that abstract operations into layers. These layers can be combined into input and output configurations in highly versatile and cus-tomizable ways. The simplest architecture, which is the one we implement below, is a sequential model, which consists of one input and one output layer, with a "stack" of layers. It is possible to define more complex models with multiple inputs and outputs, as well as the branching of layers to build very sophisticated neural network pipelines. These models are called func-tional API and subclassing API, but would not be conducive to this example.

The example model consists of Dense layers and a Dropout layer, which are arranged in sequence. Densely connected layers contain a specified num-ber of neurons with an appropriate activation function, shown in the example below. Each neuron performs the calculation outlined in Eq. (1), with σ defining the activation. Modern neural networks rarely implement `sigmoid` and `tanh` activations anymore. Their activation characteristic leads them to lose information for large positive and negative values of the input, commonly called saturation (Hochreiter, Bengio, Frasconi, Schmidhuber, et al., 2001). This saturation of neurons prevented good deep neural network performance until new nonlinear activation functions took their place (Xu, Wang, Chen, & Li, 2015). The activation function Rectified linear unit (ReLU) is generally credited with facilitating the development of very deep neural networks, due to their nonsaturating properties (Hahnloser, Sarpeshkar, Mahowald, Douglas, & Seung, 2000). It sets all negative values to zero and provides a linear response for positive values, as seen in Eq. (7) and Fig. 9. Since it is inception, many more rectifiers with different properties have been introduced.

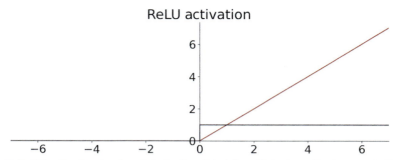

Fig. 9 ReLU activation (*red*) and derivative (*blue*) for efficient gradient computation.

$$\sigma(a) = max(0, a) \tag{7}$$

The other activation function used in the example is the "softmax" function on the output layer. This activation is commonly used for classification tasks, as it normalizes all activations at all outputs to one. It achieves this by applying the exponential function to each of the outputs in $\vec{a}$ for class C and dividing that value by the sum of all exponentials:

$$\sigma(\vec{a}) = \frac{e^{a_j}}{\sum_{p}^{C} e^{a_p}} \tag{8}$$

The example additionally uses a Dropout layer, which is a common layer used for regularization of the network by randomly setting a specified percentage of nodes to zero for each iteration. Neural networks are particularly prone to overfitting, which is counteracted by various regularization techniques that also include input data augmentation, noise injection, $\mathcal{L}_1$ and $\mathcal{L}_2$ constraints, or early-stopping of the training loop (Goodfellow, Bengio, & Courville, 2016). Modern deep learning systems may even leverage noisy student–teacher networks for regularization (Xie, Hovy, Luong, & Le, 2019).

```
import tensorflow as tf
model = tf.keras.models.Sequential([
tf.keras.layers.Dense(32, activation='relu'),
tf.keras.layers.Dropout(.3),
tf.keras.layers.Dense(16, activation='relu'),
tf.keras.layers.Dense(2, activation='softmax')])
```

These sequential models are also used for simple image classification models using CNNs. Instead of Dense layers, these are built up with convolutional layers, which are readily available in 1D, 2D, and 3D as Conv1D, Conv2D, and Conv3D, respectively. A two-dimensional CNN learns a so-called filter f for the $n \times m$-dimensional image G, expressed as:

$$G^*(x,y) = \sum_{i=1}^{n} \sum_{j=1}^{m} f(i,j) \cdot G(x-i+c, \ y-j+c), \tag{9}$$

resulting in the central result G^* around the central coordinate c. In CNNs each layer learns several of these filters f, usually following by a down-sampling operation in n and m to compress the spatial information. This serves as a forcing function to learn increasingly abstract representations in subsequent convolutional layers (Fig. 10).

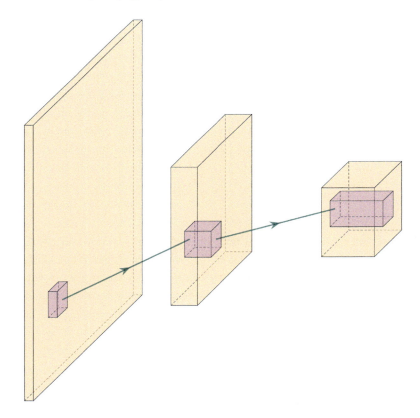

Fig. 10 Three layer convolutional network. The input image (*yellow*) is convolved with several filters or kernel matrices (*purple*). Commonly, the convolution is used to down-sample an image in the spatial dimension, while expanding the dimension of the filter response, hence expanding in "thickness" in the schematic. The filters are learned in the machine learning optimization loop. The shared weights within a filter improve efficiency of the network over classic dense networks.

This sequential example model of densely connected layers with a single input, 32, 16, and two neurons contains a total of 754 trainable weights. Initially, each of these weights is set to a pseudo-random value, which is often drawn from a distribution beneficial to fast training. Consequently, the data are passed through the network, and the result is numerically compared to the expected values. This form of training is defined as supervised training and error-correcting learning, which is a form of Hebbian learning. Other forms of learning exist and are employed in machine learning, e.g., competitive learning in self-organizing maps.

$$MAE = |y_j - o_j| \tag{10}$$

$$MSE = (y_j - o_j)^2 \tag{11}$$

In regression problems the error is often calculated using the mean absolute error (MAE) or mean squared error (MSE), the $\mathcal{L}_1$ shown in Eq. (10) and the $\mathcal{L}_2$ norm shown in Eq. (11), respectively. Classification problems form a special type of problem that can leverage a different kind of loss called cross-entropy (CE). The cross-entropy is dependent on the true label y and the prediction in the output layer.

$$CE = - \sum_j^C y_j \log(o_j) \tag{12}$$

Many machine learning data sets have one true label $y_{true} = 1$ for class $C_{j=true}$, leaving all other $y_j = 0$. This makes the sum over all labels obsolete. It is debatable how much binary labels reflects reality, but it simplifies Eq. (12) to minimizing the (negative) logarithm of the neural network output o_j, also known as negative log-likelihood:

$$CE = - \log(o_j) \tag{13}$$

Technically, the data we generated is a binary classification problem, and this means we could use the sigmoid activation function in the last layer and optimize a binary CE. This can speed up computation, but in this example, an approach is shown that works for many other problems and can therefore be applied to the readers data.

```
model.compile(optimizer='adam', # Often 'adam' or 'sgd' are good
              loss='sparse_categorical_crossentropy',
              metrics=['accuracy']) # Monitor other metrics
```

Large neural networks can be extremely costly to train with significant developments in 2019/2020 reporting multibillion parameter language models (Google, OpenAI) trained on massive hardware infrastructure for weeks with a single epoch taking several hours. This calls for validation on unseen data after every epoch of the training run. Therefore, neural networks, like all machine learning models, are commonly trained with two hold-out sets, a validation and a final test set. The validation set can be provided or be defined as a percentage of the training data, as shown below. In the example, 10% of the training data are held out for validation after every epoch, reducing the training data set from 3750 to 3375 individual samples.

```
model.fit(X_train,
          y_train,
          validation_split=.1,
          epochs=100)
>>> [...]
    Epoch 100/100
    3375/3375 [==============================] - 0s 66us/sample
    loss: 0.1567 - accuracy: 0.9401 -
    val_loss: 0.1731 - val_accuracy: 0.9359
```

Neural networks are trained with variations of stochastic gradient descent (SGD), an incremental version of the classic steepest descent algorithm. We use the Adam optimizer, a variation of SGD that converges fast, but a full explanation would go beyond the scope of this chapter. The gist of the Adam optimizer is that it maintains a per-parameter learning rate of the first statistical moment (mean). This is beneficial for sparse problems and the second moment (uncentered variance), which is beneficial for noisy and nonstationary problems (Kingma & Ba, 2014). The main alternative to Adam is SGD with Nesterov momentum (Sutskever, Martens, Dahl, & Hinton, 2013), an optimization method that models conjugate gradient methods (CG) without the heavy computation that comes with the search in CG. SGD anecdotally finds a better optimal point for neural networks than Adam but converges much slower.

In addition to the loss value, we display the accuracy metric. While accuracy should not be the sole arbiter of model performance, it gives a reasonable initial estimate, how many samples are predicted correctly with a percentage between zero and one. As opposed to scikit-learn, deep learning models are compiled after their definition to make them fit for optimization on the available hardware. Then the neural network can be fit like the SVM and random forest models before, using the X_train and y_train data.

In addition, a number of epochs can be provided to run, as well as other parameters that are left on default for the example. The amount of epochs defines how many cycles of optimization on the full training data set are performed. Conventional wisdom for neural network training is that it should always learn for more epochs than machine learning researchers estimate initially.

It can be difficult to fix all sources of randomness and stochasticity in neural networks, to make both research and examples reproducible. This example does not fix these so-called random seeds as it would detract from the example. That implies that the results for loss and accuracy will differ from the printed examples. In research fixing the seed is very important to ensure reproducibility of claims. Moreover, to avoid bad practices or so-called lucky seeds, a statistical analysis of multiple fixed seeds is good practice to report results in any machine learning model.

```
model.evaluate(X_test, y_test)
>>> 1250/1250 [==============================] - 0s 93us/sample
    loss: 0.1998 - accuracy: 0.9360
    [0.19976349686831235, 0.936]
```

In the example before, the SVM and random forest classifier were scored on unseen data. This is equally important for neural networks. Neural networks are prone to overfit, which we try to circumvent by regularizing the weights and by evaluating the final network on an unseen test set. The prediction on the test set is very close to the last epoch in the training loop, which is a good indicator that this neural network generalizes to unseen data. Moreover, the loss curves in Fig. 11 do not converge too fast, while converging. However, it appears that the network would overfit if we let training continue. The exemplary decision boundary in Fig. 12 very closely models the local distribution of the data, which is true for the entire decision volume (Dramsch, 2020a).

These examples illustrate the open-source revolution in machine learning software. The consolidated API and utility functions make it seem trivial to apply various machine learning algorithms to scientific data. This can be seen in the recent explosion of publications of applied machine learning in geoscience. The need to be able to implement algorithms has been replaced by merely installing a package and calling `model.fit(X, y)`. These developments call for strong validation requirements of models to ensure valid, reproducible, and scientific results. Without this careful validation these modern day tools can be severely misused to oversell results and even come to incorrect conclusions.

In aggregate, modern day neural networks benefit from the development of nonsaturating nonlinear activation functions. The advancements of stochastic gradient descent with Nesterov momentum and the Adam

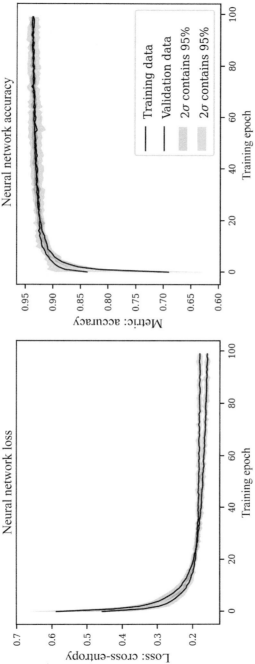

Fig. 11 Loss and accuracy of example neural network on ten random initializations. Training for 100 epochs with the shaded area showing the 95% confidence intervals of the loss and metric. Analyzing loss curves is important to evaluate overfitting. The training loss decreasing, while validation loss is close to plateauing is a sign of overfitting. Generally, it can be seen that the model converged and is only making marginal gains with the risk of overfitting.

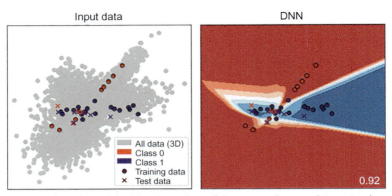

Fig. 12 Central 2D slice of decision boundary of deep neural network in trained on data with three informative features. The 3D volume is available in Dramsch (2020a).

optimizer (following AdaGrad and RMSProp) was essential faster training of deep neural networks. The leverage of graphics hardware available in most high-end desktop computers that is specialized for linear algebra computation, further reduced training times. Finally, open-source software that is well-maintained, tested, and documented with a consistent API made both shallow and deep machine learning accessible to nonexperts.

2.5 Neural network architectures

In deep learning, implementation of models is commonly more complicated than understanding the underlying algorithm. Modern deep learning makes use of various recent developments that can be beneficial to the data set it is applied to, without specific implementation details results are often not reproducible. However, the machine learning community has a firm grounding in openness and sharing, which is seen in both publications and code. New developments are commonly published alongside their open-source code, and frequently with the trained networks on standard benchmark data sets. This facilitates thorough inspection and transferring the new insights to applied tasks such as geoscience. In the following, some relevant neural network architectures and their application are explored.

2.6 Convolutional neural network architectures

The first model to discuss is the VGG-16 model, a 16-layer deep convolutional neural network (Simonyan & Zisserman, 2014) represented in Fig. 13[c]. This network was an attempt at building even deeper networks

---
[c] Visualization library (Iqbal, 2018).

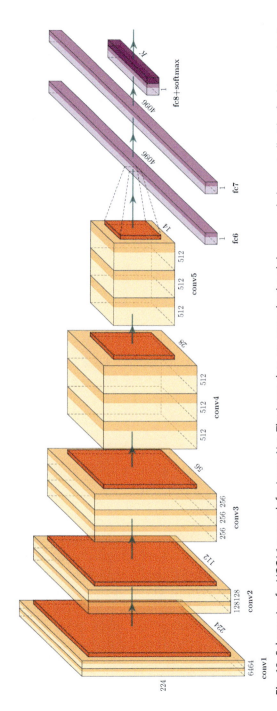

Fig. 13 Schematic of a VGG16 network for ImageNet. The input data are convolved and down-sampled repeatedly. The final image classification is performed by flattening the image and feeding it to a classic feed-forward densely connected neural network. The 1000 output nodes for the 1000 ImageNet classes are normalized by a final softmax layer (cf. Eq. 8).

and uses small 3×3 convolutional filters in the network, called f in Eq. (9). This small filter-size was sufficient to build powerful models that abstract the information from layer to deeper layer, which is easy to visualize and generalize well. The trained model on natural images also transfers well to other domains like seismic interpretation (Dramsch & Lüthje, 2018). Later, the concept of Network-in-Network was introduced, which suggested defined subnetworks or blocks in the larger network structure (Lin, Chen, & Yan, 2013). The ResNet architecture uses this concept of blocks to define residual blocks. These use a shortcut around a convolutional block (He, Zhang, Ren, & Sun, 2016) to achieve neural networks with up to 152 layers that still generalize well. ResNets and residual blocks, in particular, are very popular in modern architectures including the shortcuts or skip connections they popularized, to address the following problem:

When deeper networks start converging, a degradation problem has been exposed: with the network depth increasing, accuracy gets saturated (which might be unsurprising) and then degrades rapidly. Unexpectedly, such degradation is not caused by overfitting, and adding more layers to a suitably deep model leads to higher training error.

He et al. (2016)

The developments and successes in image classification on benchmark competitions like ImageNet and Pascal-VOC inspired applications in automatic seismic interpretation. These networks are usually single image classifiers using convolutional neural networks (CNNs). The first application of a convolutional neural network to seismic data used a relatively small deep CNN for salt identification (Waldeland & Solberg, 2017). The open-source software "MaLenoV" implemented a single image classification network, which was the earliest freely available implementation of deep learning for seismic interpretation (Ildstad & Bormann, n.d.). Dramsch and Lüthje (2018) applied pretrained VGG-16 and ResNet50 single image seismic interpretation. Recent successful applications build upon pretrained prebuilt architectures to implement into more sophisticated deep learning systems, e.g., semantic segmentation. Semantic segmentation is important in seismic interpretation. This is already a narrow field of application of machine learning and it can be observed that many early applications focus on subsections of seismic interpretation utilizing these prebuilt architectures such as salt detection (Di, Wang, & AlRegib, 2018; Gramstad & Nickel, 2018; Waldeland, Jensen, Gelius, & Solberg, 2018), fault interpretation (Araya-Polo et al., 2017; Guitton, 2018; Purves, Alaei, & Larsen, 2018), facies classification (Chevitarese, Szwarcman, Silva, & Brazil, 2018; Dramsch & Lüthje, 2018), and horizon picking (Wu & Zhang, 2018). In comparison, this is however, already a broader application than prior machine learning approaches for seismic interpretation that utilized very specific seismic attributes as input to self-organizing maps (SOM) for, e.g., sweet spot identification (Guo, Zhang, Lin, & Liu, 2017; Roden & Chen, 2017; Zhao, Li, & Marfurt, 2017).

In geoscience single image classification, as presented in the ImageNet challenge, is less relevant than other applications like image segmentation and time series classification. The developments and insights resulting from the ImageNet challenge were, however, transferred to network architectures that have relevance in machine learning for science. Fully convolutional networks are a way to better achieve image segmentation. A particularly useful implementation, the U-net, was first introduced in biomedical image segmentation, a discipline notorious for small datasets (Ronneberger, Fischer, & Brox, 2015). The U-net architecture shown in Fig. 14 utilizes several shortcuts in an encoder-decoder architecture to achieve stable segmentation results. Shortcuts (or skip connections) are a way in neural networks to combine the original information and the processed information, usually through concatenation or addition. In ResNet blocks this concept is extended to an extreme, where every block in the architecture contains a shortcut between the input and output, as seen in Fig. 15. These blocks are universally used in many architectures to implement deeper networks, i.e., ResNet-152 with 60 million parameters, with fewer parameters than previous architectures like VGG-16 with 138 million parameters. Essentially, enabling models that are ten times as deep with less than half the parameters, and significantly better accuracy on image benchmark problems.

In 2018 the seismic contractor TGS made a seismic interpretation challenge available on the data science competition platform Kaggle. Successful participants in the competition combined ResNet architectures with the Unet architecture as their base architecture and modified these with state-of-the-art image segmentation applications (Babakhin, Sanakoyeu, & Kitamura, 2019). Moreover, Dramsch and Lüthje (2018) showed that transferring networks trained on large bodies of natural images to seismic data yields good results on small datasets, which was further confirmed in this competition. The learnings from the TGS Salt Identification challenge have been incorporated in production scale models that perform human-like salt interpretation (Sen, Kainkaryam, Ong, & Sharma, 2020). In broader geoscience, U-nets have been used to model global water storage using GRAVE satellite data (Sun et al., 2019), landslide prediction

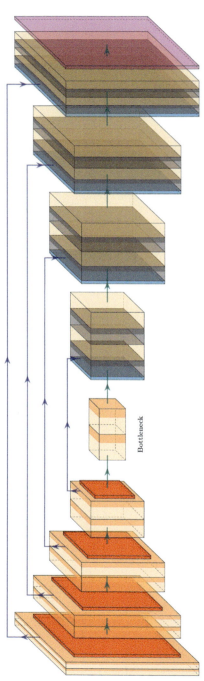

Fig. 14 Schematic of Unet architecture. Convolutional layers are followed by a down-sampling operation in the encoder. The central bottleneck contains a compressed representation of the input data. The decoder contains upsampling operations followed by convolutions. The last layer is commonly a softmax layer to provide classes. Equally sized layers are connected via shortcut connections.

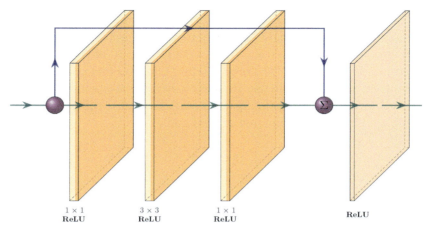

Fig. 15 Schematic of a ResNet block. The block contains a 1 × 1, 3 × 3, and 1 × 1 convolution with ReLU activation. The output is concatenated with the input and passed through another ReLU activation function.

(Hajimoradlou, Roberti, & Poole, 2019), and earthquake arrival time picking (Zhu & Beroza, 2018). A more classical approach identifies subsea scale worms in hydrothermal vents (Shashidhara, Scott, & Marburg, 2020), whereas Dramsch, Christensen, MacBeth, & Lüthje (2019) includes a U-net in a larger system for unsupervised 3D timeshift extraction from 4D seismic.

This modularity of neural networks can be seen all throughout the research and application of deep learning. New insights can be incorporated into existing architectures to enhance their predictive power. This can be in the form of swapping out the activation function σ or including new layers for improvements, e.g., regularization with batch normalization (Ioffe & Szegedy, 2015). The U-net architecture originally is relatively shallow, but was modified to contain a modified ResNet for the Kaggle salt identification challenge instead (Babakhin et al., 2019). Overall, serving as examples for the flexibility of neural networks.

2.7 Generative adversarial networks

Generative adversarial networks (GAN) take composition of neural network to another level, where two networks are trained in aggregate to get a desired result. In GANs, a generator network G and a discriminator network D work against each other in the training loop (Goodfellow et al., 2014).

The generator G is set up to generate samples from an input, these were often natural images in early GANs, but has now progressed to anything from time series (Engel et al., 2019) to high-energy physics simulation (Paganini, de Oliveira, & Nachman, 2018). The discriminator network D attempts to distinguish whether the sample is generated from G, i.e., fake or a real image from the training data. Mathematically, this defines a min max game for the value function V of G and D

$$\min_G \max_D V(D, G) = \mathbb{E}_{x \sim p_{data}(x)}[\log D(x)] + \mathbb{E}_{z \sim p_z(z)}[\log(1 - D(G(z)))],$$

(14)

with x representing the data, z is the latent space G draws samples from, and p represents the respective probability distributions. Eventually reaching a Nash equilibrium (Nash, 1951), where neither the generator network G can produce better outputs, nor the discriminator network D can improve its capability to discern between fake and real samples.

Despite how versatile U-nets are, they still need an appropriate defined loss function and labels to build a discriminative model. GANs, however, build a generative model that approximates the training sample distribution in the Generator and a discriminative model of the discriminator modeled dynamically through adversarial training. The discriminator effectively providing an adversarial loss in a GAN. In addition to providing two models that serve different purposes, learning the training sample distribution with an adversarial loss makes GANs one of the most versatile models currently discovered. Mosser, Dubrule, and Blunt (2017) were applied GANs early on to geoscience, modeling 3D porous media at the pore scale with a deep convolutional GAN. The authors extended this approach to conditional simulations of oolithic digital rock (Mosser, Dubrule, & Blunt, 2018 a). Early applications of GANs also included approximating the problem of velocity inversion of seismic data (Mosser, Kimman, et al., 2018), geostatistical inversion (Laloy, Hérault, Jacques, & Linde, 2017), and generating seismograms (Krischer & Fichtner, 2017). Richardson (2018) integrate the generator of the GAN into full waveform inversion of the scalar wavefield. Alternatively, a Bayesian inversion using the generator as prior for velocity inversion was introduced in Mosser, Dubrule, and Blunt (2018 b). In geomodeling, generation of geological channel models was presented (Chan & Elsheikh, 2017), which was subsequently extended with the capability to be conditioned on physical measurements (Dupont, Zhang, Tilke, Liang, & Bailey, 2018).

Naturally, GANs were applied to the growing field of automatic seismic interpretation (Lu, Morris, Brazell, Comiskey, & Xiao, 2018).

2.8 Recurrent neural network architectures

The final type of architecture applied in geoscience is recurrent neural networks (RNN). In contrast to all previous architectures, recurrent neural networks feed back into themselves. There are many types of RNNs, Hopfield networks being one that were applied to seismic source wavelet prediction (Wang & Mendel, 1992) early on. However, LSTMs (Hochreiter & Schmidhuber, 1997) are the main application in geoscience and wider machine learning. This type of network achieves state-of-the-art performance on sequential data like language tasks and time series applications. LSTMs solve some common problems of RNNs by implementing specific gates that regulate information flow in an LSTM cell, namely, input gate, forget gate, and output gate, visualized in Fig. 16. The input gate feeds input values to the internal cell. The forget gate overwrites the previous state. Finally, the output gate regulates the direct contribution of the input value to the output

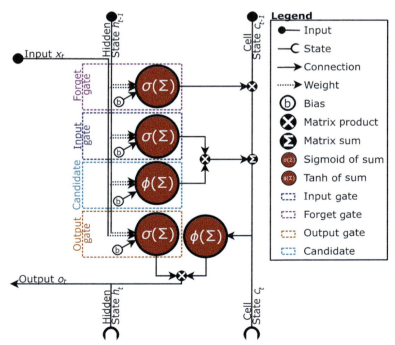

Fig. 16 Schematic of LSTM architecture. The input data are processed together with the hidden state and cell state. The LSTM avoid the exploding gradient problem by implemented a input, forget, and output gate.

value combined with the internal state of the cell. Additionally, a peephole functionality helps with the training that serves as a shortcut between inputs and gates.

A classic application of LSTMs is text analysis and natural language understanding, which has been applied to geological relation extraction from unstructured text documents (Blondelle, Juneja, Micaelli, & Neri, 2017; Luo, Zhou, Wang, Zhu, & Deng, 2017). Due to the nature of LSTMs being suited for time series data, it is has been applied to seismological event classification of volcanic activity Titos, Bueno, García, Benítez, and Ibañez (2018), multifactor landslide displacement prediction (Xie, Zhou, & Chai, 2019), and hydrological modeling (Kratzert et al., 2019). Talarico, Leão, and Grana (2019) applied LSTM to model sedimentological sequences and compared the model to baseline hidden Markov model (HMM), concluding that RNNs outperform HMMs based on first-order Markov chains, while higher order Markov chains were too complex to calibrate satisfactorily. Gated recurrent unit (GRU) (Cho et al., 2014) is another RNN developed based on the insights into LSTM, which was applied to predict petrophysical properties from seismic data (Alfarraj & AlRegib, 2018).

The scope of this review only allowed for a broad overview of types of networks that were successfully applied to geoscience. Many more specific architectures exist and are in development that provide different advantages. Siamese networks for one-shot image analysis (Koch, Zemel, & Salakhutdinov, 2015), transformer networks that largely replaced LSTM and GRU in language modeling (Vaswani et al., 2017), or attention as a general mechanism in deep neural networks (Zheng, Fu, Mei, & Luo, 2017).

Neural network architectures have been modified and applied to diverse problems in geoscience. Every architecture type is particularly suited to certain data types that are present in each field of geoscience. However, fields with data present in machine-readable format experienced accelerated adoption of machine learning tools and applications. For example, Ross, Meier, and Hauksson (2018) were able to successfully apply CNNs to seismological phase detection, relying on an extensive catalogue of hand-picked data (Ross, Meier, Hauksson, & Heaton, 2018a) and consequently generalize this work (Ross, Meier, Hauksson, & Heaton, 2018b). It has to be noted that synthetic or specifically sampled data can introduce an implicit bias into the network (Kim, Kim, Kim, Kim, & Kim, 2019; Wirgin, 2004). Nevertheless, particularly this blackbox property of machine learning model makes them versatile and powerful tools that were leveraged in every subdiscipline of the Earth sciences.

2.9 The state of ML on geoscience

Overall, geoscience and especially geophysics has followed developments in machine learning closely. Across disciplines, machine learning methods have been applied to various problems that can generally be categorized into three sections:

1. Build a surrogate ML model of a well-understood process. This model usually provides an advantage in computational cost.
2. Build an ML model of a task previously only possible with human interaction, interpretation, or knowledge and experience.
3. Build a novel ML model that performs a task that was previously not possible.

Granulometry on SEM images is an example of an application in category I, where previously sediments were hand-measured in images (Dramsch, Amour, & Lüthje, 2018). Applying large deformation diffeomorphic mapping of seismic data was computationally infeasible for matching 4D seismic data, however, made feasible by applying a U-net architecture to the problem of category II (Dramsch et al., 2019). The problem of earthquake magnitude prediction falls into category III due to the complexity of the system but was nevertheless approached with neural networks (Panakkat & Adeli, 2007).

The accessibility of tools, knowledge, and compute make this cycle of machine learning enthusiasm unique, with regard to previous decades. This unprecedented access to tools makes the application of machine learning algorithms to any problem possible, where data is available. The bibliometrics of machine learning in geoscience, shown in Fig. 17 serve as a proxy for increased access. These papers include varying degrees of depth in application and model validation. One of the primary influences for the current increase in publications are new fields such as automatic seismic interpretation, as well as, publications soliciting and encouraging machine learning publications. Computer vision models were relatively straight forward to transfer to seismic interpretation tasks, with papers in this subfield ranging from single 2D line salt identification models with limited validation to 3D multifacies interpretation with validation on a separate geographic area.

Geoscientific publishing can be challenging to navigate with respect to machine learning. While papers investigating the theoretical fundamentals of machine learning in geoscience exist, it is clear that the overwhelming majority of papers present applications of ML to geoscientific problems. It is complex to evaluate whether a paper is a case study or a methodological paper with an exemplary application to a specific data set. Despite the difficulty

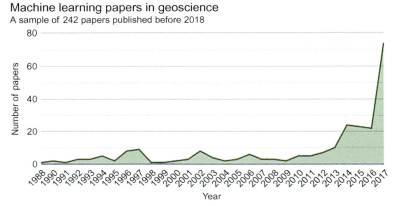

Fig. 17 Bibliometry of 242 papers in machine learning for geoscience per year. Search terms include variations of machine learning terms and geoscientific subdisciplines but exclude remote sensing and kriging.

of most thorough applications of ML, "idea papers" exist that simply present an established algorithm to a problem in geoscience without a specific implementation or addressing the possible caveats. On the flip-side, some papers apply machine learning algorithms as pure regression models without the aim to generalize the model to other data. Unfortunately, this makes meta-analysis articles difficult to impossible. This kind of meta-analysis article, is commonly done in medicine and considered a gold-standard study, and would greatly benefit the geoscientific community to determine the efficacy of algorithms on sets of similar problems.

Analogous to the medical field, obtaining accurate ground truth data, is often impossible and usually expensive. Geological ground truth data for seismic data is usually obtained through expert interpreters. Quantifying the uncertainty of these interpretations is an active field of research, which suggest a broader set of experiences and a diverse set of sources of information for interpretation facilitate correct geological interpretation between interpreters (Bond, Gibbs, Shipton, Jones, et al., 2007). Radiologists tasked to interpret X-ray images showed similar decreases in both inter- and intra-interpreter error rate with more diverse data sources Jewett et al. (1992). These uncertainties in the training labels are commonly known as "label noise" and can be a detriment to building accurate and generalizable machine learning models. A significant portion of data in geoscience, however, is not machine learning ready. Actual ground truth data from drilling reports is often locked away in running text reports, sometimes in scanned PDFs. Data are often siloed and most likely proprietary. Sometimes the amount of samples to process is so large that many insights are yet to be made from samples in core stores or the storage rooms of museums. Benchmark models are either nonexistent or made by consortia that only provide access to their members. Academic data are usually only available within academic groups for competitive advantage, respect for the amount of work, and fear of being exposed to legal and public repercussions. These problems are currently addressed by a culture change. Nevertheless, liberating data will be a significant investment, regardless of who will work on it and a slow culture change can be observed already.

Generally, machine learning has seen the fastest successes in domains where decisions are cheap (e.g., click advertising), data are readily available (e.g., online shops), and the environment is simple (e.g., games) or unconstrained (e.g., image generation). Geoscience generally is at the opposite of this spectrum. Decisions are expensive, be it drilling new wells or assessing geohazards. Data are expensive, sparse, and noisy. The environment is heterogeneous and constrained by physical limitations. Therefore, solving problems like automatic seismic interpretation see a surge of activity having fewer constraints initially. Problems like inversion have solutions that are verifiably wrong due to physics. These constraints do not prohibit machine learning applications in geoscience. However, most successes are seen in close collaboration with subject matter experts. Moreover, model explainability becomes essential in the geoscience domain. While not being a strict equivalency, simpler models are usually easier to interpret, especially regarding failure modes.

A prominent example of "excessive" (Mignan & Broccardo, 2019a) model complexity was presented in DeVries, Viégas, Wattenberg, and Meade (2018) applying deep learning to aftershock prediction. Independent data scientists identified methodological errors, including data leakage from the train set to the test set used to present results (Shah & Innig, 2019). Moreover, Mignan and Broccardo (2019b) showed that using the central physical interpretation of the deep learning model, using the von Mises yield criterion, could be used to build a surrogate logistic regression. The resulting surrogate or baseline model outperforms the deep network and overfits less. Moreover, replacing the 13,000 parameter model with the two-parameter baseline model increases calculation speed, which is essential in aftershock forecasting and disaster response.[d] More generally, this is an example where data science

---

[d] All authors point out the potential in deep and machine learning research in geoscience regardless and do not wish to stifle such research (Mignan & Broccardo, 2019 b; Shah & Innig, 2019).

practices such as model validation, baseline models, and preventing data leakage and overfitting become increasingly important when the tools of applying machine learning become readily available.

Despite potential setbacks and the field of deep learning and data science being relatively young, they can rely on mathematical and statistical foundations and make significant contributions to science and society. Machine learning systems have contributed to modeling the protein structure of the current pandemic virus COVID-19 (Jumper, Tunyasuvunakool, Kohli, Hassabis, & Team, n.d.). A deep learning computer vision system was built to stabilize food safety by identifying Cassava plant disease on offline mobile devices (Ramcharan et al., 2017, 2019). Self-driving cars have become a possibility (Bojarski et al., 2016) and natural language understanding has progressed significantly (Devlin, Chang, Lee, & Toutanova, 2018).

Geoscience is slower in the adoption of machine learning, compared to other disciplines. To be able to adapt the progress in machine learning research, many valuable data sources have to be made machine-readable. There has already been a change in making computer code open-source, which has led to collaborations and accelerating scientific progress. While specific open benchmark data sets have been tantamount to the progress in machine learning, it is questionable whether these would be beneficial to machine learning in geoscience. The problems are often very complex with nonunique explanations and solutions, which historically has led to disagreements over geophysical benchmark data sets. Open data and open-source software, however, have and will play a significant role in advancing the field. Examples of this include basic utility function to load geoscientific data (Kvalsvik & Contributors, 2019) or more specifically cross-validation functions tailored to geoscience (Uieda, 2018).

Moreover, machine learning is fundamentally conservative, training on available data. This bias of data collection will influence the ability to generate new insights in all areas of geoscience. Machine learning in geoscience may be able to generate insights and establish relationships in existing data. Entirely new insights from previously unseen or analysis of particularly complex models will still be a task performed by trained geoscientists. Transfer learning is an active field of machine learning research that geoscience can significantly benefit from. However, no significant headway has been made to transfer trained machine learning models to out-of-distribution data, i.e., data that are conceptually similar but explicitly different from the training data set. The fields of self-supervised learning, including reinforcement learning that can learn by exploration, may be able to approach some of these problems. They are, however, notoriously hard to set up and train, necessitating significant expertise in machine learning.

Large portions of publications are concerned with weakly or unconstrained predictions such as seismic interpretation and other applications that perform image recognition on SEM or core photography. These methods will continue to improve by implementing algorithmic improvements from machine learning research, specialized data augmentation strategies, and more diverse training data being available. New techniques such as multitask learning (Kendall, Gal, & Cipolla, 2018) which improved computer vision and computer linguistic models, deep Bayesian networks (Mosser, Oliveira, & Steventon, 2019) to obtain uncertainties, noisy teacher–student networks (Xie, Hovy, et al., 2019) to improve training, and transformer networks (Graves, 2012) for time series processing, will significantly improve applications in geoscience. For example, automated seismic interpretation may advance to provide reliable outputs for relatively difficult geological regimes beyond existing solutions. Success will be reliant on interdisciplinary teams that can discern why geologically specific faults are important to interpret, while others would be ignored in manual interpretations, to encode geological understanding in automatic interpretation systems.

Currently, the most successful applications of machine learning and deep learning, tie into existing workflows to automate subtasks in a grander system. These models are highly specific, and their predictive capability does not resemble an artificial intelligence or attempt to do so. Mathematical constraints and existing theory in other applied fields, especially neuroscience, were able to generate insights into deep learning and geoscience has the opportunity to develop significant contributions to the area of machine learning, considering their unique problem set of heterogeneity, varying scales and nonunique solutions. This has already taken place with the wider adoption of "kriging" or more generally Gaussian processes into machine learning. Moreover, known applications of signal theory and information theory employed in geophysics are equally applicable in machine learning, with examples utilizing complex-valued neural networks (Trabelsi et al., 2017), deep Kalman filters (Krishnan, Shalit, & Sontag, 2015), and Fourier analysis (Tancik et al., 2020). Therefore, possibly enabling additional insights, particularly when integrated with deep learning, due to its modularity and versatility.

Previous reservations about neural networks included the difficulty of implementation and susceptibility to noise in addition to computational costs. Research into updating trained models and saving the optimizer state with the model has in part alleviated the cost of retraining existing models. Moreover, fine-tuning pretrained large complex models to specific problems has proven successful in several domains. Regularization techniques and noise modeling, as well as data cleaning pipelines, can be implemented to lessen the impact of noise on machine learning models. Specific types of noise can be attenuated or even used as an additional source of information. The aforementioned concerns have mainly transitioned into a critique about overly complex models that overfit the training data and are not interpretable. Modern software makes very sophisticated machine learning models, and data pipelines available to researchers, which has, in turn, increased the importance to control for data leakage and perform thorough model validation.

Currently, machine learning for science primarily relies on the emerging field of explainability (Lundberg et al., 2018). These provide primarily post hoc explanations for predictions from models. This field is particularly important to evaluate which inputs from the data have the strongest influence on the prediction result. The major point of critique regarding post hoc explanations is that these methods attempt to explain how the algorithm reached a wrong prediction with equal confidence. Bayesian neural networks intend to address this issue by providing confidence intervals for the prediction based on prior beliefs. These neural networks intend to incorporate prior expert knowledge into neural networks, which can be beneficial in geoscientific applications, where strong priors can be necessary. Machine learning interpretability attempts to impose constraints on the machine learning models to make the model itself explainable. Closely related to these topics is the statistics field of causal inference. Causal inference attempts to model the cause of variable, instead of correlative prediction. Some methods exist that can perform causal machine learning, i.e., causal trees (Athey & Imbens, 2016). These three fields will be necessary to glean verifiable scientific insights from machine learning in geoscience. They are active fields of research and more involved to correctly apply, which often makes cooperation with a statistician necessary.

In conclusion, machine learning has had a long history in geoscience. Kriging has progressed into more general machine learning methods, and geoscience has made significant progress applying deep learning. Applying deep convolutional networks to automatic seismic interpretation has progressed these methods beyond what was possible, albeit still being an active field of research. Using modern tools, composing custom neural networks, and conventional machine learning pipelines has become increasingly trivial, enabling wide-spread applications in every subfield of geoscience. Nevertheless, it is important to acknowledge the limitations of machine learning in geoscience. Machine learning methods are often cutting edge technology, yet properly validated models take time to develop, which is often perceived as inconvenient when working in a hot scientific field. Despite being cutting edge, it is important to acknowledge that none of these applications are fully automated, as would be suggested by the lure of artificial intelligence. Nevertheless, within applied geoscience, significant new insights have been presented. Applications in geoscience are using machine learning as a utility for data preprocessing, implementing previous insights beyond the theory and synthetic cases, or the model itself enabling unprecedented applications in geoscience. Overall, applied machine learning has matured into an established tool in computational geoscience and has the potential to provide further insights into the theory of geoscience itself.

References

Abadi, M., Agarwal, A., Barham, P., Brevdo, E., Chen, Z., Citro, C., … Zheng, X. (2015). In *TensorFlow: Large-scale machine learning on heterogeneous systems.* http://tensorflow.org/ Software available from tensorflow.org.

Agterberg, F. P. (1966). Markov schemes for multivariate well data. In *Proceedings, symposium on applications of computers and operations research in the mineral industries, Pennsylvania State University, State College, Pennsylvania: Vol. 2.* (pp. X1–X18).

Aizerman, M. A. (1964). Theoretical foundations of the potential function method in pattern recognition learning. *Automation and Remote Control, 25,* 821–837.

Alfarraj, M., & AlRegib, G. (2018). Petrophysical property estimation from seismic data using recurrent neural networks. *SEG Technical Program Expanded Abstracts 2018,* 2141–2146.

Anifowose, F., Ayadiuno, C., & Rashedian, F. (2017). Carbonate reservoir cementation factor modeling using wireline logs and artificial intelligence methodology. In *79th EAGE conference and exhibition 2017-workshops.*

Araya-Polo, M., Dahlke, T., Frogner, C., Zhang, C., Poggio, T., & Hohl, D. (2017). Automated fault detection without seismic processing. *The Leading Edge, 36*(3), 208–214.

Athey, S., & Imbens, G. (2016). Recursive partitioning for heterogeneous causal effects. *Proceedings of the National Academy of Sciences, 113*(27), 7353–7360.

Babakhin, Y., Sanakoyeu, A., & Kitamura, H. (2019). Semi-supervised segmentation of salt bodies in seismic images using an ensemble of convolutional neural networks. In *German conference on pattern recognition (GCPR).*

Ballabio, C., & Sterlacchini, S. (2012). Support vector machines for landslide susceptibility mapping: The Staffora River Basin Case Study, Italy. *Mathematical Geosciences, 44*(1), 47–70. https://doi.org/10.1007/s11004-011-9379-9.

Bayes, T. (1763). LII. An essay towards solving a problem in the doctrine of chances. By the late Rev. Mr. Bayes, FRS communicated by Mr. Price, in a letter to John Canton, AMFR S. *Philosophical Transactions of the Royal Society of London, 53,* 370–418.

Belson, W. A. (1959). Matching and prediction on the principle of biological classification. *Journal of the Royal Statistical Society: Series C (Applied Statistics), 8*(2), 65–75.

Bergen, K. J., Johnson, P. A., Maarten, V., & Beroza, G. C. (2019). Machine learning for data-driven discovery in solid earth geoscience. *Science, 363*(6433), eaau0323.

Bestagini, P., Lipari, V., & Tubaro, S. (2017). A machine learning approach to facies classification using well logs. In *SEG technical program expanded abstracts 2017*. Society of Exploration Geophysicists, pp. 2137–2142. https://doi.org/10.1190/segam2017-17729805.1.

Beyreuther, M., & Wassermann, J. (2008). Continuous earthquake detection and classification using discrete Hidden Markov models. *Geophysical Journal International, 175*(3), 1055–1066. https://doi.org/10.1111/j.1365-246X.2008.03921.x.

Bicego, M., Acosta-Muñoz, C., & Orozco-Alzate, M. (2013). Classification of seismic volcanic signals using Hidden-Markov-model-based generative embeddings. *IEEE Transactions on Geoscience and Remote Sensing, 51*(6), 3400–3409. https://doi.org/10.1109/TGRS.2012.2220370.

Blondelle, H., Juneja, A., Micaelli, J., & Neri, P. (2017). *Machine learning can extract the information needed for modelling and data analysing from unstructured documents. In 79th EAGE conference and exhibition 2017-workshops.* earthdoc.org. http://www.earthdoc.org/publication/publicationdetails/?publication=89273.

Blouin, M., Caté, A., Perozzi, L., & Gloaguen, E. (2017). *Automated facies prediction in drillholes using machine learning. In 79th EAGE conference and exhibition 2017-workshops.* earthdoc.org. http://www.earthdoc.org/publication/publicationdetails/?publication=89276.

Bojarski, M., Del Testa, D., Dworakowski, D., Firner, B., Flepp, B., Goyal , P., et al. (2016). End to end learning for self-driving cars. *arXiv preprint arXiv:1604.07316*.

Bond, C. E., Gibbs, A. D., Shipton, Z. K., & Jones, S. (2007). What do you think this is? "Conceptual uncertainty" in geoscience interpretation. *GSA Today, 17*(11), 4.

Breiman, L. (2001). Random forests. *Machine learning, 45*(1), 5–32.

Bryson, A. E. (1961). A gradient method for optimizing multi-stage allocation processes. In *Proc Harvard Univ symposium on digital computers and their applications*, Vol. 72.

Buitinck, L., Louppe, G., Blondel, M., Pedregosa, F., Mueller, A., Grisel, O., … Varoquaux, G. (2013). API design for machine learning software: Experiences from the scikit-learn project. In *ECML PKDD workshop: Languages for data mining and machine learning*, pp. 108–122.

Cao, J., & Roy, B. (2017). Time-lapse reservoir property change estimation from seismic using machine learning. *The Leading Edge, 36*(3), 234–238. https://doi.org/10.1190/tle36030234.1.

Caté, A., Perozzi, L., Gloaguen, E., & Blouin, M. (2017). Machine learning as a tool for geologists. *The Leading Edge, 36*(3), 215–219. https://doi.org/10.1190/tle36030215.1.

Caté, A., Schetselaar, E., Mercier-Langevin, P., & Ross, P.-S. (2018). Classification of lithostratigraphic and alteration units from drillhole lithogeochemical data using machine learning: A case study from the Lalor volcanogenic massive sulphide deposit, Snow Lake, Manitoba, Canada. *Journal of Geochemical Exploration, 188*, 216–228. https://www.sciencedirect.com/science/article/pii/S0375674217305083.

Chaki, S., Routray, A., & Mohanty, W. K. (2018). Well-Log and seismic data integration for reservoir characterization: A signal processing and machine-learning perspective. *IEEE Signal Processing Magazine, 35*(2), 72–81. https://doi.org/10.1109/MSP.2017.2776602.

Chan, S., & Elsheikh, A. H. (2017). Parametrization and generation of geological models with generative adversarial networks. *arXiv preprint arXiv:1708.01810*.

Chang, C.-C., & Lin, C.-J. (2011). LIBSVM: A library for support vector machines. *ACM Transactions on Intelligent Systems and Technology, 2*(3), 27:1–27:27. https://doi.org/10.1145/1961189.1961199.

Chen, T., & Guestrin, C. (2016). XGBoost: A scalable tree boosting system. In *Proceedings of the 22nd ACM SIGKDD international conference on knowledge discovery and data mining.* New York, NY, USA: ACM, pp. 785–794. https://doi.org/10.1145/2939672.2939785.

Chevitarese, D., Szwarcman, D., Silva, R. M. D., & Brazil, E. V. (2018). Seismic facies segmentation using deep learning. In *ACE 2018 annual convention & exhibition*. searchanddiscovery.com.

Chiles, J. P., & Chauvet, P. (1975). Kriging: A method for cartography of the sea floor. *The International Hydrographic Review*.

Chilès, J.-P., & Desassis, N. (2018). Fifty years of kriging. In *Handbook of mathematical geosciences*. Springer, pp. 589–612.

Ching, T., Himmelstein, D. S., Beaulieu-Jones, B. K., Kalinin, A. A., Do, B. T., Way, G. P., ... Greene, C. S. (2018). Opportunities and obstacles for deep learning in biology and medicine. *Journal of the Royal Society Interface*, *15*(141). https://doi.org/10.1098/rsif.2017.0387.

Cho, K., Van Merriënboer, B., Gulcehre, C., Bahdanau, D., Bougares, F., Schwenk, H., & Bengio, Y. (2014). Learning phrase representations using RNN encoder-decoder for statistical machine translation. *arXiv preprint arXiv:1406.1078*.

Ciresan, D. C., Meier, U., Masci, J., Gambardella, L. M., & Schmidhuber, J. (2011). Flexible, high performance convolutional neural networks for image classification. In *Twenty-second international joint conference on artificial intelligence*.

Collobert, R., Bengio, S., & Mariéthoz, J. (2002). *Torch: A modular machine learning software library*.

Cortes, C., & Vapnik, V. (1995). Support-vector networks. *Machine Learning, 20*(3), 273–297.

Cover, T., & Hart, P. (1967). Nearest neighbor pattern classification. *IEEE Transactions on Information Theory, 13*(1), 21–27.

Cracknell, M. J., & Reading, A. M. (2013). The upside of uncertainty: Identification of lithology contact zones from airborne geophysics and satellite data using random forests and support vector machines. *Geophysics, 78*(3), WB113–WB126.

Cressie, N. (1990). The origins of kriging. *Mathematical Geology, 22*(3), 239–252.

Dammeier, F., Moore, J. R., Hammer, C., Haslinger, F., & Loew, S. (2016). Automatic detection of alpine rockslides in continuous seismic data using hidden Markov models. *Journal of Geophysical Research: Earth Surface, 121*(2), 351–371. https://doi.org/10.1002/2015JF003647.

Dechter, R. (1986). *Learning while searching in constraint-satisfaction problems*. University of California, Computer Science Department.

Delhomme, J. P. (1978). Kriging in the hydrosciences. *Advances in Water Resources, 1*, 251–266.

Deng, J., Dong, W., Socher, R., Li, L.-J., Li, K., & Fei-Fei, L. (2009). Imagenet: A large-scale hierarchical image database. In *2009 IEEE conference on computer vision and pattern recognition*, pp. 248–255.

Devlin, J., Chang, M.-W., Lee, K., & Toutanova, K. (2018). Bert: Pretraining of deep bidirectional transformers for language understanding. *arXiv preprint arXiv:1810.04805*.

DeVries, P. M. R., Viégas, F., Wattenberg, M., & Meade, B. J. (2018). Deep learning of aftershock patterns following large earthquakes. *Nature, 560*(7720), 632.

Di, H., Shafiq, M., & AlRegib, G. (2017a). *Multi-attribute k-means cluster analysis for salt boundary detection*. In *79th EAGE conference and exhibition*. earthdoc.org. *http://www.earthdoc.org/publication/publicationdetails/?publication=88632*.

Di, H., Shafiq, M. A., & AlRegib, G. (2017b). Seismic-fault detection based on multiattribute support vector machine analysis. In *SEG technical program expanded abstracts 2017*. Society of Exploration Geophysicists, pp. 2039–2044.

Di, H., Wang, Z., & AlRegib, G. (2018). Deep convolutional neural networks for seismic salt-body delineation. In *AAPG annual convention & exhibition*. searchanddiscovery.com.

Dodge, D. A., & Harris, D. B. (2016). Large-scale test of dynamic correlation processors: Implications for correlation-based seismic pipelines. *Bulletin of the Seismological Society of America*. https://pubs.geoscienceworld.org/ssa/bssa/article-abstract/106/2/435/332173.

Dowla, F. U., Taylor, S. R., & Anderson, R. W. (1990). Seismic discrimination with artificial neural networks: Preliminary results with regional spectral data. *Bulletin of the Seismological Society of America*, *80*(5), 1346–1373. https://pubs.geoscienceworld.org/ssa/bssa/article-abstract/80/5/1346/119382.

Zuo, R., Xiong, Y., Wang, J., & Carranza, E. J. M. (2019). Deep learning and its application in geochemical mapping. *Earth-Science Reviews*, *192*, 1–14.

Dramsch, J. (2019). Machine learning in 4D seismic data analysis: Deep neural networks in geophysics (PhD thesis).

Dramsch, J. S. (2020a). In *3D decision volume of SVM, random forest, and deep neural network*. figshare. https://doi.org/10.6084/m9.figshare.12640226.v1.

Dramsch, J. S. (2020b). *Code for 70 years of machine learning in geoscience in review*. figshare. https://doi.org/10.6084/m9.figshare.12666140.v1.

Dramsch, J. S., Amour, F., & Lüthje, M. (2018). Gaussian mixture models for robust unsupervised scanning-electron microscopy image segmentation of North Sea Chalk. In *First EAGE/PESGB workshop machine learning*. EAGE Publications BV. https://doi.org/10.3997/2214-4609.201803014.

Dramsch, J. S., Christensen, A. N., MacBeth, C., & Lüthje, M. (2019). Deep unsupervised 4D seismic 3D time-shift estimation with convolutional neural networks. *EarthArxiv*.

Dramsch, J. S., & Lüthje, M. (2018). Deep-learning seismic facies on state-of-the-art CNN architectures. In *Seg technical program expanded abstracts 2018*. Society of Exploration Geophysicists, pp. 2036–2040. https://doi.org/10.1190/segam2018-2996783.1.

Dreyfus, S. (1962). The numerical solution of variational problems. *Journal of Mathematical Analysis and Applications*, *5*(1), 30–45.

Dubrule, O. (1984). Comparing splines and kriging. *Computers & Geosciences*, *10*(2–3), 327–338.

Duvenaud, D. (2014). Automatic model construction with Gaussian processes (PhD thesis). University of Cambridge.

Dupont, E., Zhang, T., Tilke, P., Liang, L., & Bailey, W. (2018). Generating realistic geology conditioned on physical measurements with generative adversarial networks. *arXiv preprint arXiv:180203065*.

Engel, J., Agrawal, K. K., Chen, S., Gulrajani, I., Donahue, C., & Roberts, A. (2019). Gansynth: Adversarial neural audio synthesis. *arXiv preprint arXiv:190208710*.

Feng, X.-T., & Seto, M. (1998). Neural network dynamic modelling of rock microfracturing sequences under triaxial compressive stress conditions. *Tectonophysics*, *292*(3), 293–309. https://doi.org/10.1016/S0040-1951(98)00072-9.

Ferreira, R., Brazil, E. V., & Silva, R. (2018). Texture-based similarity graph to aid seismic interpretation. In *ACE 2018 annual*. searchanddiscovery.com.

Fukushima, K. (1980). Neocognitron: A self-organizing neural network model for a mechanism of pattern recognition unaffected by shift in position. *Biological Cybernetics*, *36*(4), 193–202.

Goodfellow, I., Bengio, Y., & Courville, A. (2016). *Deep learning*. MIT Press.

Goodfellow, I., Pouget-Abadie, J., Mirza, M., Xu, B., Warde-Farley, D., Ozair, S., … Bengio, Y. (2014). Generative adversarial nets. In *Advances in neural information processing systems*, pp. 2672–2680.

Görtler, J., Kehlbeck, R., & Deussen, O. (2019). A visual https://distill.pub/2019/visual-exploration-gaussian-processes exploration of Gaussian processes. https://distill.pub/2019/visual-exploration-gaussian-processes *Distill*. https://doi.org/10.23915/distill.00017. https://distill.pub/2019/visual-exploration-gaussian-processes.

Gramstad, O., & Nickel, M. (2018). *Automated top salt interpretation using a deep convolutional net. In 80th EAGE conference and exhibition 2018*. earthdoc.org. http://www.earthdoc.org/publication/publicationdetails/?publication=92117.

Graves, A. (2012). Sequence transduction with recurrent neural networks. *arXiv preprint arXiv:1211.3711*.

Guillen, P., Larrazabal, G., González, G., Boumber, D., & Vilalta, R. (2015). Supervised learning to detect salt body. In *SEG technical program expanded abstracts 2015*. Society of Exploration Geophysicists, pp. 1826–1829. https://doi.org/10.1190/segam2015-5931401.1.

Guitton, A. (2018). 3D convolutional neural networks for fault interpretation. In *80th EAGE conference and exhibition 2018*. earthdoc.org. http://www.earthdoc.org/publication/publicationdetails/?publication=92118.

Guo, R., Zhang, Y. S., Lin, H., & Liu, W. (2017). Sweet spot interpretation from multiple attributes: Machine learning and neural networks technologies. In *First EAGE/AMGP/AMGE Latin*. earthdoc.org. http://www.earthdoc.org/publication/publicationdetails/?publication=90731.

Gupta, I., Rai, C., Sondergeld, C., & Devegowda, D. (2018). Rock typing in the upper Devonian-lower Mississippian woodford shale formation, Oklahoma, USA. *Interpretation, 6*(1), SC55–SC66. https://doi.org/10.1190/INT-2017-0015.1.

Hahnloser, R. H. R., Sarpeshkar, R., Mahowald, M. A., Douglas, R. J., & Seung, H. S. (2000). Digital selection and analogue amplification coexist in a cortex-inspired silicon circuit. *Nature, 405*(6789), 947–951.

Hajimoradlou, A., Roberti, G., & Poole, D. (2019). Predicting landslides using contour aligning convolutional neural networks. *arXiv: Computer Vision and Pattern Recognition*.

Hale, D. (2013). Methods to compute fault images, extract fault surfaces, and estimate fault throws from 3D seismic images. *Geophysics, 78*(2), O33–O43.

Hall, B. (2016). Facies classification using machine learning. *The Leading Edge, 35*(10), 906–909. https://doi.org/10.1190/tle35100906.1.

Hall, M., & Hall, B. (2017). Distributed collaborative prediction: Results of the machine learning contest. *The Leading Edge, 36*(3), 267–269. https://doi.org/10.1190/tle36030267.1.

He, K., Zhang, X., Ren, S., & Sun, J. (2016). Deep residual learning for image recognition. In *Proceedings of the IEEE conference on computer vision and pattern recognition*. pp. 770–778.

Hermes, L., Frieauff, D., Puzicha, J., & Buhmann, J. M. (1999). Support vector machines for land usage classification in Landsat TM imagery. In *IEEE 1999 international geoscience and remote sensing symposium. IGARSS'99 (Cat. No. 99CH36293)* (Vol 1, pp. 348–350).

Ho, T. K. (1995). Random decision forests. In *Proceedings of 3rd international conference on document analysis and recognition* (Vol. 1, pp. 278–282).

Hochreiter, S., Bengio, Y., Frasconi, P., & Schmidhuber, J. (2001). *Gradient flow in recurrent nets: The difficulty of learning long-term dependencies*. A field guide to dynamical recurrent neural networks. IEEE Press.

Hochreiter, S., & Schmidhuber, J. (1997). Long short-term memory. *Neural Computation, 9*(8), 1735–1780.

Hopfield, J. J. (1982). Neural networks and physical systems with emergent collective computational abilities. *Proceedings of the National Academy of Sciences, 79*(8), 2554–2558.

Hornik, K., Stinchcombe, M., & White, H. (1989). Multilayer feedforward networks are universal approximators. *Neural Networks, 2*(5), 359–366. https://doi.org/10.1016/0893-6080(89)90020-8.

Huang, K. Y., Chang, W. R. I., & Yen, H. T. (1990). *Self-organizing neural network for picking seismic horizons. In Seg technical program expanded*. library.seg.org. https://library.seg.org/doi/pdf/10.1190/1.1890183.

Huijbregts, C., & Matheron, G. (1970). Universal kriging (an optimal method for estimating and contouring in trend surface analysis): 9th Intern. In *Sym on decision making in the mineral industries (proceedings to be published by Canadian Inst Mining), Montreal*.

Hulbert, C., Rouet-Leduc, B., Ren, C. X., Riviere, J., Bolton, D. C., Marone, C., & Johnson, P. A. (2018). Estimating the physical state of a laboratory slow slipping fault from seismic signals. http://arxiv.org/abs/1801.07806.

Ildstad, C. R., & Bormann, P. (n.d.). MalenoV_nD (MAchine LEarNing of Voxels). Retrieved from https://github.com/bolgebrygg/MalenoV.

Ioffe, S., & Szegedy, C. (2015). Batch normalization: Accelerating deep network training by reducing internal covariate shift. *arXiv preprint arXiv:1502.03167*.

Iqbal, H. (2018). In *HarisIqbal88/PlotNeuralNet v100*. https://doi.org/10.5281/zenodo.2526396.

Jeong, J., Park, E., Han, W. S., & Kim, K.-Y. (2014). A novel data assimilation methodology for predicting lithology based on sequence labeling algorithms. *Journal of Geophysical Research: Solid Earth, 119*(10), 7503–7520. https://doi.org/10.1002/2014JB011279.

Jewett, M. A. S., Bombardier, C., Caron, D., Ryan, M. R., Gray, R. R., Louis, E. L. S., … Psihramis, K. E. (1992). Potential for inter-observer and intra-observer variability in x-ray review to establish stone-free rates after lithotripsy. *The Journal of Urology, 147*(3), 559–562.

Jumper, J., Tunyasuvunakool, K., Kohli, P., Hassabis, D., & Team, A. (n.d.). Computational predictions of protein structures associated with COVID-19 (Tech. Rep.). Retrieved from https://deepmind.com/research/open-source/computational-predictions-of-protein-structures-associated-with-COVID-19.

Kadurin, A., Nikolenko, S., Khrabrov, K., Aliper, A., & Zhavoronkov, A. (2017). druGAN: An advanced generative adversarial autoencoder model for de novo generation of new molecules with desired molecular properties in silico. *Molecular Pharmaceutics, 14*(9), 3098–3104. https://doi.org/10.1021/acs.molpharmaceut.7b00346.

Karra, S., O'Malley, D., Hyman, J. D., Viswanathan, H. S., & Srinivasan, G. (2018). Modeling flow and transport in fracture networks using graphs. *Physical Review E, 97*(3-1), 033304. https://doi.org/10.1103/PhysRevE.97.033304.

Kelley, H. J. (1960). Gradient theory of optimal flight paths. *ARS Journal, 30*(10), 947–954.

Kendall, A., Gal, Y., & Cipolla, R. (2018). Multi-task learning using uncertainty to weigh losses for scene geometry and semantics. In *Proceedings of the IEEE conference on computer vision and pattern recognition*, pp. 7482–7491.

Khoshnevis, N., & Taborda, R. (2018). Prioritizing groundmotion validation metrics using semisupervised and supervised learning. *Bulletin of the Seismological Society of America*. https://pubs.geoscienceworld.org/ssa/bssa/article-abstract/108/4/2248/536309.

Kim, B., Kim, H., Kim, K., Kim, S., & Kim, J. (2019). Learning not to learn: Training deep neural networks with biased data. In *Proceedings of the IEEE conference on computer vision and pattern recognition*. pp. 9012–9020.

Kingma, D. P., & Ba, J. (2014). Adam: A method for stochastic optimization. *arXiv:14126980*.

Koch, G., Zemel, R., & Salakhutdinov, R. (2015). Siamese neural networks for one-shot image recognition. In *ICML deep learning workshop*, Vol. 2.

Kolmogorov, A. N. (1939). Sur l'interpolation et extrapolation des suites stationnaires. *Comptes Rendus de l'Académie des Sciences, 208*, 2043–2045.

Kong, Q., Trugman, D. T., Ross, Z. E., Bianco, M. J., Meade, B. J., & Gerstoft, P. (2019). Machine learning in seismology: Turning data into insights. *Seismological Research Letters, 90*(1), 3–14.

Kratzert, F., Klotz, D., Shalev, G., Klambauer, G., Hochreiter, S., & Nearing, G. (2019). Benchmarking a catchment-aware long short-term memory network (LSTM) for large-scale hydrological modeling. *arXiv preprint arXiv:1907.08456*.

Krige, D. G. (1951). A statistical approach to some mine valuation and allied problems on the Witwatersrand (PhD thesis). Johannesburg.

Krischer, L., & Fichtner, A. (2017). Generating seismograms with deep neural networks. *AGU Fall Meeting Abstracts*, *2017*, S41D-03.

Krishnan, R. G., Shalit, U., & Sontag, D. (2015). Deep Kalman filters. *arXiv preprint arXiv:1511.05121*.

Krizhevsky, A., Sutskever, I., & Hinton, G. E. (2012). Imagenet classification with deep convolutional neural networks. In *Advances in neural information processing systems*, pp. 1097–1105.

Krumbein, W. C., & Dacey, M. F. (1969). Markov chains and embedded Markov chains in geology. *Journal of the International Association for Mathematical Geology*, *1*(1), 79–96.

Kuehn, N. M., & Riggelsen, C. (2011). Modeling the joint probability of earthquake, site, and ground-motion parameters using Bayesian networks. *Bulletin of the Seismological Society of America*, *101*(1), 235–249. https://pubs.geoscienceworld.org/ssa/bssa/article-abstract/101/1/235/349494.

Kuzma, H. A. (2003). A support vector machine for avo interpretation. In *SEG technical program expanded abstracts 2003*. Society of Exploration Geophysicists, pp. 181–184.

Kvalsvik, J., et al. (2019). In *SegyIO*. https://github.com/equinor/segyio/.

Laloy, E., Hérault, R., Jacques, D., & Linde, N. (2017). *Efficient training-image based geostatistical simulation and inversion using a spatial generative adversarial neural network*. arXiv preprint arXiv:1708.04975.

Lary, D. J., Alavi, A. H., Gandomi, A. H., & Walker, A. L. (2016). Machine learning in geosciences and remote sensing. *Geoscience Frontiers*, *7*(1), 3–10. https://doi.org/10. 1016/j.gsf.2015.07.003.

LeCun, Y., Bengio, Y., & Hinton, G. (2015). Deep learning. *Nature*, *521*(7553), 436–444.

Legendre, A. M. (1805). *Nouvelles méthodes pour la détermination des orbites des comètes*. F. Didot.

Li, J., & Castagna, J. (2004). Support vector machine (SVM) pattern recognition to AVO classification. *Geophysical Research Letters*, *31*(2), 948. https://doi.org/10.1029/2003GL018299.

Linnainmaa, S. (1970). The representation of the cumulative rounding error of an algorithm as a Taylor expansion of the local rounding errors (Master's thesis, in Finnish). Univ. Helsinki.

Lin, M., Chen, Q., & Yan, S. (2013). Network in network. *arXiv preprint arXiv:1312.4400*.

Liu, Y., Chen, Z., Wang, L., Zhang, Y., Liu, Z., & Shuai, Y. (2015). Quantitative seismic interpretations to detect biogenic gas accumulations: A case study from Qaidam Basin, China. *Bulletin of Canadian Petroleum Geology*, *63*(1), 108–121. https://doi.org/10.2113/ gscpgbull.63.1.108.

Lu, P., Morris, M., Brazell, S., Comiskey, C., & Xiao, Y. (2018). Using generative adversarial networks to improve deep-learning fault interpretation networks. *The Leading Edge*, *37*(8), 578–583.

Lundberg, S. M., Nair, B., Vavilala, M. S., Horibe, M., Eisses, M. J., Adams, T., et al. (2018). Explainable machine-learning predictions for the prevention of hypoxaemia during surgery. *Nature Biomedical Engineering*, *2*(10), 749–760.

Luo, X., Zhou, W., Wang, W., Zhu, Y., & Deng, J. (2017). Attention-based relation extraction with bidirectional gated recurrent unit and highway network in the analysis of geological data. *IEEE Access*, *6*, 5705–5715.

Ma, J., Jiang, Z., Tian, Q., & Couples, G. D. (2012). Classification of digital rocks by machine learning. In *ECMOR XIII-13th European*. earthdoc.org. http://www.earthdoc.org/ publication/publicationdetails/?publication=62262.

Maggi, A., Ferrazzini, V., Hibert, C., Beauducel, F., Boissier, P., & Amemoutou, A. (2017). Implementation of a multistation approach for automated event classification at Piton de la Fournaise Volcano. *Seismological Research Letters*, *88*(3), 878–891. https://doi.org/ 10.1785/0220160189. https://pubs.geoscienceworld.org/ssa/srl/article-abstract/88/3/ 878/284054.

Malfante, M., Mura, M. D., Metaxian, J., Mars, J. I., Macedo, O., & Inza, A. (2018). Machine learning for volcano-seismic Signals: Challenges and perspectives. *IEEE Signal Processing Magazine, 35*(2), 20–30. https://doi.org/10.1109/MSP.2017.2779166.

Mardan, A., & Javaherian, A. (2017). *Channel characterization using support vector machine. In 79th EAGE conference.* earthdoc.org. http://www.earthdoc.org/publication/publicationdetails/?publication=89283.

Marjanović, M., Kovačević, M., Bajat, B., & Voženílek, V. (2011). Landslide susceptibility assessment using SVM machine learning algorithm. *Engineering Geology, 123*(3), 225–234. https://doi.org/10.1016/j.enggeo.2011.09.006.

Markov, A. A. (1906). Rasprostranenie zakona bol'shih chisel na velichiny, zavisyaschie drug ot druga. *Izvestiya Fiziko-matematicheskogo obschestva pri Kazanskom universitete, 15*(135–156), 18.

Markov, A. A. (1971). Extension of the limit theorems of probability theory to a sum of variables connected in a Chain. *Dynamic Probabilistic Systems, 1,* 552–577. Reprint in English of Markov, 1906.

Martinelli, G., Eidsvik, J., & Sinding-Larsen, R. (2013). Building Bayesian networks from basin-modelling scenarios for improved geological decision making. *Petroleum, 19,* 289–304. http://pg.lyellcollection.org/content/early/2013/06/24/petgeo2012-057.abstract.

Masotti, M., Campanini, R., Mazzacurati, L., Falsaperla, S., Langer, H., & Spampinato, S. (2008). TREMOrEC: A software utility for automatic classification of volcanic tremor. *Geochemistry, Geophysics, Geosysystem, 9*(4), 4007. https://doi.org/10.1029/2007GC001860.

Masotti, M., Falsaperla, S., Langer, H., Spampinato, S., & Campanini, R. (2006). Application of support vector machine to the classification of volcanic tremor at Etna, Italy. *Geophysical Research Letters, 33*(20), 113. https://doi.org/10.1029/2006GL027441.

Matalas, N. C. (1967). Mathematical assessment of synthetic hydrology. *Water Resources Research, 3*(4), 937–945.

Matheron, G. (1963). Principles of geostatistics. *Economic Geology, 58*(8), 1246–1266.

Matheron, G. (1981). *Splines and kriging; their formal equivalence.* Syracuse, New York: Syracuse University Geology Contribution D. F. Merriam ed.

McCormack, M. (1991). Neural computing in geophysics. *The Leading Edge, 10*(1), 11–15. https://doi.org/10.1190/1.1436771.

Mignan, A., & Broccardo, M. (2019a). A deeper look into 'deep learning of aftershock patterns following large earthquakes': Illustrating first principles in neural network physical interpretability. In *International work-conference on artificial neural networks.* pp. 3–14.

Mignan, A., & Broccardo, M. (2019b). One neuron versus deep learning in aftershock prediction. *Nature, 574*(7776), E1–E3.

Mitchell, T. M. (1997). *Machine learning.* New York: McGraw-Hill.

Mjolsness, E., & DeCoste, D. (2001). Machine learning for science: State of the art and future prospects. *Science, 293*(5537), 2051–2055. https://doi.org/10.1126/science.293.5537.2051.

Mosser, L., Dubrule, O., & Blunt, M. J. (2017). Reconstruction of three-dimensional porous media using generative adversarial neural networks. *Physical Review E, 96*(4-1), 043309. https://doi.org/10.1103/PhysRevE.96.043309.

Mosser, L., Dubrule, O., & Blunt, M. J. (2018a). Conditioning of three-dimensional generative adversarial networks for pore and reservoirscale models. http://arxiv.org/abs/1802.05622.

Mosser, L., Dubrule, O., & Blunt, M. J. (2018b). Stochastic seismic waveform inversion using generative adversarial networks as a geological prior. http://arxiv.org/abs/1806.03720.

Mosser, L., Kimman, W., Dramsch, J. S., Purves, S., De la Fuente Briceño, A., & Ganssle, G. (2018). Rapid seismic domain transfer: Seismic velocity inversion and modeling using deep generative neural networks. In *80th EAGE conference and exhibition 2018, June*. EAGE Publications BV. https://doi.org/10.3997/2214-4609.201800734.

Mosser, L., Oliveira, R., & Steventon, M. (2019). Probabilistic seismic interpretation using Bayesian neural networks. In *81st EAGE conference and exhibition 2019. Vol. 2019.* (pp. 1–5).

Nash, J. (1951). Non-cooperative games. *Annals of Mathematics, 54*(2), 286–295.

Neal, R. M. (1996). *Bayesian learning for neural networks*. Springer Science & Business Media.

Newendorp, P. D. (1976). *Decision analysis for petroleum exploration*. Tulsa, OK: Penn Well Books.

Ochoa, L. H., Niño, L. F., & Vargas, C. A. (2018). Fast magnitude determination using a single seismological station record implementing machine learning techniques. *Geodesy and Geodynamics, 9*(1), 34–41. https://doi.org/10.1016/j.geog.2017.03.010.

Ohrnberger, M. (2001). researchgate.net. Accessed: 2018-12-17.

Paganini, M., de Oliveira, L., & Nachman, B. (2018). CaloGAN: Simulating 3D high energy particle showers in multilayer electromagnetic calorimeters with generative adversarial networks. *Physical Review D, 97*(1), 014021.

Panakkat, A., & Adeli, H. (2007). Neural network models for earthquake magnitude prediction using multiple seismicity indicators. *International Journal of Neural Systems, 17*(01), 13–33.

Pasolli, E., Melgani, F., & Donelli, M. (2009). Automatic analysis of GPR images: A pattern-recognition approach. *IEEE Transactions on Geoscience and Remote Sensing, 47*(7), 2206–2217.

Paszke, A., Gross, S., Chintala, S., Chanan, G., Yang, E., DeVito, Z., ... Lerer, A. (2017). Automatic differentiation in PyTorch. In *Nips autodiff workshop*.

Pearl , J. (2012). The do-calculus revisited. *arXiv preprint arXiv:1210.4852.*

Pedregosa, F., Varoquaux, G., Gramfort, A., Michel, V., Thirion, B., Grisel, O., ... Duchesnay, E. (2011). Scikit-learn: Machine learning in Python. *Journal of Machine Learning Research, 12*, 2825–2830.

Poulton, M., Sternberg, B., & Glass, C. (1992). Location of subsurface targets in geophysical data using neural networks. *Geophysics, 57*(12), 1534–1544. https://doi.org/10.1190/1.1443221.

Preston, F. W., & Henderson, J. (1964). *Fourier series characterization of cyclic sediments for stratigraphic correlation*. Kansas Geological Survey.

Purves, S., Alaei, B., & Larsen, E. (2018). Bootstrapping Machine-Learning based seismic fault interpretation. In *ACE 2018 annual convention & exhibition*. searchanddiscovery.com. http://www.searchanddiscovery.com/abstracts/html/2018/ace2018/abstracts/2856016.html

Ramcharan, A., Baranowski, K., McCloskey, P., Ahmed, B., Legg, J., & Hughes, D. P. (2017). Deep learning for image-based cassava disease detection. *Frontiers in Plant Science, 8*, 1852.

Ramcharan, A., McCloskey, P., Baranowski, K., Mbilinyi, N., Mrisho, L., Ndalahwa, M., ... Hughes, D. P. (2019). A mobile-based deep learning model for cassava disease diagnosis. *Frontiers in Plant Science, 10*, 272.

Rasmussen, C. E. (2003). Gaussian processes in machine learning. In *Summer school on machine learning*, pp. 63–71.

Recht, B., Roelofs, R., Schmidt, L., & Shankar, V. (2019). Do imagenet classifiers generalize to imagenet? *arXiv preprint arXiv:1902.10811.*

Reddy, R. K. T., & Bonham-Carter, G. F. (1991). A decision-tree approach to mineral potential mapping in snow lake area, Manitoba. *Canadian Journal of Remote Sensing, 17*(2), 191–200. https://doi.org/10.1080/07038992.1991.10855292.

Richardson, A. (2018). Seismic Full-Waveform inversion using deep learning tools and techniques. http://arxiv.org/abs/1801.07232.

Roden, R., & Chen, C. W. (2017). Interpretation of DHI characteristics with machine learning. *First Break*. http://www.earthdoc.org/publication/publicationdetails/?publication=88069.

Rolnick, D., Donti, P. L., Kaack, L. H., Kochanski, K., Lacoste, A., Sankaran , K., et al. (2019). Tackling climate change with machine learning. *arXiv preprint arXiv:1906.05433*.

Ronneberger, O., Fischer, P., & Brox, T. (2015). U-net: Convolutional networks for biomedical image segmentation. In *International conference on medical image computing and computer-assisted intervention*, pp. 234–241.

Rosenblatt, F. (1958). The perceptron: A probabilistic model for information storage and organization in the brain. *Psychological Review*, 65(6), 386.

Ross, Z. E., Meier, M.-A., Hauksson, E., & Heaton, T. H. (2018a). Generalized seismic phase detection with deep learning. http://arxiv.org/abs/1805.01075.

Ross, Z. E., Meier, M. A., & Hauksson, E. (2018). P-wave arrival picking and first-motion polarity determination with deep learning. *Journal of Geophysical Research: Solid Earth*, 123, 5120–5129. https://agupubs.onlinelibrary.wiley.com/doi/abs/10.1029/2017JB015251.

Ross, Z. E., Meier, M.-A., Hauksson, E., & Heaton, T. H. (2018b). Generalized seismic phase detection with deep learning. *Bulletin of the Seismological Society of America*, 108(5A), 2894–2901.

Röth, G., & Tarantola, A. (1994). Neural networks and inversion of seismic data. *Journal of Geophysical Research: Solid Earth*, 99(B4), 6753. https://doi.org/10.1029/93JB01563.

Rouet-Leduc, B., Hulbert, C., & Bolton, D. C. (2018). Estimating fault friction from seismic signals in the laboratory. *Geophysical Research Letters*, 45, 1321–1329. https://onlinelibrary.wiley.com/doi/abs/10.1002/2017GL076708.

Rouet-Leduc, B., Hulbert, C., & Lubbers, N. (2017). Machine learning predicts laboratory earthquakes. *Geophysical Research Letters*, 44, 9276–92826. https://onlinelibrary.wiley.com/doi/abs/10.1002/2017GL074677.

Rumelhart, D. E., Hinton, G. E., & Williams, R. J. (1988). Learning representations by back-propagating errors. *Cognitive Modeling*, 5(3), 1.

Russakovsky, O., Deng, J., Su, H., Krause, J., Satheesh, S., Ma, S., ... Fei-Fei, L. (2015). ImageNet large scale visual recognition challenge. *International Journal of Computer Vision (IJCV)*, 115(3), 211–252. https://doi.org/10.1007/s11263-015-0816-y.

Russell, S. J., & Norvig, P. (2010). *Artificial intelligence—A modern approach third international edition*. Pearson Education. http://vig.pearsoned.com/store/product/1,1207,store-12521_isbn-0136042597,00.html.

Samuel, A. L. (1959). Some studies in machine learning using the game of checkers. *IBM Journal of Research and Development*, 3(3), 210–229.

Saporetti, C. M., da Fonseca, L. G., Pereira, E., & de Oliveira, L. C. (2018). Machine learning approaches for petrographic classification of carbonate-siliciclastic rocks using well logs and textural information. *Journal of Applied Geophysics*, 155, 217–225. https://doi.org/10.1016/j.jappgeo.2018.06.012.

Schütt, K. T., Arbabzadah, F., Chmiela, S., Müller, K. R., & Tkatchenko, A. (2017). Quantum-chemical insights from deep tensor neural networks. *Nature Communication*, 8, 13890. https://doi.org/10.1038/ncomms13890.

Schwarzacher, W. (1972). The semi-Markov process as a general sedimentation model. In *Mathematical models of sedimentary processes*. Springer, pp. 247–268.

Sen, S., Kainkaryam, S., Ong, C., & Sharma, A. (2020). SaltNet: A production-scale deep learning pipeline for automated salt model building. *The Leading Edge*, 39(3), 195–203.

Shah, R., & Innig, L. (2019). In *Aftershock issues*. https://github.com/rajshah4/aftershocks_issues.

Shashidhara, B. M., Scott, M., & Marburg, A. (2020). Instance segmentation of benthic scale worms at a hydrothermal site. In *The ieee winter conference on applications of computer vision*, pp. 1314–1323.

Shen, C. (2018). A transdisciplinary review of deep learning research and its relevance for water resources scientists. *Water Resources Research*, *54*(11), 8558–8593.

Shen, D., Wu, G., & Suk, H.-I. (2017). Deep learning in medical image analysis. *Annual Review of Biomedical Engineering*, *19*, 221–248. https://doi.org/10.1146/annurev-bioeng-071516-044442.

Simonyan, K., & Zisserman, A. (2014). Very deep convolutional networks for large-scale image recognition. *arXiv preprint arXiv:1409.1556*.

Sun, A. Y., Scanlon, B. R., Zhang, Z., Walling, D., Bhanja, S. N., Mukherjee, A., & Zhong, Z. (2019). Combining physically based modeling and deep learning for fusing GRACE satellite data: Can we learn from mismatch? *Water Resources Research*, *55*(2), 1179–1195.

Sutskever, I., Martens, J., Dahl, G., & Hinton, G. (2013). On the importance of initialization and momentum in deep learning. In S. Dasgupta & D. McAllester (Eds.), *Proceedings of the 30th international conference on machine learning, 17–19 June*. (Vol. 28, pp. 1139–1147). Atlanta, Georgia, USA: PMLR. http://proceedings.mlr.press/v28/sutskever13.html.

Talarico, E., Leão, W., & Grana, D. (2019). Comparison of recursive neural network and Markov chain models in facies inversion. In *Petroleum geostatistics 2019*. (Vol. 2019, pp. 1–5).

Tancik, M., Srinivasan, P. P., Mildenhall, B., Fridovich-Keil, S., Raghavan, N., Singhal, U., … Ng, R. (2020). Fourier features let networks learn high frequency functions in low dimensional domains. *arXiv preprint arXiv:2006.10739*.

Theano Development Team. (2016). Theano: A Python framework for fast computation of mathematical expressions. *arXiv e-prints*, *1605.02688*. http://arxiv.org/abs/1605.02688.

Titos, M., Bueno, A., García, L., Benítez, M. C., & Ibañez, J. (2018). Detection and classification of continuous volcano-seismic signals with recurrent neural networks. *IEEE Transactions on Geoscience and Remote Sensing*, *57*(4), 1936–1948.

Trabelsi, C., Bilaniuk, O., Zhang, Y., Serdyuk, D., Subramanian, S., Santos, J. F., … Pal, C. J. (2017). Deep complex networks. *arXiv preprint arXiv:1705.09792*.

Turing, A. M. (1950). I.—Computing machinery and intelligence. *Mind*, *LIX*(236), 433–460. https://doi.org/10.1093/mind/LIX.236.433.

Uieda, L. (2018). Verde: Processing and gridding spatial data using Green's functions. *Journal of Open Source Software*, *3*(29), 957. https://doi.org/10.21105/joss.00957.

Valentine, A. P., & Kalnins, L. M. (2016). An introduction to learning algorithms and potential applications in geomorphometry and earth surface dynamics. *Earth Surface Dynamics*, *4*, 445–460.

Valera, M., Guo, Z., Kelly, P., Matz, S., Cantu, V. A., Percus, A. G., … Viswanathan, H. S. (2017). Machine learning for graphbased representations of three-dimensional discrete fracture networks. http://arxiv.org/abs/1705.09866.

van der Baan, M., & Jutten, C. (2000). Neural networks in geophysical applications. *Geophysics*, *65*(4), 1032–1047. https://doi.org/10.1190/1.1444797.

Vaswani, A., Shazeer, N., Parmar, N., Uszkoreit, J., Jones, L., Gomez, A. N., & Polosukhin, I. (2017). Attention is all you need. In *Advances in neural information processing systems*, pp. 5998–6008.

Waldeland, A., Jensen, A., Gelius, L., & Solberg, A. (2018). Convolutional neural networks for automated seismic interpretation. *The Leading Edge*, *37*(7), 529–537. https://doi.org/10.1190/tle37070529.1.

Waldeland, A. U., & Solberg, A. (2017). Salt classification using deep learning. *79th EAGE conference and exhibition*. http://www.earthdoc.org/publication/publicationdetails/?publication=88635.

Wang, H., Wellmann, J. F., Li, Z., Wang, X., & Liang, R. Y. (2017). A segmentation approach for stochastic geological modeling using Hidden Markov random fields. *Mathematical Geosciences, 49*(2), 145–177. https://doi.org/10.1007/s11004-016-9663-9.

Wang, K., Lomask, J., & Segovia, F. (2017). Automatic, geologic layer-constrained well-seismic tie through blocked dynamic warping. *Interpretation, 5*(3), SJ81–SJ90. https://doi.org/10.1190/INT-2016-0160.1.

Wang, L. X., & Mendel, J. M. (1992). Adaptive minimum prediction-error deconvolution and source wavelet estimation using Hopfield neural networks. *Geophysics, 57*(5), 670–679. https://library.seg.org/doi/abs/10.1190/1.1443281.

Wang, Z., Di, H., Shafiq, M. A., Alaudah, Y., & AlRegib, G. (2018). Successful leveraging of image processing and machine learning in seismic structural interpretation: A review. *The Leading Edge, 37*(6), 451–461.

Watkins, C. J. C. H. (1989). *Learning from delayed rewards.* Cambridge: King's College.

Wei, S., Yonglin, O., Qingcai, Z., & Jiaqiang, H. (2018). *Unsupervised machine learning: K-means clustering velocity semblance Auto-Picking. In 80th EAGE conference.* earthdoc.org. http://www.earthdoc.org/publication/publicationdetails/?publication=92299

Wickman, F. E. (1968). Repose period patterns of volcanoes. V. General discussion and a tentative stochastic model. *Arkiv for Mineralogi Och Geologi, 4*(5), 351.

Williams, C. K. I. (1998). Prediction with Gaussian processes: From linear regression to linear prediction and beyond. In M. I. Jordan (Ed.), *Learning in graphical models* (pp. 599–621). Springer.

Williams, C. K. I., & Rasmussen, C. E. (2006). *Gaussian processes for machine learning (Vol. 2) (No. 3).* Cambridge, MA: MIT Press.

Wirgin, A. (2004). The inverse crime. *arXiv preprint math-ph/0401050.*

Witten, I. H., Frank, E., & Hall, M. A. (2005). *Practical machine learning tools and techniques.* Morgan Kaufmann. p. 578.

Wu, H., & Zhang, B. (2018). A deep convolutional encoderdecoder neural network in assisting seismic horizon tracking. http://arxiv.org/abs/1804.06814.

Xie, P., Zhou, A., & Chai, B. (2019). The application of long short-term memory (lstm) method on displacement prediction of multifactor-induced landslides. *IEEE Access, 7,* 54305–54311.

Xie, Q., Hovy, E., Luong, M.-T., & Le, Q. V. (2019). Self-training with Noisy Student improves ImageNet classification. *arXiv preprint arXiv:191104252.*

Xie, X., Qin, H., Yu, C., & Liu, L. (2013). An automatic recognition algorithm for GPR images of RC structure voids. *Journal of Applied Geophysics, 99,* 125–134. https://doi.org/10.1016/j.jappgeo.2013.02.016.

Xu, B., Wang, N., Chen, T., & Li, M. (2015). Empirical evaluation of rectified activations in convolutional network. *arXiv preprint arXiv:1505.00853.*

Zhang, Y., & Paulson, K. V. (1997). Magnetotelluric inversion using regularized Hopfield neural networks. *Geophysical Prospecting, 45*(5), 725–743. http://www.earthdoc.org/publication/publicationdetails/?publication=33881.

Zhao, T., Li, F., & Marfurt, K. (2017). Constraining self-organizing map facies analysis with stratigraphy: An approach to increase the credibility in automatic seismic facies classification. *Interpretation, 5*(2), T163–T171. https://doi.org/10.1190/INT-2016-0132.1.

Zhao, X., & Mendel, J. M. (1988). *Minimum-variance deconvolution using artificial neural networks. In library.seg.org.* https://library.seg.org/doi/pdf/10.1190/1.1892433.

Zhao, Z., & Gross, L. (2017). Using supervised machine learning to distinguish microseismic from noise events. In *SEG technical program expanded abstracts 2017*. Society of Exploration Geophysicists, pp. 2918–2923. https://doi.org/10.1190/segam2017-17727697.1.

Zheng, H., Fu, J., Mei, T., & Luo, J. (2017). Learning multi-attention convolutional neural network for fine-grained image recognition. In *Proceedings of the IEEE international conference on computer vision*, pp. 5209–5217.

Zhu, W., & Beroza, G. C. (2018). PhaseNet: A deep-neural-network-based seismic arrival time picking method. http://arxiv.org/abs/1803.03211.

CHAPTER TWO

Machine learning and fault rupture: A review

Christopher X. Ren[a,†], Claudia Hulbert[b,†], Paul A. Johnson[c], and Bertrand Rouet-Leduc[c,*]

[a]Intelligence and Space Research Division, Los Alamos National Laboratory, Los Alamos, NM, United States
[b]Laboratoire de Géologie, Département de Géosciences, École Normale Supérieure, PSL Université, Paris, France
[c]Geophysics Group, Los Alamos National Laboratory, Los Alamos, NM, United States
*Corresponding author: e-mail address: bertrandrl@lanl.gov

Contents

1. Introduction 57
2. Machine learning: A shallow dive 61
   2.1 Learning tasks, performance, and experience 61
   2.2 Learning capacity 63
   2.3 Geophysical data 65
3. Laboratory studies 66
   3.1 Laboratory geodesy 67
   3.2 Laboratory seismology 71
4. Field studies 75
   4.1 Techniques 78
   4.2 Applications 91
5. Conclusion 95
Acknowledgments 97
References 97

1. Introduction

The study of material failure and rupture in geophysics is an extremely broad field (Scholz, 2019), involving observation and analysis of geophysical data from failure simulations at laboratory and field scale (Dorostkar & Carmeliet, 2018; Dorostkar, Guyer, Johnson, Marone, & Carmeliet, 2017, 2018; Gao et al., 2018; Gao, Guyer, Rougier, Ren, & Johnson, 2019;

[†] These two authors contributed equally.

Ren et al., 2019; Richards-Dinger & Dieterich, 2012; Thomas & Bhat, 2018), laboratory experiments (Beeler, 2004, 2006; Bolton et al., 2019; Corbi et al., 2013; Hulbert et al., 2019; Johnson et al., 2016; Kenigsberg, Riviére, Marone, & Saffer, 2019; Leeman, Saffer, Scuderi, & Marone, 2016b; Logan, 1975; Lubbers et al., 2018; Marone, 1998a; McLaskey, 2019; Rosakis, Kanamori, & Xia, 2006; Rosenau, Corbi, & Dominguez, 2017; Rouet-Leduc et al., 2018, 2017; Ryan, Riviére, & Marone, 2018; Xia, Rosakis, & Kanamori, 2005), and in the solid Earth (Brodsky et al., 2020; Collettini, Niemeijer, Viti, & Marone, 2009; Ross, Meier, & Hauksson, 2018; Ross, Meier, Hauksson, & Heaton, 2018a; Ross, Trugman, Hauksson, & Shearer, 2019; Rouet-Leduc, Hulbert, McBrearty, & Johnson, 2020a; Scholz, 2019; Trippetta et al., 2019; Trugman & Ross, 2019; Trugman, Ross, & Johnson, 2020). Despite the vast apparent scope of this field, one approach to distill its essence is based on the following question: given a seismic signal received by a sensor or a set of sensors, what information can be gleaned concerning the process generating the signal? Often these signals are noisy, numerous, and can be high-dimensional in nature meaning it is nontrivial to extract meaningful information from them (Bergen, Johnson, de Hoop, & Beroza, 2019; Kong et al., 2019). It then follows that the study of fault slip, whether in simulated granular fault gouge, or in situ, is a rich area of application for machine learning (ML).

Solid Earth geoscience and seismology are historically data-rich areas of research, drawing insights from seismic, electromagetic, magenotelluric, gravity datasets, among others. Indeed, some applications of what might be considered "modern" ML techniques originated from the fields of geostatistics and geophysics. For example, Gaussian process regression was initially developed from an effort to estimate the distribution of gold from samples from boreholes (Matheron, 1962) in the 1960s. Similarly, least absolute shrinkage and selection operator (LASSO) regression, was initially introduced in the geophysics literature for the linear inversion of band-limited reflection seismograms in 1986 (Santosa & Symes, 1986), before being independently discovered and further developed in the statistics community in 1996 (Tibshirani, 1996). Currently these fields are undergoing rapid and substantial changes with respect to the sheer volume of data available to the public: Kong et al. noted the volume of seismic waveform data available from the Incorporated Research Institutions for Seismology (IRIS) is growing at an exponential rate and that as of September 2018

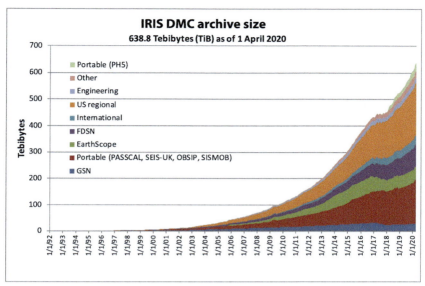

Fig. 1 IRIS Data Management Center archive size since its inception. The archive's size has been doubling every 3 years since the 2000s. *Courtesy of IRIS.*

the total size of the IRIS Data Management Center archive was 496.8 TB (Kong et al., 2019). As of May 1st 2020, the archive was 644.5 TiB (708 TB), and has been doubling roughly every 3 years for the past 20 years, as shown in Fig. 1. This rate of data increase is expected to further accelerate due to novel developments such as large-N deployments and distributed acoustic sensing (DAS). A similar dramatic increase in acquired data has been taking place in laboratory experiments of faulting. The case of the Rock and Sediments Mechanics Lab at Penn State University is telling: recent advances in experimental setups, related in particular to new high-throughput acoustic data acquisition systems have led to a surge in experimental data volume, with a doubling roughly every year for the last 5 years, as shown in Fig. 2. This explosive increase in available data presents many challenges: increasing volumes can lead to prohibitive access and processing times, however, it also provides exciting new opportunities as ML algorithms thrive in problem settings with large volumes of data. The objective of learning algorithms is to extract patterns from training data which *generalize to unseen data* (Goodfellow, Bengio, & Courville, 2016), rather than simply fit data. Thus, providing learning

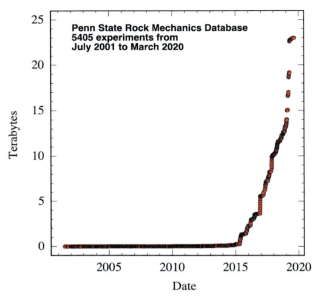

Fig. 2 Experimental data stored at the Rock Mechanics Lab at Penn State. The database's size has been doubling every year since 2015. *Courtesy of Chris Marone, Penn State University.*

algorithms with more high quality data typically enables them to extract relevant patterns more accurately.

This work will review recent advances in the application of ML in studying seismic signals originating from fault slip, rupture, and failure. We will provide an overview of the manners in which ML has been applied to advance research in studying fault slip in laboratory experiments and in the solid Earth. In Section 2 we provide a brief overview of the principles of ML, and specificities in the application of ML to geophysical data. Section 3 we provide an overview of the applications of ML to the analysis of data generated from laboratory studies. Following this, Section 4 reviews advances in the application of ML to the analysis of data in Solid Earth, including seismological analysis such as catalog building, phase picking and polarity determination and event location. Section 4 also covers the use of ML in the detection of geodetic deformation by InSAR. The latter part of Section 4 provides an overview of the uses of ML in specific applications such as early warning systems, the study of induced seismicity and earthquake forecasting. Finally, we conclude with thoughts on the future direction of ML in the field of geophysics.

2. Machine learning: A shallow dive

Machine learning (ML) and deep learning are scientific disciplines in their own right, and readers interested in a deeper dive will find many reviews and books dedicated to the subject (Abu–Mostafa, Magdon-Ismail, & Lin, 2012a; Bishop, 2006; Goodfellow et al., 2016). In this section we will provide a brief overview of ML. Previous reviews have already covered many of the types of algorithms used in ML and their properties (Bergen, Johnson, Hoop, & Beroza, 2019; Kong et al., 2019), so we will instead cover overarching definitions and concepts which unify these different algorithms.

2.1 Learning tasks, performance, and experience

The oft–quoted and rather succinct definition of a learning algorithm by Mitchell et al. (1997) goes as follows: "

"A computer program is said to learn from experience E with respect to some class of tasks T and performance measure P, if its performance at tasks in T, as measured by P, improves with experience E."

Typically, the types of tasks T are grouped into two main categories: classification and regression. In a classification task, the objective of an algorithm is to specify which predefined categories a given input belongs to, and outputs a categorical value. For a given regression task, the objective is to predict a numerical value given some input. The main differences between these two types of tasks is the format of the output.

It is interesting to note that often the difference between these two types of learning algorithms is simply the type of output: the "machinery" behind the two may be the same. Nowhere is this more obvious than with logistic regression, often one of the first "classification" algorithms taught in introductory ML classes. Logistic regression is itself a regression model which estimates the *probability of class membership* as a multilinear function of the input features. Only when combined with a decision rule, or a threshold, can we claim to be performing a classification task using logistic regression. A simple example can be seen in Fig. 3, which shows how both the logistic regression model and a decision rule (set at 0.5 here) can be combined to create a classifier.

We note that there are many other categories of tasks not covered by this introduction, such as: anomaly detection, data synthesis and imputation,

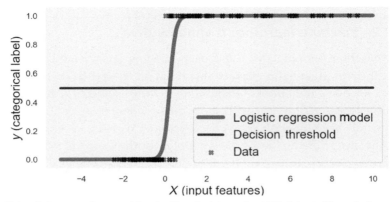

Fig. 3 Logistic regression combined with a decision rule at 0.5. *Adapted from Pedregosa, F., Varoquaux, G., Gramfort, A., Michel, V., Thirion, B., Grisel, O., ... Duchesnay, E. (2011). Scikit-learn: Machine Learning in Python. Journal of Machine Learning Research, 12, 2825–2830.*

ranking, density estimation among others. For a comprehensive overview of these we refer the reader to Goodfellow et al. (2016).

In the context of this review, the task **T** is typically one of detection (e.g., does this continuous seismic signal contain volcanic or nonvolcanic tremor, or an earthquake?), prediction (e.g., how close is our system to failing, given a portion of some continuous signal?).

The performance measure **P** in this context is a quantitative measure of the ability of the learning algorithm, and is task-specific (e.g., what is the difference between the predicted failure point and true failure point in terms of time). The performance metric during the training procedure is measured through the value of the loss function. The act of choosing an appropriate loss is typically quite an involved one, as it can influence model performance, generalization, computational cost and sometimes whether the learning algorithm converges to a solution at all. A simple example is shown in Fig. 4, where we fit two linear models, one with a least-squares loss and one with a Huber loss (Huber, 1964) to a simple linear dataset with outliers. Clearly in this case the effect of the Huber loss is to render the linear model more robust to the outliers. In testing, the key performance measure is the evaluation metric applied to the test-set: since we are most interested in the ability of the learning algorithm to generalize to unseen data, a portion of the available data is reserved for the testing procedure. The trained algorithm is evaluated on this "unseen" separate dataset, which was not involved in the training procedure. The evaluation metric is usually, but not necessarily, the same as the loss function.

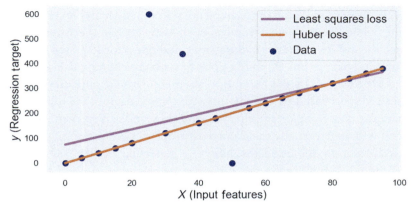

Fig. 4 Linear models fit to data with l2 loss (*fuchsia*) and Huber loss (*orange line*).

Finally the experience **E** provided to the learning algorithm allows us to draw one of the primary distinctions between types of learning algorithms: **supervised learning** algorithms are provided with a dataset containing input features where each example is associated with a target (the y-axis in Figs. 3 and 4). The term "supervised" originates from the idea that by providing the learning algorithm with a target we are essentially instructing it to find patterns related to this target (Goodfellow et al., 2016).

On the other hand, in an **unsupervised learning** setting the learning algorithm is provided with the input features, but no label or target. The goal of the learning procedures is then to learn useful patterns from the dataset such as dividing the dataset into similar groups (clustering), or learning the generating probability distribution behind the dataset (density estimation, denoising, synthesis) (Goodfellow et al., 2016).

2.2 Learning capacity

We noted in Section 2.1 that the central aim of ML is to learn a model which can generalize effectively to unseen data (i.e., hypothesis testing). During the learning process, an ML algorithm has access to some training data, on which it must attempt to reduce some error measure, known as the training error. Machine learning goes beyond simple curve fitting by assessing models based on performance on datasets not used for fitting/training: validation and testing sets.

The values of the training and test error and their relationship are linked to two central issues in ML: underfitting, occurring when the model is incapable of minimizing the training error to an satisfactory level and overfitting, when the training error has been reduced but the model cannot reduce the gap between the training error and test error (e.g., the model fits the training dataset but exhibits a poor fit with the training dataset). These issues are governed by the ML model's representational **capacity**, or complexity. Models with low capacity may struggle to capture the full spectrum of patterns in a complex training set, and thus underfit to them. Likewise models with high capacity can memorize properties of the training set which are irrelevant to the model's ability to generalize to the test set, thus overfitting. Fig. 5 illustrates this principle: we take the simple task of fitting

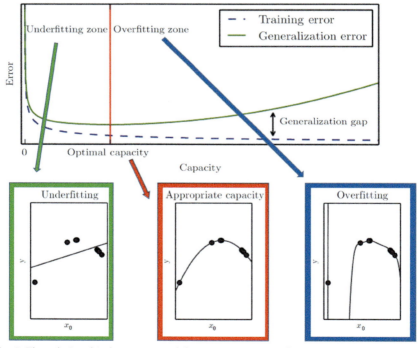

Fig. 5 The relationship between model capacity (x-axis) and error (y-axis). In the underfitting region shown at *left* (*green frame*), both training and test error decrease with increasing capacity. Once the capacity of the model exceeds the complexity of the data, the likelihood of overfitting, and thus increasing the test or generalization error increases (*blue frame*). The optimal model (*red frame*) is well matched to the complexity of the task, and will generalize to unseen data drawn from the same distribution as the training data. *Adapted from Goodfellow, I., Bengio, Y., & Courville, A. (2016). Deep learning. MIT Press.*

a dataset generated by a quadratic function: the linear model (green frame) does not have the capacity to fit the data, whereas the high-degree polynomial model (blue frame), with more parameters than training examples, overfits due to the fact is has the representational capacity to pick many potential solutions to the problem, reducing the chances of selecting the correct solution. The model at optimal capacity (red frame), is well matched to the complexity of the task, and will generalize well to unseen data drawn from the same distribution as the training data.

We note that in the case of scientific research, the models and results published can be sometimes in the *overfitting* regime. This is likely to be due to several reasons: one aspect is inherent to the scientific publishing process whereby less-overfit results are less likely to be considered novel or spectacular and thus less likely to be published journal articles. Another is related to data sparsity: though the overall volume of available geophysical data may be increasing, when compared to fields such as computer vision or natural language processing the amount of *labeled* data available for specific tasks is generally still small (examples of labeled data are earthquakes, low frequency earthquakes or tremor that have been identified and verified by an analyst). Last but not least, the test set will rarely be kept strictly independent from the learning phase, as researchers will go back to tuning their models **after** having seen their results on the test, effectively resulting in "data snooping" and some overfitting (Abu-Mostafa, Magdon-Ismail, & Lin, 2012b). Going back the drawing board after having looked at model performance on the test set effectively makes the test set really a validation set. In most cases one should then expect performance to drop on a dataset that was actually kept completely separate. However, even slightly overfit models will contain useful insights into the physical processes they are applied to if interpreted with caution.

2.3 Geophysical data

An assumption often made in the ML literature is that data samples are i.i.d., independently and identically distributed (i.e., the value from a sample has no influence on the other samples, and every sample is drawn from the same distribution). This i.i.d. assumption is more often than not violated in geophysical data. Time series data, *as well as features derived from it*, are autocorrelated, and the underlying physical process typically evolves over time. Geophysical data are often neither independently distributed, nor identically distributed. Autocorrelation in particular is extremely common in geophysical data analysis, and therefore such data should be handled with care. For example if one is analyzing data from a laboratory experiment: measurements (e.g., shear strain) at a time $t + 1$ are strongly dependent on the state of the experiment at time t, and both measurements are correlated. Similarly, field measurements of geophysical phenomena (e.g., seismicity or GPS displacement) will be strongly correlated depending on their proximity in time. A fundamental corollary to the autocorrelation of geophysical data, is that a contiguous train-test split is required (Bergmeir & Benítez, 2012) to ensure independent hypothesis testing, e.g., the first n years of data are used to train the model and the following n years are used to test it (e.g., McBrearty, Delorey, & Johnson, 2019). Moreover, when dealing with geophysical data, strategies to ensure stationarity can be necessary, for example by differentiating the data over time (i.e., trying to relate rates of change of properties to each other instead of the properties themselves (Rouet-Leduc, Hulbert, & Johnson, 2019)).

3. Laboratory studies

Laboratory friction experiments have served as a framework for the study of earthquake physics since the original experiments of stick–slip on a fault analog by Brace and Byerlee (1966). The theory of earthquakes as frictional instabilities and fracture propagation with associated elastic radiation stems from such experiments (Brace & Byerlee, 1966; Scholz, 2019). The favored phenomenological theory describing fault behavior is rate and state friction (Dieterich, 1979; Marone, 1998b; Ruina, 1983), although there are a host of others (e.g., Daub, Shelly, Guyer, & Johnson, 2011). Rate and state friction has been placed in a more physically-based context by using Arrhenius theory as a basis by Rice, Lapusta, and Kunnath (2001).

Just as in field seismology and geodesy, laboratory experiments are seeing an explosion in terms of volume of experimental data recorded, as noted previously. Experimental setups are improving with an increasing number of acquisition sensors, while these sensors are themselves rapidly improving in acquisition speed, sometimes resulting in terabytes of data for a single experiment (Bolton et al., 2019), making the field ripe for the use of ML. As distributed acoustic sensors (Parker, Shatalin, & Farhadiroushan, 2014) are incorporated into laboratory shear experiments, data volumes are expected to grow even further.

There has been a recent focus on using laboratory earthquake experiments to tackle unsolved mysteries of earthquake physics: earthquake precursors and their utility in earthquake forecasting (Johnson et al., 2013a), earthquake nucleation (Latour, Schubnel, Nielsen, Madariaga, & Vinciguerra, 2013), and the interplay between slow earthquakes and dynamic rupture (Latour et al., 2013; Leeman, Saffer, Scuderi, & Marone, 2016a), just to name a few. In this section we will review applications of ML that are helping to answer these fundamental questions in the context of laboratory friction experiments.

3.1 Laboratory geodesy

In this section we will highlight recent advances in applying ML techniques to analyzing what we term "laboratory geodesy" experiments. While point strain sensors have been employed in shear experiments for years using strain gauges or laser measurements, advances in high-resolution strain monitoring enabled by techniques such as digital image correlation (Adam et al., 2005) are advancing the field of analogue seismotectonic modeling by allowing for the measurement of small-scale deformation on the order of magnitude of displacements related to single earthquakes (Rosenau et al., 2017). We note that digital image techniques including digital image correlation have been demonstrated as effective techniques for the study of nucleation and phenomena such as supershear rupture by a number of groups using photoelastic materials, e.g., Xia et al., 2005 as well as in photoelastic experiments using "granular" discs (Daniels, Kollmer, & Puckett, 2016; Majmudar & Behringer, 2005).

Seismotectonic scale models represent a new form of laboratory experiments, possessing several desirable features in terms of rupture and earthquake modeling: coseismic dynamic weakening which may occur in nature for reasons such as frictional melting or thermal pressurization can be mimicked as these models feature nonlinear frictional material properties (Rosenau et al., 2017). Laboratory seismotectonic models can also feature properly scaled elasticity, allowing for depth-dependent pressurization of the "faults" in a more realistic manner.

Currently, all of the efforts in applying ML to the analysis of seismotectonic scale models make use of data from a viscoelastic gelatine wedge experiment initially developed by Corbi et al. (2013) to study subduction megathrust earthquakes. In this setup, a gelatin wedge is underthrust at a constant rate by a dipping, rigid plate. This experiment is meant to be the analogue of a subduction zone, as shown in Fig. 6A. The interface between the gelatin wedge and the rigid plate is where the analog megathrust earthquake develops. Two velocity weakening patches

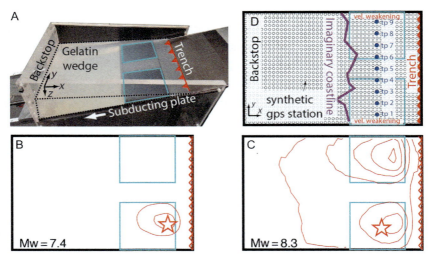

Fig. 6 (A) Experimental setup of a viscoelastic model of megathrust earthquakes. *Cyan rectangles* delineate the velocity weakening patches, with examples of ruptures involving (B) one or (C) two patches characterized in terms of coseismic displacement shown as the *red contour*, with the stars showing the epicenters. (D) Monitoring performed using image cross correlation, each *circle* represents an discretized area used as a "synthetic" GPS station. *Adapted from Corbi, F., Sandri, L., Bedford, J., Funiciello, F., Brizzi, S., Rosenau, M., & Lallemand, S. (2019). Machine learning can predict the timing and size of analog earthquakes. Geophysical Research Letters, 46(3), 1303–1311.*

(asperities modeled using gelatin on sandpaper) separated by a velocity strengthening patch (a barrier modeled using gelatin on plastic contact) are embedded in this analog megathrust. The viscoelastic seismotectonic model undergoes stick-slip cycles characterized by stress build-up punctuated by the nucleation of frictional instabilities (earthquakes) propagating at the interface between the gelatin and the subducting plate (Corbi et al., 2019). The velocity field is extracted by image cross correlation between consecutive images: this field is discretized into windows which in this case provide analogues of GPS stations above the model surface. Corbi et al. (2019) refer to these discrete windows derived from image cross correlation as "synthetic GPS stations."

Corbi et al. utilized a Gradient Boosted Regression Trees algorithm (Friedman, 2001) in an attempt to predict the time to failure for the next cycle.

The input ML features used are derived from the velocity field measured at the synthetic GPS stations, and are statistical features such as the variance, kurtosis and skewness of the velocity field, as well as correlations between subsequent frames and measures of cumulative displacement (see Supplementary Information of Corbi et al. (2019) for more details). The authors use information from previous cycles as input into the model; this is a hyperparameter that can be optimized: the authors reported using 10 previous cycles gives the best model performance, shown in Fig. 7A, with a coefficient of correlation $R = 0.3$ (Corbi et al., 2019). The authors then further refined their model by using as label the time to failure at specific locations, denoted by the nine blue points in Fig. 6D, with the results of these different models shown in Fig. 7B. It is stated in Corbi et al. (2019) that the data from the points located above the velocity weakening regions allowed for the learning of a model with $0.7 < R < 0.8$, even when using shorter training windows of 5 cycles. Analyzing the spatio-temporal distribution of the time to failure predictions allowed for the prediction of not only the timing and location of upcoming ruptures but also the size by examining how many adjacent points were expected to rupture simultaneously (Corbi et al., 2019). The most predictive features were determined to be measures related to cumulative displacement at specific points, indicating the learning algorithm is heavily informed by the relative loading history of these positions.

A further study was conducted by Corbi et al. on the same dataset in Corbi et al. (2020). Rather than framing the problem as a regression task (Corbi et al., 2019), the authors instead chose to train an ML algorithm to determine whether the deformation observed was characteristic of a time window immediately preceding slip onset or not: a binary classification problem (Corbi et al., 2020). The authors also modified the "positions" of the synthetic GPS network in the experiment, excluding data originating from the equivalent of the "offshore" seismogenic zone in an attempt to mimic limitations in geodetic coverage in real subduction zones.

The features used in this study consisted of the displacement measured at a given synthetic GPS station. The only features used for this study were thus displacements parallel and orthogonal to the "trench" (or by association, the coastline), as these were determined to be the variables most predictive of slip (Corbi et al., 2019). The target variable was derived from the same nine locations as in Corbi et al. (2019), where the slip was identified at times where the displacement at these locations exceeded a threshold. For each event, the category of "alarm" was assigned to the Δt seconds prior to the onset of this displacement rate. The output thus consists of nine

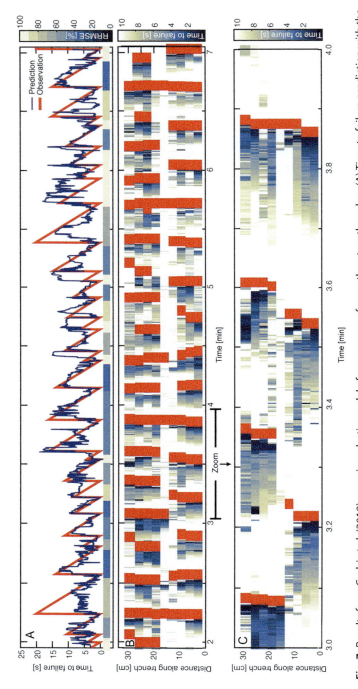

Fig. 7 Results from Corbi et al. (2019) on a viscoelastic model of a sequence of megathrust earthquakes. (A) Time to failure prediction with the cycle specific performance shown along the *color-coded horizontal line*, (B) Spatio-temporal evolution of the predicted time to failure for the nine points shown in Fig. 6D, with the distribution of the earthquakes highlighted by the *red squares*, (C) 1 min zoom of the period shown in (B).

response vectors, with each element of each vector indicating whether a given time step is part of the "alarm" category or not.

The authors use an ML algorithm known as Random Undersampling Boosting (RUSBoost) (Seiffert, Khoshgoftaar, Van Hulse, & Napolitano, 2008), an algorithm which combines random undersampling of the most prominent class with adaptive boosting (Freund, Schapire, & Abe, 1999). This is an effective method in dealing with the imbalanced nature of the dataset: there are simply more negative examples in the dataset than positive "alarm" examples (depending on the Δt the alarms represent 3%–27% of the total observations).

Since in this case the features are essentially locations in the experimental set up, the authors utilize sequential feature selection to determine informative regions or stations in the set up. Relevant features selected by this process are thus deemed informative to the model, while redundant or uninformative features which were discarded by the feature selection process are not. The authors find that regions adjacent to the two velocity weakening zones are the most informative in terms of monitoring the state of the system.

3.2 Laboratory seismology

In this section, we will focus on applications of ML to "laboratory seismology." In these experiments, an analogue model (a model of the fault block variety) is used to generate acoustic emissions, which are collected by accelerometers, strain gauges or acoustic sensors.

The first application of ML to the analysis of failure in a laboratory system was performed by Rouet-Leduc et al. (2017). This system was the basis of many further ML studies (Bolton et al., 2019; Hulbert et al., 2019; Jasperson, Bolton, Johnson, Marone, & de Hoop, 2019; Lubbers et al., 2018; Rouet-Leduc et al., 2018), as well as the basis for a Kaggle ML competition that drew more than 4500 teams (LANL Earthquake Prediction | Kaggle, n.d.), and as such merits some description in the context of this review. The experimental apparatus consists of a biaxial shearing device: a dual fault-configuration containing fault gouge material is driven at constant normal load and a constant velocity. The driving piston accelerates during slip. Simulated faults fail periodically throughout repetitive loading and failure cycles, with the goal of mimicking loading and failure on a fault patch in Earth. Acoustic emission (AE) produced by the shearing layers is recorded using an accelerometer, and the shear stress resulting from the

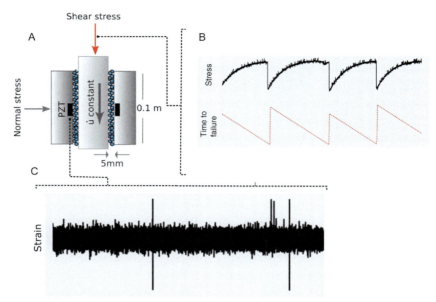

Fig. 8 (A) Biaxial shearing apparatus (B) recorded shear stress and derived time-to-failure (C) recorded strain by the accelerometers. *Adapted from Rouet-Leduc, B., Hulbert, C., Lubbers, N., Barros, K., Humphreys, C. J., & Johnson, P. A. (2017). Machine learning predicts laboratory earthquakes. Geophysical Research Letters. https://doi.org/10.1002/2017GL074677; Rouet-Leduc, B., Hulbert, C., Bolton, D. C., Ren, C. X., Riviere, J., Marone, C., … Johnson, P. A. (2018). Estimating fault friction from seismic signals in the laboratory. Geophysical Research Letters, 45(3), 1321–1329.*

driving block, shearing rate, gouge layer thickness, friction and applied load are monitored and recorded, as shown in Fig. 8.

This biaxial shearing system has been thoroughly studied and characterized in the past in many studies (Johnson et al., 2013a; Marone, 1998a; Niemeijer, Marone, & Elsworth, 2010; Scuderi, Marone, Tinti, Stefano, & Collettini, 2016). Stick-slip frictional failure causes the shearing block to displace, while the gouge material undergoes dilation and strengthening. Characteristics of the critical-stress regime exhibited by the material as it approaches failure, include minute shear failures emitting impulsive bursts of AE, as shown by Johnson et al. (2013a). The gradual progression of the system toward instability culminates in a "laboratory earthquake," or stick-slip failure characterized by rapid displacement of the shearing block, precipitous decreases in friction and shear stress, and compaction of the gouge layers.

In the initial study by Rouet-Leduc et al., ML was applied in an exploratory manner to identify relevant statistical features derived from the dynamic strain acoustic emission signal. Approximately 100 statistical features (mean, kurtosis, variance, etc.) were computed from moving time windows over the signal shown in Fig. 8C, which were then used to predict time remaining until the next stick–slip failure (red trace in Fig. 8B). In this study, an ML algorithm known as Random Forest (RF) (Breiman, 2001), an ensemble method based on decision trees (Breiman, 1984), was used to model this relationship. The learning algorithm identifies a relationship between a given time window of the acoustic emissions and the time remaining until the next stick–slip failure, as evidenced by the high R^2 value, or goodness of fit of the model.

The key takeaway from this experiment is that the time to failure in this system can be derived with fairly good accuracy even in the initial stages of the slip cycle, indicating a continuous progression toward failure. It is also important to note that even large fluctuations in slip cycle periodicity, such as the drastically shorter cycles shown in Fig. 9E, were accurately modeled by the RF algorithm.

This initial success indicating that the state of the laboratory fault can be derived from the time windowed features prompted a further study with the aim of explicitly predicting the frictional state, or shear stress of the fault using a similar approach (Rouet-Leduc et al., 2018). Feature importance derived from the model revealed that the variance of the acoustic signal in a given time window was the most predictive feature in determining the instantaneous friction of the fault system. Even under varying normal loads, the biaxial apparatus demonstrated a predictable relationship between frictional state (shear stress) and the seismic power (variance of the signal), with the observation that scaling the seismic power by the cube of the normal load revealed a single, empirical "friction law" for all normal loads.

Lubbers et al. demonstrated a different methodology employing the same data used in Rouet-Leduc et al. (2017) but deriving input features from event catalogs built from acoustic emission rather than directly from the continuous signal (Lubbers et al., 2018). In this study the features were chosen to capture cumulative statistics of event counts and amplitudes for a given time window. The authors demonstrated the ability of a RF model trained on the features to predict shear stress to a high degree of accuracy, as well as time *since* failure, with a lower performance for time to failure. The results suggest that models using cataloged-based features can reach a performance similar to models built on features from the continuous

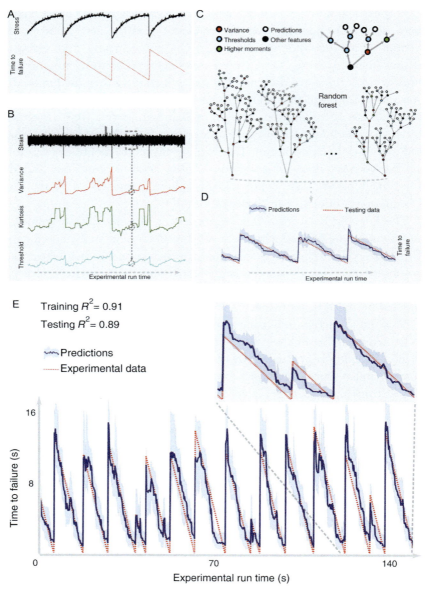

Fig. 9 Biaxial experimental results. (A) Shear stress and time to failure (*labels*), (B) acoustic emission and derived features (C) RF model schematic: the model averages the predictions of multiple decision trees, where each tree makes a prediction (*white node*) given an example based on a series of decisions made on the features associated with that example (*colored nodes*) (D) and (E) test-set performance of the trained RF model for the time to failure prediction task. *Adapted from Rouet-Leduc, B., Hulbert, C., Lubbers, N., Barros, K., Humphreys, C. J., & Johnson, P. A. (2017). Machine learning predicts laboratory earthquakes. Geophysical Research Letters. https://doi.org/10.1002/2017GL074677.*

signal, but only in the presence of an extremely complete catalog, something currently not possible in Earth. The performance rapidly deteriorates when the smallest events are progressively removed (Lubbers et al., 2018).

Hulbert et al. (2019) used a similar approach to analyze an experiment with aperiodic slip cycles consisting of both short and long slip durations, showing good performance in predicting not only the instantaneous shear stress of the system, but also fault displacement, fault gouge layer thickness, slip duration, and event magnitude. Signals originating from both slow and fast earthquakes behave in the same way when attempting to deduce the state of the fault. The level of acoustic power early on in the slip cycle is found to be correlated to the duration and the stress drop of the impending slip event.

Bolton et al. (2019) further analyzed the data used in Rouet-Leduc et al. (2017), Hulbert et al. (2019), and Lubbers et al. (2018), using unsupervised learning to study the evolution of features of the seismic data throughout the laboratory earthquake cycle, with a focus on identifying precursors from their anomalous distribution.

Jasperson et al. (2019) analyzed the same data set. They applied a Conscience Self-Organizing Map to perform topologically ordered vector quantization based on seismic waveform properties. The resulting map was used to interactively cluster catalogued events according to "damage mechanism" of the granular fault gouge material, a term borrowed from the domain of nondestructive evaluation. They applied an event-based long short term memory (LSTM) network to test the predictive power of each cluster. By tracking cumulative waveform features over the seismic cycle, the network forecast the time-to-failure of the fault.

4. Field studies

This section will review recent progress at the interface between ML and geophysics for the observation of the earthquake cycle, with a focus on novel applications of ML to better detect and characterize earthquakes, precursors to earthquakes (foreshocks, slow slip), and aftershocks.

Most of the recent advances at this interface between ML and geophysics have been made in seismology, with the goal of improving the detection of earthquakes of small magnitude, with broad-ranging applications: (i) for the study of earthquake dynamics—clusters in time, e.g., earthquake swarms, clusters in space, e.g., finer fault structures, (ii) for the study of foreshocks—elevated seismicity prior to main earthquakes that may indicate the nucleation of the main shock, (iii) for the study of aftershocks. These advances have relied for the most part on deep learning techniques, which perform exceedingly well on complex problems containing unstructured data such as seismic data. Another avenue of research at the interface between ML and geophysics that has seen a number of studies in recent years has been the search for predictability in earthquake catalogs. Finally, we will review the early works emerging from a field of study that is poised for dramatic advances: the interface between ML and geodesy, interferometric synthetic aperture radar (InSAR) in particular.

The stakes are high in terms of the detection of precursory signals, as these may lead to an improved understanding of earthquake nucleation, one of the most challenging and elusive phenomena in geoscience. This challenge is characterized most notably by the discrepancy between laboratory studies, where clear precursors are generally observed [foreshocks (Johnson et al., 2013b), aseismic slip (Dieterich, 1978; Latour et al., 2013)] as discussed in Section 3, and field earthquakes where precursors are only sometimes observed (Beckouche & Ma, 2014; Bouchon, Durand, Marsan, Karabulut, & Schmittbuhl, 2013; Bouchon et al., 2013) for reasons that may be related to earthquake catalog fidelity, or to the fact that not all earthquakes exhibit foreshocks.

In a broad sense, two end-members of earthquake nucleation exist in the literature (Beroza & Ellsworth, 1996; McLaskey, 2019). The cascade model of earthquake nucleation posits that a spontaneous rupture triggers an avalanche of foreshocks that leads to the main shock, with no prior activity. The preslip model of earthquake nucleation posits on the contrary that earthquakes are triggered by previous aseismic slip, of which foreshocks are a byproduct (see Fig. 10).

These two models have dramatic implications for the observation of earthquake nucleation, and open very different avenues for their detection using ML. The cascade model only leaves the improvement of earthquake catalogs, the current focus of the majority of ML research in geophysics (Kong et al., 2019), as an avenue for improving earthquake nucleation detection. The preslip model, on the other hand, leaves hope for a variety of possible improvements in the observation of earthquake nucleation through ML. The preslip model's hypothesis of foreshock and main shock triggering by aseismic slip adds slow slip to the list of potential observable earthquake dynamics. As a result, in addition to improved foreshock detection from improved earthquakes catalogs, improvements in a variety of geodetic observations of slow slip (GRACE, GPS, InSAR, etc.) stemming from ML may improve the detection of earthquake nucleation along with analysis of the continuous seismic signals emitted by the fault.

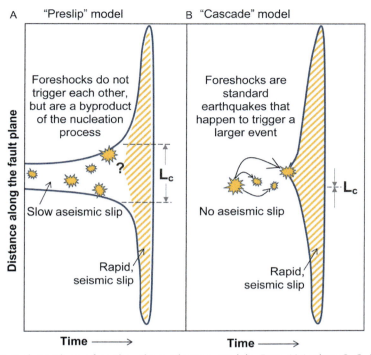

Fig. 10 End-members of earthquake nucleation models. *From McLaskey, G. C. (2019). Earthquake initiation from laboratory observations and implications for foreshocks. Journal of Geophysical Research: Solid Earth, 124(12), 12882–12904. https://doi.org/10.1029/2019JB018363.*

Recent direct (Ito et al., 2013; Ruiz et al., 2014) and indirect (Bouchon et al., 2013; Kato et al., 2012) observations of aseismic slip preceding large earthquakes hold the promise that with ML-enhanced detection of aseismic slip from novel seismic and geodetic data analysis, will come an enhanced detectability of earthquake nucleation. Likewise, recent improvements in the detection threshold of earthquake catalogs have shown that foreshocks may be more prevalent before earthquakes than previously thought (Trugman & Ross, 2019), indicating that improvements in precursor detection may also aid in detecting the elusive nucleation of earthquakes.

In the following sections of the review, in Section 4.1 we will review recent ML-driven technical advances in earthquake detection and characterization, with a focus on (i) building higher-fidelity catalogs, a requisite in particular in the quest for precursory foreshocks, and (ii) improving the detection of slow earthquakes from novel seismology and geodesy processing, a promising avenue for the detection of earthquake nucleation.

In Section 4.2, we will review few developments at the intersection between rupture physics in the earth and ML that require specific developments in addition to event detection and characterization. This work will cover induced seismicity, a topic of intense focus in recent years, including the development of ML-based techniques to better relate it to injection or extraction practices. We will also provide an overview of the specific developments made for earthquake early warning with a focus on faster and more reliable detection as well as better early-magnitude estimation. Lastly we will briefly cover earthquake forecasting, for which progress from ML techniques has arguably been lacking.

4.1 Techniques

4.1.1 Earthquake catalog building

Building catalogs with lower magnitudes of completeness has long been a major effort in seismology. More complete catalogs are useful for many applications such as the study of fault rupture, the identification of faulting networks, the estimation of seismic hazard and analyses of induced seismicity among others. They are also of major interest to identify foreshocks preceding large earthquakes.

The creation of catalogs has historically been conducted by hand in a long and tedious process involving the identification of an event and its seismic phase arrivals on seismograms, associating these phases to a particular earthquake, and pinpointing it to a location. The manual nature of this task is a hindrance in terms of scalability. With the explosion in the volume of seismic data recorded over the last decade (see Fig. 1), the need has arisen for automation of catalog building. Methods developed to combine the automatic detection of events and further verification by an analyst are now commonly used; however, because they still require human intervention, these approaches nonetheless still suffer scalability issues.

Given that they remain unchecked by analysts, fully automated methodologies for building catalogs require sufficient robustness to ensure a minimal false detection rate. In the scope of fully automated event detection, using robust confidence metrics (and ideally measuring uncertainty within a Bayesian framework) is therefore of paramount importance.

ML algorithms are particularly adapted to this problem, as they provide an opportunity for fast and automatic detection. While standard confidence metrics for the existence of an event is often based upon simple statistics such as a correlation threshold, ML algorithms provide an assessment of performance that can be evaluated (in the case of supervised learning), and in some cases can enable the estimation of uncertainty of event detection. As a result, catalog building based upon ML techniques has been actively pursued and was introduced early in the literature, starting in the early 1990s (Dowla, Taylor, & Anderson, 1990; Dysart & Pulli, 1991). In the following, we introduce the original work in this field and describe more recent developments for event detection, picking, association, and location. We also conduct a critical analysis of the shortcomings of ML algorithms with respect to catalog creation.

4.1.1.1 Earthquake detection

The first applications of ML for earthquake recognition focused on discriminating the spectra of seismic waveforms due to earthquakes from those due to explosions (Dowla et al., 1990; Dysart & Pulli, 1991). While not directly related to construction of earthquake catalogs, these analyses pioneered the detection of characteristic seismic signatures of earthquakes with neural networks. They were soon followed by studies aimed at identifying earthquakes from seismic signal and discriminating them from noise. In 1995, Wang and Teng (1995) found that neural networks trained with data from the Landers earthquake aftershocks outperformed the results of a standard threshold classification approach, STA/LTA (short term average over long term average). In 1999, Tiira (1999) reached similar conclusions when analyzing teleseism data in Finland, reporting that a neural network approach could improve the training catalog by 25%.

All the studies mentioned above relied on shallow, feed-forward neural networks (with the exception of Tiira (1999), which utilized slightly more complex architectures in the form of simple recurrent networks). They relied on hand-built features (either waveform spectra, STA/LTA functions, or both), and were trained on a small number of examples (from a few dozen to several hundred).

Most of the following studies have relied on the same tools, and were also based on relatively small datasets, with examples in the 100–1000 range (Madureira & Ruano, 2009; Mousavi, Horton, Langston, & Samei, 2016; Reynen & Audet, 2017; Ruano et al., 2014). However, the majority of these studies also incorporated ML approaches other than feed-forward networks, such as SVMs (support vector machines) (Madureira & Ruano, 2009; Ruano et al., 2014), random forests (Reynen & Audet, 2017), or logistic regressions (Reynen & Audet, 2017), which were found to systematically outperform the simple perceptron (Freund & Schapire, 1999) models. A parallel algorithmic avenue explored to detect earthquakes has relied on probabilistic graphical models other than feed-forward networks, using either hidden Markov models (Beyreuther & Wassermann, 2008) or dynamic Bayesian networks (Riggelsen & Ohrnberger, 2014). These approaches provide further benefits in associating a true Bayesian uncertainty measure to each detection, which can be used to help filter out false positives. In the two studies, the authors comment that their algorithms produced a number of false detections, but the uncertainties associated to those false detections were much higher than those associated to real earthquakes. Unsupervised algorithms based upon regrouping event waveforms by similarity have also been explored (Aguiar & Beroza, 2014; Sick, Guggenmos, & Joswig, 2015; Yoon et al., 2015).

Recently, researchers have begun to train models using larger datasets (Lomax, Michelini, & Jozinović, 2019; Mousavi, Zhu, Sheng, & Beroza, 2019a; Perol, Gharbi, & Denolle, 2018; Wiszniowski, Plesiewicz, & Trojanowski, 2014), enabling the use of more complex neural network architectures and automatic feature extraction. In particular, Mousavi et al. (2019a) used a database of 250,000 earthquakes and 250,000 noise samples in California to train a residual network based upon convolutional and recurrent units, and reported that their algorithm outperforms other approaches. Interestingly, when attempting to generalize their model from California to Arkansas, the model produced a higher rate of false positives, suggesting that complex models built upon large databases can have difficulties generalizing to other areas.

Overall, the analysis of the existing literature shows that earthquake detection with ML algorithms clearly outperform a standard a STA/LTA approach, in particular due to a lower rate of false positives as shown in Fig. 11. The comparison with template matching algorithms is less clearcut. Some definitive advantages of ML earthquake detection over template matching are (i) reduced computation time (once the model is trained), and (ii) the fact that events not included in the template list can be detected (depending on model generalization). However, ML algorithms may generate more false positives than template matching. Further developments of Bayesian deep networks trained on large datasets to associate clear uncertainties to each detection, such as that proposed by Gu, Marzouk, and Toksoz (2019), would be a straightforward improvement of the state of the art.

Machine learning and fault rupture 81

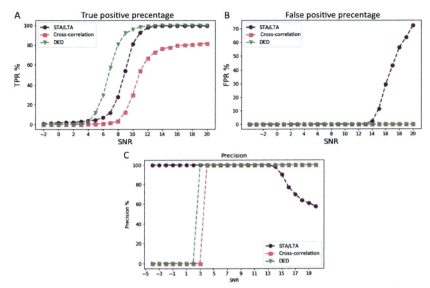

Fig. 11 Comparison between neural network detection (DED), STA/LTA, and cross-correlation on semi-synthetic data, at different signal-to-noise ratios. *From Mousavi, S. M., Zhu, W., Sheng, Y., & Beroza, G. C. (2019a). CRED: A deep residual network of convolutional and recurrent units for earthquake signal detection. Scientific Reports, 9(1), 1–14. https://doi.org/10.1038/s41598-019-45748-1.*

The comparison between template matching and neural network detection using convolutional networks is of particular interest. Because convolutional layers perform a cross-correlation operation, template matching corresponds effectively to a single 1D convolutional layer using one kernel of known weights for each template. In contrast, across the convolutional layers of a neural network, because the pooling operations change the resolution of the filtered input waveform, multiple (smaller) templates at different resolutions are learned.

Another useful comparison relates to the generalization of earthquake detection algorithms. STA/LTA algorithms are obviously straightforward to generalize to other areas and/or other seismic stations, but tend to generate a high rate of false positives. On the contrary, template matching algorithms do not transfer well, and can only detect events for which templates have been identified by analysts. ML algorithms are promising in terms of building robust models that generalize due to their ability to learn templates and identify events different from those included in training. However, the current state of the art does not seem to include models generalizing well to regions outside the training area. This is problematic as it precludes the use of such models for the analysis of regions of interest that are poorly instrumented, or where only a small number of events were cataloged. A likely future area of research may be focused on improving the generalization of these models.

4.1.1.2 Phase picking and polarity determination

Once an earthquake is detected, the arrival times of P- and S-waves have to be picked on seismograms in order to allow for the determination the hypocenter; the polarity of the P-wave is required to determine an event's focal mechanism. The identification of phase picks, as well as polarity, has been actively pursued with ML recently.

The first applications of ML for seismic picking was focused on recovering first arrivals from seismic reflection data, in the early 1990s (McCormack, Zaucha, & Dushek, 1993; Murat & Rudman, 1992). ML estimation of earthquake P- and S-wave arrivals was further investigated in 1995 by Dai and MacBeth (1995), where a shallow feed-forward neural network (in combination with a discriminant function) was tasked with recovering arrivals from the vector modulus of the three components of seismic waveforms. A similar analysis by the same authors (Dai & MacBeth, 1997) extended the approach to arrival detections from a single station component. In both studies, a model trained on a very small number of examples (nine pairs of seismic arrivals and noise) was able to reach a relatively high picking accuracy on hundreds of earthquakes. Shallow perceptron architectures were also applied for the same purpose by Wang and Teng (1997) and Zhao and Takano (1999). Gentili and Michelini (2006) found that picking from a neural tree was robust to errors from noise data and outperforms a standard STA/LTA algorithm. Chen (2017) also found that unsupervised algorithms outperform STA/LTA picking for small events.

In the late 2010s models were developed using much larger datasets which, combined with increases in computational power, enabled the use of more complex neural network architectures and automated feature extraction for phase picking. Zhu and Beroza (2019) developed a U-Net type network for picking estimations, trained on over 600,000 earthquakes in Northern California. The network takes as input the three components of seismic waveforms, and outputs either a P arrival, S arrival, or noise class. Wang, Li, Kemao, Di, and Zhao (2019) trained two separate convolutional networks to pick P and S arrival times, on a database of over 700,000 picks

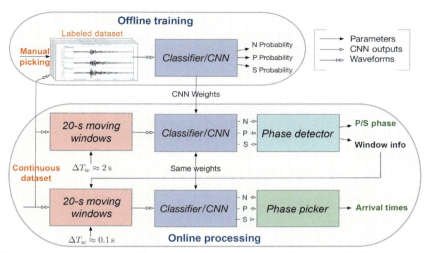

Fig. 12 Schematic of an offline and online picking classifier. N stands for noise, and P and S represent the pick arrivals of the P and S waves. *From Zhu, L., Peng, Z., McClellan, J., Li, C., Yao, D., Li, Z., & Fang, L. (2019). Deep learning for seismic phase detection and picking in the aftershock zone of 2008 Mw7.9 Wenchuan Earthquake. Physics of the Earth and Planetary Interiors, 293, 106261. https://doi.org/10.1016/j.pepi.2019.05.004.*

in Japan. Zhu et al. (2019) also utilized on a CNN to study the aftershock sequences of the 2008 M W 7.9 Wenchuan Earthquake. For all studies, results in testing were found to outperform existing methods. An online version of their algorithm, identifying picks on continuous waveforms, was also proposed. A sketch of the associated classification pipeline is shown in Fig. 12.

Besides picking arrival times, a few studies attempt to estimate P-wave first motion polarity. Ross, Meier, Hauksson, and Heaton (2018b) trained two different CNNs on 18.2 million manually picked seismograms in Southern California: one to identify P-wave arrival time, and a second to determine the polarity of the first motion. Hara, Fukahata, and Iio (2019) similarly relied on a CNN classifier to determine first-motion polarity from earthquakes in Western Japan.

Interestingly, models for phase picking appear to generalize better than models trained to detect earthquakes. Two recent studies report that their algorithms generalize well to regions outside of the training area, either without finetuning (Wang et al., 2019), or with minimal finetuning (Zhu et al., 2019). While this improved generalization could be attributed to larger training sets or model specificities, it might be the case that earthquake detection is intrinsically a more challenging problem in terms of generalization due to models indirectly learning specific Green's functions.

4.1.1.3 Event association

Once waveform picks are identified, they have to be associated to a given earthquake; the timing of arrivals appearing on multiple stations is analyzed to identify patterns characteristic of an earthquake. Standard methods mostly rely on travel time information for determining associations, which can be challenging in the presence of false arrivals, or when the earthquake rate is very high as shown in Fig. 13. ML algorithms have only very recently been applied to this problem.

The main approach employed in the literature is to associate several picks to an event by classification (McBrearty, Delorey, & Johnson, 2019; Ross, Yue, Meier, Hauksson, & Heaton, 2019). McBrearty, Delorey, and Johnson (2019) captained the use of convolutional neural networks to recognize whether or not a pair of seismic waveforms originated from the same earthquake. In Ross, Yue, et al. (2019), phases, travel times, and station

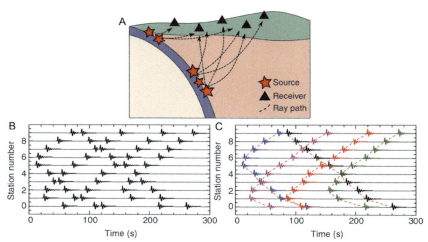

Fig. 13 Illustration of the earthquake association problem. In particular, in the presence of dense seismicity, associating the earthquake arrivals recorded at different sensors to a given earthquake (subplots (A) and (B)) can be challenging. (C) A schematic of the associated arrivals (each *color corresponds* to an earthquake). *From McBrearty, I. W., Gomberg, J., Delorey, A. A., & Johnson, P. A. (2019). Earthquake arrival association with backprojection and graph. Bulletin of the Seismological Society of America, 109(6), 2510–2531. https://doi.org/10.1785/0120190081.*

locations are fed as input to a recurrent network; the network then estimates whether picks within a moving window belong to the same event.

Another approach has focused on the use of clustering instead of classification for associating picks. In McBrearty, Gomberg, Delorey, and Johnson (2019) a backprojection algorithm is tasked with identifying candidate sources. A spectral clustering algorithm, combined with an optimization routine, is then used to associate the arrivals to the smallest number of sources possible that still match the data. The application of the algorithm in Northern Chile allows the authors outperform existing catalogs by nearly an order of magnitude.

4.1.1.4 Event location

Automatic earthquake location with ML has been actively explored in recent years. Once detected, earthquakes need to be located, which usually involves the back-propagation of seismic waves toward a source, or their association to the location of a given template. ML algorithms provide a promising avenue to refine and accelerate the location process. Two approaches have been explored in the literature: (i) multistation methods that are likely to favor event move-out patterns for event location similar to existing seismology approaches; (ii) single-station location, which employs the ability of algorithms to recognize the path of the wave's propagation encoded in seismic waveforms.

Among multistation location techniques, past attempts to locate seismic sources have relied upon clustering algorithms and neural networks. Riahi and Gerstoft (2017) use a graph-based clustering approach to regroup sensors affected by a common source; the source's location can then be inferred from the area spanned by the clusters. Trugman and Shearer (2017) use hierarchical clustering for earthquake relocation. The clustering algorithm takes as input differential travel times, cross-correlation values, and starting locations; events grouped within similar clusters are then simultaneously relocated. This approach is also able to return estimates of location errors. Kriegerowski, Petersen, Vasyura-Bathke, and Ohrnberger (2019) trained a neural network to locate earthquakes from multiple stations during an earthquake swarm in West Bohemia. The network takes as input the orthogonal components of all stations, and returns the depth, east and north source coordinates. The model is trained with events from a double-difference relocated catalog. Zhang et al. (2020) rely on a fully convolutional network to estimate a 3D image of the earthquake location, from data recorded at multiple stations. While location errors are small for earthquakes greater than magnitude 2, smaller events were reported to be more challenging to locate using this approach and associated with larger errors.

Studies of earthquake location from single stations all rely on neural networks, either in the classification or regression setting. Perol et al. (2018) and Lomax et al. (2019) trained neural networks to output classes that correspond to geographical areas associated to source locations. In Perol et al. (2018), a convolutional neural network was tasked with regrouping earthquakes within 6 geographic clusters. However, while the accuracy of the model's locations is reasonably good (74.5%), this metric drops drastically with increasing number of classes—potentially due to the small number of training samples. Lomax et al. (2019) employed a similar idea, but with a larger number of classes; however, the model returns high error rates. Mousavi and Beroza (2020) used an innovative approach for earthquake location: two Bayesian neural networks were trained separately, tasked with estimating respectively the epicentral distance and the P travel time associated to an event. A parallel neural network was then used to estimate the back-azimuth and its uncertainty. By regrouping the results of the three models, an event location is proposed along with an associated uncertainty. Epicenter, origin time, and depth errors are small in this approach, and because the models were trained on globally distributed events, this approach is found to generalize to different areas.

Location algorithms based on a single station are of particular interest in this field. First, the fact that models are able to locate events from single stations shows that they can reconstruct the wave propagation path from seismic data, thereby learning from the physics of the medium. Second, such models are portable, in the sense that they can be applied to areas outside of the training region. A model trained on several stations from a specific network will not be useful on data from other stations, which makes single-station models potentially more generalizable. Single-station models can also be useful for locating earthquakes that are recorded on a very few number of stations.

4.1.2 Seismic waveform denoising and enhancing

A very promising, recent use of ML in seismology is related to the ability of several algorithms to process or denoise data, thereby helping identify signals of interest, such as the detection of small events in seismology. Denoised data can afterwards be combined with standard seismology analysis or other ML algorithms for event detection and location.

The ability of many algorithms to reduce the dimension of the data and build a sparse representation from original signals makes ML tools particularly suited for the task of denoising seismic signals. In the literature, two main approaches have been used for such analysis of seismic waveforms: dictionary learning and autoencoders. Dictionary learning has been applied in particular to the denoising of seismic signals. In Beckouche and Ma (2014) and Chen and Guestrin (2016), the authors show that dictionary learning algorithms outperform several other approaches for this task (such as wavelets, curvelets, etc,), and result in higher signal-to-noise ratios.

In one of the first application of autoencoders in seismology Valentine and Trampert (2012) show that seismic waveforms can be successfully encoded and decoded, therefore learning an effective low-dimension representation of seismic signals. This procedure is applied to the identification of seismograms of good quality versus those of bad quality, with the goal of building a selection criteria of retaining only waveforms that have high signal-to-noise ratios, leading to more robust analyses. A recent study published by Zhu et al. (2019) similarly employs autoencoders with the objective of separating seismic signals from noise. In this work, an autoencoder trained on a large number of earthquake waveforms with high signal-to-noise ratios that are overlayed with noise waveforms was tasked with separating the signal from the noise. Once trained, the autoencoder was able to produce masks corresponding, respectively, to the signal and to the noise. These masks were used to clean new waveform data; on a few examples, retaining only the masked signal is shown to drastically improve the STA/LTA functions associated to earthquakes (Fig. 14), a demonstration of the potential of combining ML-based waveform denoising with standard seismology tools. A different approach for the same problem has been developed by Rouet-Leduc, Hulbert, McBrearty, and Johnson (2020b), who showed that neural network interpretation techniques can be used to denoise tremor waveforms and produce much cleaner recovered signals of interest.

A different and interesting application aimed at recovering seismic signal hidden in noise was proposed by Sun and Demanet (2020). The authors rely on a convolutional neural network to extrapolate missing low frequencies in synthetic seismic records. Inputs to the model were bandlimited recordings of seismograms, with low frequencies removed. The model is tasked with recovering the missing low frequencies from these inputs. The model was able to estimate the unobserved low-frequencies, both in terms of phase and in amplitude, reasonably well.

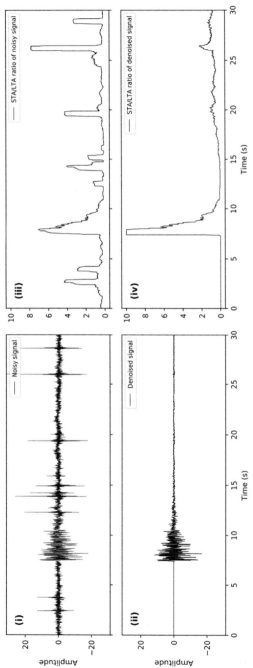

Fig. 14 Denoising of seismic waveforms with an auto-encoder. Improvement of STA/LTA characteristic function after denoising. (i) Example of a noisy waveform. (ii) Denoised waveform using a deep learning auto-encoder, (iii) and (iv) corresponding STA/LTA characteristic functions. From Zhu, L., Peng, Z., McClellan, J., Li, C., Yao, D., Li, Z., & Fang, L. (2019). Deep learning for seismic phase detection and picking in the aftershock zone of 2008 Mw7.9 Wenchuan Earthquake. *Physics of the Earth and Planetary Interiors*, 293, 106261. https://doi.org/10.1016/j.pepi.2019.05.004.

Further developments of ML algorithms to denoise or enhance the quality of seismic data may be of considerable interest in seismology. These approaches are particularly promising for the identification of small events hidden in the noise. They can be used to leverage standard seismology tools while limiting the risks of false detection, and might help to increase waveform cross-correlation levels.

4.1.3 Tectonic tremor detection

Tectonic tremor, along with GPS measures of deformation, is the most direct evidence for aseismic slip (Rogers & Dragert, 2003). As such, progress driven by ML in the detection of tectonic tremor is poised to advance the understanding of slow earthquakes and their interplay with regular earthquakes.

Recent advances in tectonic tremor detection using ML fall into two categories: developments of models to estimate tectonic tremor intensity and slow slip displacement rate from features of continuous seismic data, and developments of models to directly detect tectonic tremor.

In terms of the first category mentioned above: Rouet-Leduc et al. (2019) demonstrated that a ML model can be trained to accurately track slow slip rate on Vancouver Island, using features of the continuous seismic noise that are characteristic of tectonic tremor. Hulbert, Rouet-Leduc, Jolivet, and Johnson (2020) showed that slow earthquakes under Vancouver Island are likely preceded by a months-long nucleation phase, during which seismic features characteristic of tremor rise exponentially.

In terms of the second category: we note here that direct detection of tectonic tremor is a task perfectly suited for deep neural networks, which deal particularly well with rich unstructured data (e.g., text, images, videos, sound). Indeed, this has been demonstrated by Nakano, Sugiyama, Hori, Kuwatani, and Tsuboi (2019) that a CNN could reliably discriminate between earthquakes and tectonic tremor using a catalog of tremors recorded near the Nankai trough in Japan. Rouet-Leduc, Hulbert, et al. (2020a) showed that a CNN trained to distinguish tectonic tremor from background noise could be used to extract the tremor signals to improve event detection, using seismic records from Vancouver Island.

4.1.4 Fault slip inversion

Fault slip inversion relies on seismic and/or geodetic observations to determine earthquake source properties. Determining source properties is of importance for a variety of problems, in particular for early warning applications (see Section 4.2.1). Slip distribution also has implications for the distribution of subsequent aftershocks (e.g., Rietbrock et al., 2012), as well as the general understanding of slip processes on faults (e.g., Rousset et al., 2016).

Early applications of neural networks for slip inversion were developed by Käufl, Valentine, O'Toole, and Trampert (2014). The authors trained a number of small neural networks to invert for slip properties from synthetic geodetic data, and successfully applied their models to real data from the 2010 M_w 10 El Mayor Cupah earthquake, using the predictions from their ensemble of neural networks as a proxy for an a posteriori distribution of the source parameters.

Slip inversion is a particularly difficult problem for ML algorithms, for two reasons: (i) inversion is a generally nonunique problem, and therefore the estimation of fault slip properties requires probabilistic approaches, which inhibits the available tools; (ii) slip inversion often relies on very heterogeneous sources of data (e.g., seismic and GPS data), where deep learning does not perform well in general. Recent advances in probabilistic deep learning (Bingham et al., 2019) may solve this apparent incompatibility.

4.1.5 Automatic detection of geodetic deformation

Interferometric synthetic aperture radar (InSAR) is a powerful geodetic technique developed in the early 1990s (Massonnet, Briole, & Arnaud, 1995; Massonnet et al., 1993) for measuring ground surface displacements. InSAR has been successfully applied to monitor large displacements due to earthquakes (Massonnet et al., 1993; Peltzer, Crampé, & King, 1999; Simons, Fialko, & Rivera, 2002), as well as smaller displacements related to interseismic deformation (Jolivet et al., 2015, 2012; Peltzer, Crampé, Hensley, & Rosen, 2001; Wright, Parsons, England, & Fielding, 2004), slow moving landslides (Hilley, Bürgmann, Ferretti, Novali, & Rocca, 2004), and slow slip events (Cavalié et al., 2013; Jolivet et al., 2013). However, in general these measurements have been successful only through painstaking manual exploration of the deformation data by InSAR experts.

In the past low satellite revisit rates typically resulted in large time delays between measurements. The launch of the Sentinel-1 satellite constellation in 2014 has been revolutionary in that it provides systematic radar mapping of all actively deforming regions in the world with a 6-day return period. This wealth of data represents an opportunity as well as a challenge. InSAR data processing is not fully automatic, with unwrapping errors in particular often requiring manual correction, although rapid progress is being made toward full InSAR unwrapping automation (Benoit, Pinel-Puysségur, Jolivet, & Lasserre, 2020). Furthermore, the analysis of InSAR deformation data requires expert interpretation, and usually requires a priori knowledge on the analyzed area. The launch of additional InSAR satellites with even shorter return period (NASA's NISAR mission in

2022) will prove even more challenging in terms of data volumes and computational and time costs. Automation of InSAR processing and analysis has been identified as one of the main challenges of the field of InSAR geodesy (Lohman, 2019), and is poised to become a rapidly growing avenue of research, that may enable the automatic detection of small or slow deformation on faults.

Anantrasirichai, Biggs, Albino, Hill, and Bull (2018), Anantrasirichai, Biggs, Albino, and Bull (2019b), and Anantrasirichai, Biggs, Albino, and Bull (2019a) began developing deep learning tools for InSAR analysis by finetuning the AlexNet model (Krizhevsky, Sutskever, & Hinton, 2012) on manually labeled wrapped InSAR data as well as synthetic data. Their models flag volcano deformation (see Fig. 15), demonstrating a 3 cm detection capability.

Rouet-Leduc, Dalaison, Hulbert, Johnson, and Jolivet (2020) built a deep autoencoder to output cumulative ground deformation from unwrapped InSAR time series, with a focus on fault deformation. Trained on synthetic data, the authors demonstrate the recovery of millimeter scale deformation on real data from creeping faults as well as small earthquakes.

4.2 Applications

4.2.1 Early warning

Real-time alerts of destructive earthquakes while they are in progress are based on Earthquake Early Warning (EEW) systems. The very proposition of earthquake early warning (EEW)—a rapid assessment of immediate seismic hazard—demands the use of algorithmic solutions. As a result, EEW has seen considerable use of ML methods as early as the 2000s. These systems rely on the fact that (i) damaging S-waves follow identifiable P-waves and (ii) seismic waves propagate at a few km/s, much slower than speed-of-light telecommunication messages (seismic stations closer to the earthquake source can therefore be used to alert locales further away).

Böse, Wenzel, and Erdik (2008) designed a series of small, fully connected neural networks on features of seismic recordings following the onset of P-wave arrivals to estimate location and magnitude of earthquakes, on simulated data. Ochoa, Niño, and Vargas (2018) used a similar approach on real data, feeding features of the waveform that follows the P arrival to a support vector machine to estimate earthquake magnitude.

Ongoing efforts by Kong, Allen, Schreier, and Kwon (2016) to build a very large array of EEW detectors using a smartphone application (ShakeAlert), make use of a neural network applied to features of the phone's accelerometer data to estimate if the phone is recording an

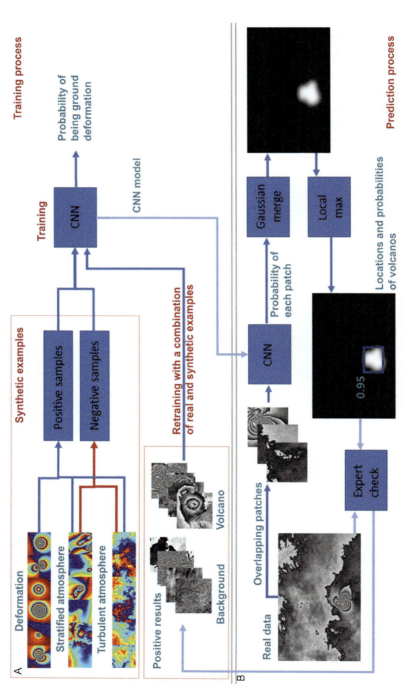

Fig. 15 Deformation detection in a wrapped InSAR interferogram. Anantrasirichai et al. finetuned the AlexNet model on synthetic data and on real data to classify whether or not a portion of wrapped InSAR interferograms contains ground deformation. *From Anantrasirichai, N., Biggs, J., Albino, F., & Bull, D. (2019b). A deep learning approach to detecting volcano deformation from satellite imagery using synthetic datasets. Remote Sensing of Environment, 230, 111179. https://doi.org/10.1016/J.RSE.2019.04.032.*

earthquake or some other motion. With the same goal of reducing false alarm rates from false detections, Li, Meier, Hauksson, Zhan, and Andrews (2018) trained a generative adversarial network (GAN), consisting a generative network and a discriminator network, (Goodfellow & Young, 2014; Schmidhuber, 1992), and use the discriminator network as a feature extractor for a random forest classifier tasked with telling earthquakes from impulsive noise on short waveforms from ShakeAlert data.

Meier, Heaton, and Clinton (2015) developed a Bayesian algorithm that starts from single station analysis for the very early onset of the detection on nearby stations, and evolves into a more accurate multistation analysis. Trained on a dataset of 60,000+ waveforms, their algorithm can provide a timely assessment of earthquake magnitude for areas close to the epicenter, where damage is greater. Their ML approach of training and testing an empirical model also demonstrates that information about earthquake magnitude saturates at magnitudes that depend on seismic record length, but are in general about magnitude 7–7.5.

4.2.2 Induced seismicity

A major effort in earthquake detection and location with ML algorithms has been focused on induced seismicity, either in geothermal fields (Beyreuther, Hammer, Wassermann, Ohrnberger, & Megies, 2012), in mines (Mousavi et al., 2016), or in gas (Gu et al., 2019) and oil (Mousavi et al., 2019a; Perol et al., 2018; Reynen & Audet, 2017; Zhang et al., 2020; Zhu et al., 2019) extraction fields (in particular in Oklahoma). These studies aim at building automatic and more complete catalogs, and were described in Section 4.1.

In this section we will therefore focus on another application of ML algorithms: the analysis of existing earthquake catalogs to better understand the connection between injected fluids and induced seismicity. Two studies are of particular interest, both relying on probabilistic graphical models to extract patterns from catalog data.

Holtzman, Paté, Paisley, Waldhauser, and Repetto (2018) analyzed the waveforms of 46,000 cataloged earthquakes in the Geysers geothermal field, in order to identify subtle changes in seismic activity. By combining non-negative matrix factorization and hidden Markov models, the authors were able to extract low-dimension representations of the earthquake seismic waveforms, which were then fed to a clustering algorithm. The resulting clusters did not reflect the geographical proximity of earthquakes, and showed instead clear temporal patterns that appear to be connected to injection rates. Interestingly, not all clusters correspond to higher seismic rates at times of high water injection; in some cases, clusters are correlated with the minimum injected rates. An interpretation of the results is that the model is able to identify changes in faulting mechanisms, and that the spectral-temporal patterns identified may correspond to changes in earthquake source physics, possibly due to changes in the thermo-mechanical conditions of the reservoir. As a result these observations may have applications for operational monitoring, aid in mitigating the risk of dangerous earthquakes, and enable the optimization of production efficiency from geothermal fields. Hincks, Aspinall, Cooke, and Gernon (2018) also relied on probabilistic graphical models to analyze the connections between geological characteristics, well operations, and induced seismicity in Oklahoma. They provided well characteristics (injection volume, pressure, etc.), geological features, as well as seismic moment release (calculated from cataloged earthquakes) to a Bayesian network. The model was then tasked with modeling the joint conditional dependencies between these different variables. Interestingly, the injection depth with respect to the crystalline basement was found to correlate with seismic moment release much more strongly than the total injected volume. Therefore, lithologic and fault network characteristics were found to play a more fundamental role with respect to induced seismicity than simple injection volumes: deeper fluid injection may favor migration into the basement, and favor the reactivation of faults as others have observed, e.g., Walsh and Zoback (2015). This analysis has important implications for operation management and regulation; in particular, it suggests that injecting water at lower depths (further away from the crystalline basement) may significantly reduce annual seismic moment release.

Both studies show that ML algorithms are promising tools for identifying patterns between known earthquakes and injection data, potentially leading to a better understanding of the physics of injection systems. Such findings may be useful for the operational management of geothermal reservoirs and extraction fields, by helping mitigate the risks of potentially damaging seismicity, and/or help to increase production. The combination of the aforementioned creation of denser catalogs, and similar analyses that link seismicity to injection operations may help to better characterize injection systems in the future.

4.2.3 Earthquake catalog forecasting

Mignan and Broccardo (2020) wrote a detailed analysis of the existing neural network-based earthquake forecasting literature (77 articles from 1994 to 2019). They concluded that no new insights on earthquake predictability stemmed from these works, and demonstrated that simpler models

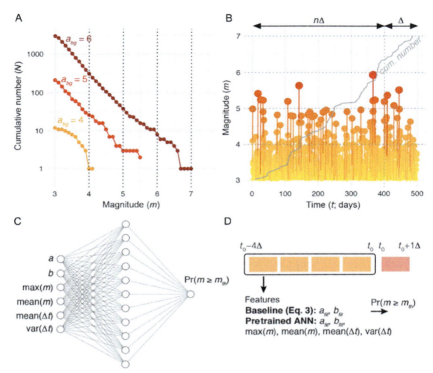

Fig. 16 Neural network forecasting of earthquakes: no improvement over classical statistical seismology so far. Figure from Mignan and Broccardo (2020) showing the approach of the authors to compare neural network-based earthquake predictions with classical statistical seismology laws as baseline. (A) The Gutenberg-Richter law as a baseline, (B) earthquake time series from the ETAS model, (C) a neural network model for earthquake prediction, (D) earthquake prediction in a window of time based on previous history.

based on classical empirical laws from statistical seismology offer similar or superior predictive power, perhaps indicating that prediction from earthquake catalogs is still a distant goal. In particular, the authors discuss the lack of baseline in the earthquake prediction literature, and show that on synthetic time series of earthquakes (ETAS model, Ogata, 1988) a neural network model does not outperform predicting mainshock magnitude in a window using the Gutenberg-Richter law (see Fig. 16).

5. Conclusion

In this review, we have attempted to present the current state of the art at the interface between fault physics and ML. From our analysis, two trends emerge. On the one hand, the field of seismology embraced ML very early on: seismology has historically been a highly data intensive discipline, and on occasions even preceded the field of ML in the development of certain algorithms. On the other hand, the field of fault physics (the why and how of rupture nucleation and propagation) has historically been much less data intensive. Earlier studies focused on the understanding of a few isolated slip events, whether in the laboratory or in the field. The shift toward ML and data science is only a very recent development, most notably for laboratory studies, for which large scale acquisition of seismic data has only recently become a reality.

As the amount of geophysical data continues increasing at ever faster rates, it also increases in complexity. Laboratory studies are recording increasing amounts of data from a variety of mechanical and seismic data from increasingly dense arrays of sensors, challenging models (such as neural networks) that can struggle with structured, heterogeneous data. The coverage of seismic networks in the field is also improving, challenging the generalization ability of models trained on particular regions, while also opening up the possibility of detecting events below the noise level of any single station. Distributed acoustic sensing (DAS) has recently started to be used for the study of earthquakes (Lindsey et al., 2017), and is poised to enable seismology at high frequency, large distances, and very fine spatial sampling (on the order of meters)—resolutions unattainable with classical seismic sensors. As a counterpart, DAS are also likely to generate staggering amounts of data. Similarly, the sheer volume of currently available InSAR data already provides a challenge for existing processing methods and algorithms which suffer from scalability issues and can require manual intervention. Next generation InSAR satellites will soon launch and dramatically increase the amount of acquisitions, making the automation of InSAR data analysis an even more pressing challenge.

As the stage is set for geophysical data to become ever harder to analyze using traditional methods, ML and data science techniques may well become fundamental tools of the trade in the geosciences. However, the growing use of modern machine learning algorithms does not come without risks, and the community will have exercise caution with its many pitfalls. These pitfalls are beyond the scope of this review but merit a few words. First, minute technical mishaps can have dramatic consequences: chief among them is a mixing of training and testing sets. This error can take many forms, and when dealing with time series, as is more often than not the case when dealing with geophysical data, randomly partitioning the data into training and testing sets instead of splits appropriate for time series will usually yield a model that makes accurate predictions simply due to the autocorrelation of the data. Second, even when trained and used correctly, machine learning models have inherent weaknesses (Marcus, 2018): models may not transfer well (Kansky et al., 2017) to even apparently similar problems (a model trained to detect earthquakes may not work on waveforms from different seismometers, for example). Related to this, deep learning models in particular can be very sensitive to subtle changes in the data to which they are applied, explaining the efficacy of adversarial attacks where minute but targeted perturbations can dramatically change a network's response (Goodfellow, Shlens, & Szegedy, 2015). Furthermore, unlike physics models, many machine learning models are unable to extrapolate. Where some classical machine learning models (decision trees, Gaussian processes with standard kernels) will consistently predict within the observed empirical distribution of the output when tasked with extrapolating, deep learning models can display erratic behavior in this regard.

Still, as field and laboratory studies of earthquakes and earthquake nucleation processes incorporate ML in the search for ever finer observations and slip characterization, new information may be revealed concerning fault behavior. This could improve our understanding of the accuracy of earthquake nucleation models analogs. In particular, the discrepancy between the abundance of earthquake precursors (seismic and aseismic) in the lab and their apparent lack in the field may be reconciled as the detection limit of observation methods is pushed down with the incorporation of ML algorithms.

Acknowledgments

C.X.R. and P.A.J. acknowledge support of this work by the US Department of Energy, Office of Science, Office of Basic Energy Sciences, Chemical Sciences, Geosciences, and Biosciences Division under grant 89233218CNA000001 with the Los Alamos National Laboratory. C.H. was supported by a joint research laboratory effort in the framework of the CEA-ENS Yves Rocard LRC (France), and by the European Research Council (ERC) under the European Union's Horizon 2020 research and innovation program (Geo-4D project, grant agreement 758210). B.R.L was funded by LANL institutional support (LDRD).

References

Abu-Mostafa, Y. S., Magdon-Ismail, M., & Lin, H.-T. (2012a). *Learning from data*. (Vol. 4). AMLBook.

Abu-Mostafa, Y. S., Magdon-Ismail, M., & Lin, H. T. (2012b). *Learning from data: A short course*. AMLBook. https://doi.org/10.1108/17538271011063889.

Adam, J., Urai, J. L., Wieneke, B., Oncken, O., Pfeiffer, K., Kukowski, N., ... Schmatz, J. (2005). Shear localisation and strain distribution during tectonic faultingnew insights from granular-flow experiments and high-resolution optical image correlation techniques. *Journal of Structural Geology, 27*(2), 283–301.

Aguiar, A. C., & Beroza, G. C. (2014). PageRank for earthquakes. *Seismological Research Letters, 85*(2), 344–350. https://doi.org/10.1785/0220130162.

Anantrasirichai, N., Biggs, J., Albino, F., Hill, P., & Bull, D. (2018). Application of machine learning to classification of volcanic deformation in routinely generated InSAR data. *Journal of Geophysical Research: Solid Earth, 123*(8), 6592–6606.

Anantrasirichai, N., Biggs, J., Albino, F., & Bull, D. (2019a). The application of convolutional neural networks to detect slow, sustained deformation in InSAR time series. *Geophysical Research Letters, 46*(21), 11850–11858. https://doi.org/10.1029/2019GL084993.

Anantrasirichai, N., Biggs, J., Albino, F., & Bull, D. (2019b). A deep learning approach to detecting volcano deformation from satellite imagery using synthetic datasets. *Remote Sensing of Environment, 230*, 111179. https://doi.org/10.1016/J.RSE.2019.04.032.

Beckouche, S., & Ma, J. (2014). Simultaneous dictionary learning and denoising for seismic data. *Geophysics, 79*(3), A27–A31. https://doi.org/10.1190/GEO2013-0382.1.

Beeler, N. (2004). Review of the physical basis of laboratory-derived relations for brittle failure and their implications for earthquake occurrence and earthquake nucleation. *Pure and Applied Geophysics, 161*, 1853–1876. https://doi.org/10.1007/s00024-004-2536-z.

Beeler, N. (2006). Inferring earthquake source properties from laboratory observations and the scope of lab contributions to source physics. *Washington DC American Geophysical Union Geophysical Monograph Series, 170*, 99–117. https://doi.org/10.1029/170GM12.

Benoit, A., Pinel-Puysségur, B., Jolivet, R., & Lasserre, C. (2020). CorPhU: An algorithm based on phase closure for the correction of unwrapping errors in SAR interferometry. *Geophysical Journal International, 221*(3), 1959–1970. https://doi.org/10.1093/gji/ggaa120.

Bergen, K. J., Johnson, P. A., de Hoop, M. V., & Beroza, G. C. (2019). Machine learning for data-driven discovery in solid Earth geoscience. *Science, 363*(6433), eaau0323. https://doi.org/10.1126/science.aau0323.

Bergen, K. J., Johnson, P. A., Hoop, M. V. d., & Beroza, G. C. (2019). Machine learning for data-driven discovery in solid Earth geoscience. *Science, 363*(6433), 1–10. https://doi.org/10.1126/science.aau0323.

Bergmeir, C., & Benítez, J. M. (2012). On the use of cross-validation for time series predictor evaluation. *Information Sciences, 191*, 192–213. https://doi.org/10.1016/j.ins.2011.12.028.

Beroza, G. C., & Ellsworth, W. L. (1996). Properties of the seismic nucleation phase. *Tectonophysics, 261*(1–3 Spec. Iss.), 209–227. https://doi.org/10.1016/0040-1951(96)00067-4.

Beyreuther, M., Hammer, C., Wassermann, J., Ohrnberger, M., & Megies, T. (2012). Constructing a hidden Markov model based earthquake detector: Application to induced seismicity. *Geophysical Journal International, 189*(1), 602–610. https://doi.org/10.1111/j.1365-246X.2012.05361.x.

Beyreuther, M., & Wassermann, J. (2008). Continuous earthquake detection and classification using discrete hidden Markov models. *Geophysical Journal International, 175*(3), 1055–1066. https://doi.org/10.1111/j.1365-246X.2008.03921.x.

Bingham, E., Chen, J. P., Jankowiak, M., Obermeyer, F., Pradhan, N., Karaletsos, T., ... Goodman, N. D. (2019). Pyro: Deep universal probabilistic programming. *Journal of Machine Learning Research, 20*(1), 973–978.

Bishop, C. M. (2006). *Pattern recognition and machine learning.* Springer.

Bolton, D. C., Shokouhi, P., Rouet-Leduc, B., Hulbert, C., Riviére, J., Marone, C., & Johnson, P. A. (2019). Characterizing acoustic signals and searching for precursors during the laboratory seismic cycle using unsupervised machine learning. *Seismological Research Letters*, *90*(3), 1088–1098. https://doi.org/10.1785/0220180367.

Böse, M., Wenzel, F., & Erdik, M. (2008). PreSEIS: a neural network-based approach to earthquake early warning for finite faults. *Bulletin of the Seismological Society of America*, *98*(1), 366–382. https://doi.org/10.1785/0120070002.

Bouchon, M., Durand, V., Marsan, D., Karabulut, H., & Schmittbuhl, J. (2013). The long precursory phase of most large interplate earthquakes. *Nature Geoscience*, *6*(4), 299–302. https://doi.org/10.1038/ngeo1770.

Brace, W. F., & Byerlee, J. D. (1966). Stick-slip as a mechanism for earthquakes. *Science*, *153*(3739), 990–992. https://doi.org/10.1126/science.153.3739.990.

Breiman, L. (1984). *Classification and regression trees*. New York: Routledge. https://doi.org/10.1201/9781315139470.

Breiman, L. (2001). Random forests. *Machine Learning*, *45*(1), 5–32.

Brodsky, E. E., Mori, J. J., Anderson, L., Chester, F. M., Conin, M., Dunham, E. M., … Yang, T. (2020). The state of stress on the fault before, during, and after a major earthquake. *Annual Review of Earth and Planetary Sciences*, *48*(1), 40–74. https://doi.org/10.1146/annurev-earth-053018-060507.

Cavalié, O., Pathier, E., Radiguet, M., Vergnolle, M., Cotte, N., Walpersdorf, A., … Cotton, F. (2013). Slow slip event in the Mexican subduction zone: Evidence of shallower slip in the Guerrero seismic gap for the 2006 event revealed by the joint inversion of InSAR and GPS data. *Earth and Planetary Science Letters*, *367*, 52–60. https://doi.org/10.1016/j.epsl.2013.02.020.

Chen, T., & Guestrin, C. (2016). Xgboost: A scalable tree boosting system. In *Proceedings of the 22nd ACM SIGKDD international conference on knowledge discovery and data mining*. pp. 785–794.

Chen, Y. (2017). Automatic microseismic event picking via unsupervised machine learning. *Geophysical Journal International*, *212*(1), 88–9102. https://doi.org/10.1093/gji/ggx420.

Collettini, C., Niemeijer, A., Viti, C., & Marone, C. (2009). Fault zone fabric and fault weakness. *Nature*, *462*, 907–910. https://doi.org/10.1038/nature08585.

Corbi, F., Bedford, J., Sandri, L., Funiciello, F., Gualandi, A., & Rosenau, M. (2020). Predicting imminence of analog megathrust earthquakes with machine learning: Implications for monitoring subduction zones. *Geophysical Research Letters*, *47*(7), e2019GL086615.

Corbi, F., Funiciello, F., Moroni, M., Van Dinther, Y., Mai, P. M., Dalguer, L. A., & Faccenna, C. (2013). The seismic cycle at subduction thrusts: 1. Insights from laboratory models. *Journal of Geophysical Research: Solid Earth*, *118*(4), 1483–1501.

Corbi, F., Sandri, L., Bedford, J., Funiciello, F., Brizzi, S., Rosenau, M., & Lallemand, S. (2019). Machine learning can predict the timing and size of analog earthquakes. *Geophysical Research Letters*, *46*(3), 1303–1311.

Dai, H., & MacBeth, C. (1995). Automatic picking of seismic arrivals in local earthquake data using an artificial neural network. *Geophysical Journal International*, *120*(3), 758–774. https://doi.org/10.1111/j.1365-246X.1995.tb01851.x.

Dai, H., & MacBeth, C. (1997). The application of back-propagation neural network to automatic picking seismic arrivals from single-component recordings. *Journal of Geophysical Research: Solid Earth*, *102*(B7), 15105–15113. https://doi.org/10.1029/97jb00625.

Daniels, K., Kollmer, J., & Puckett, J. (2016). Photoelastic force measurements in granular materials. *Review of Scientific Instruments, 88*, 051808-01–051808-13. https://doi.org/10.1063/1.4983049.

Daub, E., Shelly, D., Guyer, R., & Johnson, P. (2011). Brittle and ductile friction and the physics of tectonic tremor. *Geophysical Research Letters*, *38*, 1–4. https://doi.org/10.1029/2011GL046866.

Dieterich, J. H. (1978). Preseismic fault slip and earthquake prediction. *Journal of Geophysical Research*, *83*(B8), 3940. https://doi.org/10.1029/jb083ib08p03940.

Dieterich, J. H. (1979). Modeling of rock friction: 1. Experimental results and constitutive equations. *Journal of Geophysical Research: Solid Earth*, *84*(B5), 2161–2168.

Dorostkar, O., & Carmeliet, J. (2018). Potential energy as metric for understanding stickslip dynamics in sheared granular fault gouge: A coupled CFDDEM study. *Rock Mechanics and Rock Engineering*, *51*(10), 3281–3294. https://doi.org/10.1007/s00603-018-1457-6.

Dorostkar, O., Guyer, R. A., Johnson, P. A., Marone, C., & Carmeliet, J. (2017). On the role of fluids in stick-slip dynamics of saturated granular fault gouge using a coupled computational fluid dynamics-discrete element approach. *Journal of Geophysical Research: Solid Earth*, *122*(5), 3689–3700. https://doi.org/10.1002/2017JB014099.

Dorostkar, O., Guyer, R. A., Johnson, P. A., Marone, C., & Carmeliet, J. (2018). Cohesion-induced stabilization in stick-slip dynamics of weakly wet, sheared granular fault gouge. *Journal of Geophysical Research: Solid Earth*, *123*, 2115–2126. https://doi.org/10.1002/2017JB015171.

Dowla, F. U., Taylor, S. R., & Anderson, R. W. (1990). Seismic discrimination with artificial neural networks: Preliminary results with regional spectral data. *Bulletin—Seismological Society of America*, *80*(5), 1346–1373.

Dysart, P. S., & Pulli, J. J. (1991). Regional seismic event classification at the NORESS array: Seismological measurements and the use of trained neural networks. *Bulletin—Seismological Society of America*, *80*(6 B), 1910–1933.

Freund, Y., & Schapire, R. (1999). Large margin classification using the perceptron algorithm. *Machine Learning*, *37*, 277–296. https://doi.org/10.1023/A:1007662407062.

Freund, Y., Schapire, R., & Abe, N. (1999). A short introduction to boosting. *Journal-Japanese Society For Artificial Intelligence*, *14*(771–780), 1612.

Friedman, J. H. (2001). Greedy function approximation: A gradient boosting machine. *Annals of Statistics*, *29*, 1189–1232.

Gao, K., Euser, B. J., Rougier, E., Guyer, R. A., Lei, Z., Knight, E. E., … Johnson, P. A. (2018). Modeling of stick-slip behavior in sheared granular fault gouge using the combined finite-discrete element method. *Journal of Geophysical Research: Solid Earth*, *123*(7), 5774–5792. https://doi.org/10.1029/2018JB015668.

Gao, K., Guyer, R., Rougier, E., Ren, C. X., & Johnson, P. A. (2019). From stress chains to acoustic emission. *Physical Review Letters*, *123*(4), 048003. https://doi.org/10.1103/PhysRevLett.123.048003.

Gentili, S., & Michelini, A. (2006). Automatic picking of P and S phases using a neural tree. *Journal of Seismology*, *10*(1), 39–63. https://doi.org/10.1007/s10950-006-2296-6.

Goodfellow, I., Bengio, Y., & Courville, A. (2016). *Deep learning*. MIT Press.

Goodfellow, I. J., Shlens, J., & Szegedy, C. (2015). *Explaining and harnessing adversarial examples*. ICLR.

Goodfellow, S. D., & Young, R. P. (2014). A laboratory acoustic emission experiment under in situ conditions. *Geophysical Research Letters*, *41*(10), 3422–3430. https://doi.org/10.1002/2014GL059965.

Gu, C., Marzouk, Y. M., & Toksoz, M. N. (2019). Bayesian deep learning and uncertainty quantification applied to induced seismicity locations in the Groningen gas field in the Netherlands: What do we need for safe AI? In *SEG international exposition and annual meeting*. San Antonio, Texas, USA: Society of Exploration Geophysicists, pp. 6.

Hara, S., Fukahata, Y., & Iio, Y. (2019). P-wave first-motion polarity determination of waveform data in western Japan using deep learning. *Earth, Planets and Space*, *71*(1), 1–11. https://doi.org/10.1186/s40623-019-1111-x.

Hilley, G. E., Bürgmann, R., Ferretti, A., Novali, F., & Rocca, F. (2004). Dynamics of slow-moving landslides from permanent scatterer analysis. *Science*, *304*(5679), 1952–1955. https://doi.org/10.1126/science.1098821.

Hincks, T., Aspinall, W., Cooke, R., & Gernon, T. (2018). Oklahoma's induced seismicity strongly linked to wastewater injection depth. *Science, 359*(6381), 1251–1255. https://doi.org/10.1126/science.aap7911.

Holtzman, B. K., Paté, A., Paisley, J., Waldhauser, F., & Repetto, D. (2018). Machine learning reveals cyclic changes in seismic source spectra in Geysers geothermal field. *Science Advances, 4*(5), eaao2929. https://doi.org/10.1126/sciadv.aao2929.

Huber, P. J. (1964). Robust estimation of a location parameter. In *Breakthroughs in Statistics: Methodology and Distribution.* New York: Springer, pp. 492–518.

Hulbert, C., Rouet-Leduc, B., Johnson, P. A., Ren, C. X., Riviére, J., Bolton, D. C., & Marone, C. (2019). Similarity of fast and slow earthquakes illuminated by machine learning. *Nature Geoscience, 12*(1), 69–74. https://doi.org/10.1038/s41561-018-0272-8.

Hulbert, C., Rouet-Leduc, B., Jolivet, R., & Johnson, P. A. (2020). An exponential build-up in seismic energy suggests a months-long nucleation of slow slip in Cascadia. Nature Communications. *Nature Communications, 11*(1), 1–8.

Ito, Y., Hino, R., Kido, M., Fujimoto, H., Osada, Y.,Inazu, D. ... Others. (2013). Episodic slow slip events in the Japan subduction zone before the 2011 Tohoku-Oki earthquake. *Tectonophysics, 600,* 14–26.

Jasperson, H., Bolton, C., Johnson, P., Marone, C., & de Hoop, M. V. (2019). Unsupervised classification of acoustic emissions from catalogs and fault time-to-failure prediction. *arXiv preprint arXiv:1912.06087.*

Johnson, P. A., Carmeliet, J., Savage, H. M., Scuderi, M., Carpenter, B. M., Guyer, R. A., ... Marone, C. (2016). Dynamically triggered slip leading to sustained fault gouge weakening under laboratory shear conditions. *Geophysical Research Letters, 43*(4), 1559–1565. https://doi.org/10.1002/2015GL067056.

Johnson, P. A., Ferdowsi, B., Kaproth, B. M., Scuderi, M., Griffa, M., Carmeliet, J., ... Marone, C. (2013a). Acoustic emission and microslip precursors to stick-slip failure in sheared granular material. *Geophysical Research Letters, 40*(21), 5627–5631.

Johnson, P. A., Ferdowsi, B., Kaproth, B. M., Scuderi, M., Griffa, M., Carmeliet, J., ... Marone, C. (2013b). Acoustic emission and microslip precursors to stick-slip failure in sheared granular material. *Geophysical Research Letters, 40*(21), 5627–5631.

Jolivet, R., Candela, T., Lasserre, C., Renard, F., Klinger, Y., & Doin, M. P. (2015). The burst-like behavior of aseismic slip on a rough fault: The creeping section of the Haiyuan fault, China. *Bulletin of the Seismological Society of America, 105*(1), 480–488. https://doi.org/10.1785/0120140237.

Jolivet, R., Lasserre, C., Doin, M. P., Guillaso, S., Peltzer, G., Dailu, R., ... Xu, X. (2012). Shallow creep on the Haiyuan fault (Gansu, China) revealed by SAR interferometry. *Journal of Geophysical Research: Solid Earth, 117*(6), B06401. https://doi.org/10.1029/2011JB008732.

Jolivet, R., Lasserre, C., Doin, M. P., Peltzer, G., Avouac, J. P., Sun, J., & Dailu, R. (2013). Spatio-temporal evolution of aseismic slip along the Haiyuan fault, China: Implications for fault frictional properties. *Earth and Planetary Science Letters, 377-378,* 23–33. https://doi.org/10.1016/j.epsl.2013.07.020.

Kansky, K., Silver, T., Mély, D. A., Eldawy, M., Lázaro-Gredilla, M., Lou, X., ... George, D. (2017). Schema networks: Zero-shot transfer with a generative causal model of intuitive physics. *arXiv preprint arXiv:170604317.*

Kato, A., Obara, K., Igarashi, T., Tsuruoka, H., Nakagawa, S., & Hirata, N. (2012). Propagation of slow slip leading up to the 2011 Mw 9.0 Tohoku-Oki earthquake. *Science, 335*(6069), 705–708. https://doi.org/10.1126/science.1215141.

Käufl, P., Valentine, A. P., O'Toole, T. B., & Trampert, J. (2014). A framework for fast probabilistic centroid-moment-tensor determination-inversion of regional static displacement measurements. *Geophysical Journal International, 196*(3), 1676–1693. https://doi.org/10.1093/gji/ggt473.

Kenigsberg, A. R., Riviére, J., Marone, C., & Saffer, D. M. (2019). The effects of shear strain, fabric, and porosity evolution on elastic and mechanical properties of clay-rich fault gouge. *Journal of Geophysical Research: Solid Earth, 124*(11), 10968–10982. https://doi.org/10.1029/2019JB017944.

Kong, Q., Allen, R. M., Schreier, L., & Kwon, Y. W. (2016). Earth sciences: MyShake: A smartphone seismic network for earthquake early warning and beyond. *Science Advances, 2*(2), e1501055. https://doi.org/10.1126/sciadv.1501055.

Kong, Q., Trugman, D. T., Ross, Z. E., Bianco, M. J., Meade, B. J., & Gerstoft, P. (2019). Machine learning in seismology: Turning data into insights. *Seismological Research Letters, 90*(1), 3–14. https://doi.org/10.1785/0220180259.

Kriegerowski, M., Petersen, G. M., Vasyura-Bathke, H., & Ohrnberger, M. (2019). A deep convolutional neural network for localization of clustered earthquakes based on multistation full waveforms. *Seismological Research Letters, 90*(2 A), 510–516. https://doi.org/10.1785/0220180320.

Krizhevsky, A., Sutskever, I., & Hinton, G. E. (2012). ImageNet classification with deep convolutional neural networks. In *Advances in neural information processing systems*. Curran Associates, Inc., pp. 1097–1105.

LANL Earthquake Prediction | Kaggle (n.d.). https://www.kaggle.com/c/LANL-Earthquake-Prediction (2020-05-26).

Latour, S., Schubnel, A., Nielsen, S., Madariaga, R., & Vinciguerra, S. (2013). Characterization of nucleation during laboratory earthquakes. *Geophysical Research Letters, 40*(19), 5064–5069. https://doi.org/10.1002/grl.50974. https://onlinelibrary.wiley.com/doi/abs/10.1002/grl.50974..

Leeman, J. R., Saffer, D. M., Scuderi, M. M., & Marone, C. (2006a). Laboratory observations of slow earthquakes and the spectrum of tectonic fault slip modes. *Nature Communications, 7,* 11104.

Leeman, J. R., Saffer, D. M., Scuderi, M. M., & Marone, C. (2016b). Laboratory observations of slow earthquakes and the spectrum of tectonic fault slip modes. *Nature Communications, 7*(1), 1–6. https://doi.org/10.1038/ncomms11104.

Li, Z., Meier, M. A., Hauksson, E., Zhan, Z., & Andrews, J. (2018). Machine learning seismic wave discrimination: Application to earthquake early warning. *Geophysical Research Letters, 45*(10), 4773–4779. https://doi.org/10.1029/2018GL077870.

Lindsey, N. J., Martin, E. R., Dreger, D. S., Freifeld, B., Cole, S., James, S. R., ... Ajo-Franklin, J. B. (2017). Fiber-optic network observations of earthquake wavefields. *Geophysical Research Letters, 44*(23), 11,792–11,799. https://doi.org/10.1002/2017GL075722.

Logan, J. (1975). Friction in rocks. *Reviews of Geophysics and Space Physics, 13,* 358–361. https://doi.org/10.1029/RG013i003p00358.

Lohman, R. (2019). *100 years of InSAR—Setting the stage for a new era of imaging geodesy* (No. G54A-06, AGU Fall Meeting).

Lomax, A., Michelini, A., & Jozinović, D. (2019). An investigation of rapid earthquake characterization using single-station waveforms and a convolutional neural network. *Seismological Research Letters, 90*(2 A), 517–529. https://doi.org/10.1785/0220180311.

Lubbers, N., Bolton, D. C., Mohd-Yusof, J., Marone, C., Barros, K., & Johnson, P. A. (2018). Earthquake catalog-based machine learning identification of laboratory fault states and the effects of magnitude of completeness. *Geophysical Research Letters, 45*(24), 13269–13276. https://doi.org/10.1029/2018GL079712.

Madureira, G., & Ruano, A. E. (2009). A neural network seismic detector. In *Vol. 2. IFAC proceedings volumes (IFAC-PapersOnline), January*. Elsevier. https//doi.org/10.3182/20090121-3-TR-3005.00054.

Majmudar, T., & Behringer, R. (2005). Contact force measurements and stress-induced anisotropy in granular materials. *Nature, 435,* 1079–1082. https://doi.org/10.1038/nature03805.

Marcus, G. (2018). Deep learning: A critical appraisal. *arXiv preprint arXiv:180100631*.

Marone, C. (1998a). Laboratory-derived friction laws and their application to seismic faulting. *Annual Review of Earth and Planetary Sciences, 26*(1), 643–696.

Marone, C. (1998b). Laboratory-derived friction laws and their application to seismic faulting. *Annual Review of Earth and Planetary Sciences, 26*(1), 643–696.

Massonnet, D., Briole, P., & Arnaud, A. (1995). Deflation of Mount Etna monitored by spaceborne radar interferometry. *Nature, 375*(6532), 567–570. https://doi.org/10.1038/375567a0.

Massonnet, D., Rossi, M., Carmona, C., Adragna, F., Peltzer, G., Feigl, K., & Rabaute, T. (1993). The displacement field of the Landers earthquake mapped by radar interferometry. *Nature, 364*(6433), 138–142. https://doi.org/10.1038/364138a0.

Matheron, G. (1962). *Traité de géostatistique appliquée. 1 (1962).* (Vol. 1). Editions Technip.

McBrearty, I. W., Delorey, A. A., & Johnson, P. A. (2019). Pairwise association of seismic arrivals with convolutional neural networks. *Seismological Research Letters, 90*(2A), 503–509. https://doi.org/10.1785/0220180326.

McBrearty, I. W., Gomberg, J., Delorey, A. A., & Johnson, P. A. (2019). Earthquake arrival association with backprojection and graph. *Bulletin of the Seismological Society of America, 109*(6), 2510–2531. https://doi.org/10.1785/0120190081.

McCormack, M. D., Zaucha, D. E., & Dushek, D. W. (1993). First-break refraction event picking and seismic data trace editing using neural networks. *Geophysics, 58*(1), 67–78. https://doi.org/10.1190/1.1443352.

McLaskey, G. C. (2019). Earthquake initiation from laboratory observations and implications for foreshocks. *Journal of Geophysical Research: Solid Earth, 124*(12), 12882–12904. https://doi.org/10.1029/2019JB018363.

Meier, M. A., Heaton, T., & Clinton, J. (2015). The Gutenberg algorithm: Evolutionary Bayesian magnitude estimates for earthquake early warning with a filter bank. *Bulletin of the Seismological Society of America, 105*(5), 2774–2786. https://doi.org/10.1785/0120150098.

Mignan, A., & Broccardo, M. (2020). Neural network applications in earthquake prediction (1994–2019): Meta-analytic and statistical insights on their limitations. *Seismological Research Letters, 91*, 2330–2342. https://doi.org/10.1785/0220200021.

Mitchell, T. M., et al. (1997). Machine learning. *Machine learning.* New York: McGraw-Hill.

Mousavi, S. M., & Beroza, G. C (2020). Bayesian-deep-learning estimation of earthquake location from single-station observations. In *IEEE Transactions on Geoscience and Remote Sensing*. https://doi.org/10.1109/TGRS.2020.2988770.

Mousavi, S. M., Horton, S. P., Langston, C. A., & Samei, B. (2016). Seismic features and automatic discrimination of deep and shallow induced-microearthquakes using neural network and logistic regression. *Geophysical Journal International, 207*(1), 29–46. https://doi.org/10.1093/gji/ggw258.

Mousavi, S. M., Zhu, W., Sheng, Y., & Beroza, G. C. (2019a). CRED: A deep residual network of convolutional and recurrent units for earthquake signal detection. *Scientific Reports, 9*(1), 1–14. https://doi.org/10.1038/s41598-019-45748-1.

Murat, M. E., & Rudman, A. J. (1992). Automated first arrival picking: A neural network approach. *Geophysical Prospecting, 40*(6), 587–604. https://doi.org/10.1111/j.1365-2478.1992.tb00543.x.

Nakano, M., Sugiyama, D., Hori, T., Kuwatani, T., & Tsuboi, S. (2019). Discrimination of seismic signals from earthquakes and tectonic tremor by applying a convolutional neural network to running spectral images. *Seismological Research Letters, 90*(2 A), 530–538. https://doi.org/10.1785/0220180279.

Niemeijer, A., Marone, C., & Elsworth, D. (2010). Frictional strength and strain weakening in simulated fault gouge: Competition between geometrical weakening and chemical strengthening. *Journal of Geophysical Research: Solid Earth, 115*(B10), B10207.

Ochoa, L. H., Niño, L. F., & Vargas, C. A. (2018). Fast magnitude determination using a single seismological station record implementing machine learning techniques. *Geodesy and Geodynamics, 9*(1), 34–41. https://doi.org/10.1016/j.geog.2017.03.010.

Ogata, Y. (1988). Statistical models for earthquake occurrences and residual analysis for point processes. *Journal of the American Statistical Association, 83*(401), 9–27. https://doi.org/10.1080/01621459.1988.10478560.

Parker, T., Shatalin, S., & Farhadiroushan, M. (2014). Distributed acoustic sensing—A new tool for seismic applications. *First Break, 32*(2). https://doi.org/10.3997/1365-2397.2013034.

Peltzer, G., Crampé, F., Hensley, S., & Rosen, P. (2001). Transient strain accumulation and fault interaction in the Eastern California shear zone. *Geology, 29*(11), 975–978. https://doi.org/10.1130/0091-7613(2001)029<0975:TSAAFI>2.0.CO;2.

Peltzer, G., Crampé, F., & King, G. (1999). Evidence of nonlinear elasticity of the crust from the Mw7.6 Manyi (Tibet) earthquake. *Science, 286*(5438), 272–276. https://doi.org/10.1126/science.286.5438.272.

Perol, T., Gharbi, M., & Denolle, M. (2018). Convolutional neural network for earthquake detection and location. *Science Advances, 4*(2), e1700578. https://doi.org/10.1126/sciadv.1700578.

Ren, C. X., Dorostkar, O., Rouet-Leduc, B., Hulbert, C., Strebel, D., Guyer, R. A., … Carmeliet, J. (2019). Machine learning reveals the state of intermittent frictional dynamics in a sheared granular fault. *Geophysical Research Letters, 46*(13), 7395–7403.

Reynen, A., & Audet, P. (2017). Supervised machine learning on a network scale: Application to seismic event classification and detection. *Geophysical Journal International, 210*(3), 1394–1409. https://doi.org/10.1093/gji/ggx238.

Riahi, N., & Gerstoft, P. (2017). Using graph clustering to locate sources within a dense sensor array. *Signal Processing, 132*, 110–120. https://doi.org/10.1016/j.sigpro.2016.10.001.

Rice, J., Lapusta, N., & Kunnath, R. (2001). Rate and state dependent friction and the stability of sliding between elastically deformable solids. *Journal of the Mechanics and Physics of Solids, 49*, 1865–1898. https://doi.org/10.1016/S0022-5096(01)00042-4.

Richards-Dinger, K., & Dieterich, J. (2012). RSQSim earthquake simulator. *Seismological Research Letters, 83*, 983–990. https://doi.org/10.1785/0220120105.

Rietbrock, A., Ryder, I., Hayes, G., Haberland, C., Comte, D., Roecker, S., & Lyon-Caen, H. (2012). Aftershock seismicity of the 2010 Maule Mw=8.8, Chile, earthquake: Correlation between co-seismic slip models and aftershock distribution? *Geophysical Research Letters, 39*(8), L08310. https://doi.org/10.1029/2012GL051308.

Riggelsen, C., & Ohrnberger, M. (2014). A machine learning approach for improving the detection capabilities at 3C seismic stations. *Pure and Applied Geophysics, 171*(3–5), 395–411. https://doi.org/10.1007/s00024-012-0592-3.

Rogers, G., & Dragert, H. (2003). Episodic tremor and slip on the Cascadia subduction zone: The chatter of silent slip. *Science, 300*(5627), 1942–1943.

Rosakis, A. J., Kanamori, H., & Xia, K. (2006). Laboratory earthquakes. In A. Carpinteri, Y.-W. Mai, & R. O. Ritchie (Eds.), *Advances in fracture research* (pp 211–218). Springer.

Rosenau, M., Corbi, F., & Dominguez, S. (2017). Analogue earthquakes and seismic cycles: Experimental modelling across timescales. *Solid Earth, 8*(3), 597–635. https://doi.org/10.5194/se-8-597-2017.

Ross, Z. E., Meier, M.-A., & Hauksson, E. (2018). P wave arrival picking and first-motion polarity determination with deep learning. *Journal of Geophysical Research: Solid Earth, 123*(6), 5120–5129. https://doi.org/10.1029/2017JB015251.

Ross, Z. E., Meier, M.-A., Hauksson, E., & Heaton, T. H. (2018a). Generalized seismic phase detection with deep learning. *Bulletin of the Seismological Society of America, 108*(5A), 2894–2901. https://doi.org/10.1785/0120180080.

Ross, Z. E., Meier, M. A., Hauksson, E., & Heaton, T. H. (2018b). Generalized seismic phase detection with deep learning. *Bulletin of the Seismological Society of America*, *108*(5), 2894–2901. https://doi.org/10.1785/0120180080.

Ross, Z. E., Trugman, D. T., Hauksson, E., & Shearer, P. M. (2019). Searching for hidden earthquakes in Southern California. *Science*, *364*(6442), 767–771. https://doi.org/10.1126/science.aaw6888.

Ross, Z. E., Yue, Y., Meier, M. A., Hauksson, E., & Heaton, T. H. (2019). PhaseLink: A deep learning approach to seismic phase association. *Journal of Geophysical Research: Solid Earth*, *124*(1), 856–869. https://doi.org/10.1029/2018JB016674.

Rouet-Leduc, B., Dalaison, M., Hulbert, C., Johnson, P., & Jolivet, R. (2020). A deep Learning approach for detecting transient deformation in InSAR.

Rouet-Leduc, B., Hulbert, C., Bolton, D. C., Ren, C. X., Riviere, J., Marone, C., … Johnson, P. A. (2018). Estimating fault friction from seismic signals in the laboratory. *Geophysical Research Letters*, *45*(3), 1321–1329. https://doi.org/10.1002/2017GL076708.

Rouet-Leduc, B., Hulbert, C., & Johnson, P. A. (2019). Continuous chatter of the Cascadia subduction zone revealed by machine learning. *Nature Geoscience*, *12*(1), 908–1752. https://doi.org/10.1038/s41561-018-0274-6.

Rouet-Leduc, B., Hulbert, C., Lubbers, N., Barros, K., Humphreys, C. J., & Johnson, P. A. (2017). Machine learning predicts laboratory earthquakes. *Geophysical Research Letters*, *44*, 9276–9282. https://doi.org/10.1002/2017GL074677.

Rouet-Leduc, B., Hulbert, C., McBrearty, I. W., & Johnson, P. A. (2020a). Probing slow earthquakes with deep learning. *Geophysical Research Letters*, *47*(4), e2019GL085870. https://doi.org/10.1029/2019GL085870.

Rouet-Leduc, B., Hulbert, C., McBrearty, I. W., & Johnson, P. A. (2020b). Probing slow earthquakes with deep learning. *Geophysical Research Letters*, *47*(4). e2019GL085870. https://doi.org/10.1029/2019GL085870.

Rousset, B., Jolivet, R., Simons, M., Lasserre, C., Riel, B., Milillo, P., … Renard, F. (2016). An aseismic slip transient on the North Anatolian Fault. *Geophysical Research Letters*, *43*(7), 3254–3262. https://doi.org/10.1002/2016GL068250.

Ruano, A. E., Madureira, G., Barros, O., Khosravani, H. R., Ruano, M. G., & Ferreira, P. M. (2014). Seismic detection using support vector machines. *Neurocomputing*, *135*, 273–283. https://doi.org/10.1016/j.neucom.2013.12.020.

Ruina, A. (1983). Slip instability and state variable friction laws. *Journal of Geophysical Research: Solid Earth*, *88*(B12), 10359–10370. https://doi.org/10.1029/JB088iB12p10359.

Ruiz, S., Metois, M., Fuenzalida, A., Ruiz, J., Leyton, F., Grandin, R., … Campos, J. (2014). Intense foreshocks and a slow slip event preceded the 2014 Iquique Mw 8.1 earthquake. *Science*, *345*(6201), 1165–1169. https://doi.org/10.1126/science.1256074.

Ryan, K. L., Riviére, J., & Marone, C. (2018). The role of shear stress in fault healing and frictional aging. *Journal of Geophysical Research: Solid Earth*, *123*(12), 10,479–10,495. https://doi.org/10.1029/2018JB016296.

Santosa, F., & Symes, W. W. (1986). Linear inversion of band-limited reflection seismograms. *SIAM Journal on Scientific and Statistical Computing*, *7*(4), 1307–1330.

Schmidhuber, J. (1992). Learning factorial codes by predictability minimization. *Neural Computation*, *4*(6), 863–879. https://doi.org/10.1162/neco.1992.4.6.863.

Scholz, C. H. (2019). *The mechanics of earthquakes and faulting* (3rd ed.). Cambridge University Press. https://doi.org/10.1017/9781316681473.

Scuderi, M. M., Marone, C., Tinti, E., Stefano, G. D., & Collettini, C. (2016). Precursory changes in seismic velocity for the spectrum of earthquake failure modes. *Nature Geoscience*, *9*(9), 695–700. https://doi.org/10.1038/ngeo2775.

Seiffert, C., Khoshgoftaar, T. M., Van Hulse, J., & Napolitano, A. (2008). RUSBoost: Improving classification performance when training data is skewed. In *2008 19th international conference on pattern recognition*. IEEE, pp. 1–4.

Sick, B., Guggenmos, M., & Joswig, M. (2015). Chances and limits of single-station seismic event clustering by unsupervised pattern recognition. *Geophysical Journal International*, *201*(3), 1801–1813. https://doi.org/10.1093/gji/ggv126.

Simons, M., Fialko, Y., & Rivera, L. (2002). Coseismic deformation from the 1999 Mw 7.1 hector mine, California, earthquake as inferred from InSAR and GPS observations. *Bulletin of the Seismological Society of America*, *92*(4), 1390–1402. https://doi.org/10.1785/0120000933.

Sun, H., & Demanet, L. (2020). Extrapolated full waveform inversion with deep learning. *Geophysics*, *85*(3), R275–R288. https://doi.org/10.1190/geo2019-0195.1.

Thomas, M., & Bhat, H. (2018). Dynamic evolution of off-fault medium during an earthquake: A micromechanics based model. *Geophysical Journal International*, *214*, 1267–1280. https://doi.org/10.1093/GJI/GGY129.

Tibshirani, R. (1996). Regression shrinkage and selection via the lasso. *Journal of the Royal Statistical Society: Series B (Methodological)*, *58*(1), 267–288.

Tiira, T. (1999). Detecting teleseismic events using artificial neural networks. *Computers and Geosciences*, *25*(8), 929–938. https://doi.org/10.1016/S0098-3004(99)00056-4.

Trippetta, F., Petricca, P., Billi, A., Collettini, C., Cuffaro, M., Lombardi, A., …Doglioni, C. (2019). From mapped faults to fault-length earthquake magnitude (FLEM): A test on Italy with methodological implications. *Solid Earth*, *10*, 1555–1579. https://doi.org/10.5194/se-10-1555-2019.

Trugman, D. T., & Ross, Z. E. (2019). Pervasive foreshock activity across Southern California. *Geophysical Research Letters*, *46*(15), 8772–8781. https://doi.org/10.1029/2019GL083725.

Trugman, D. T., Ross, Z. E., & Johnson, P. A. (2020). Imaging stress and faulting complexity through earthquake waveform similarity. *Geophysical Research Letters*, *47*(1), e2019GL085888. https://doi.org/10.1029/2019GL085888.

Trugman, D. T., & Shearer, P. M. (2017). GrowClust: A Hierarchical clustering algorithm for relative earthquake relocation, with application to the Spanish Springs and Sheldon, Nevada, earthquake sequences. *Seismological Research Letters*, *88*(2), 379–391. https://doi.org/10.1785/0220160188.

Valentine, A. P., & Trampert, J. (2012). Data space reduction, quality assessment and searching of seismograms: Autoencoder networks for waveform data. *Geophysical Journal International*, *189*(2), 1183–1202. https://doi.org/10.1111/j.1365-246X.2012.05429.x.

Walsh, F. R., & Zoback, M. D. (2015). Oklahoma's recent earthquakes and saltwater disposal. *Science Advances*, *1*(5), e1500195. https://doi.org/10.1126/sciadv.1500195.

Wang, J., & Teng, T.-L. (1995). Artificial neural network-based seismic detector. *Bulletin— Seismological Society of America*, *85*(1), 308–319. https://doi.org/10.1016/0148-9062(96)86904-x.

Wang, J., & Teng, T.-L. (1997). Identification and picking of S phase using an artificial neural network. *Bulletin of the Seismological Society of America*, *87*(5), 1140–1149.

Wang, K., Li, Y., Kemao, Q., Di, J., & Zhao, J. (2019). One-step robust deep learning phase unwrapping. *Optics Express*, *27*(10), 15100. https://doi.org/10.1364/OE.27.015100.

Wiszniowski, J., Plesiewicz, B. M., & Trojanowski, J. (2014). Application of real time recurrent neural network for detection of small natural earthquakes in Poland. *Acta Geophysica*, *62*(3), 469–485. https://doi.org/10.2478/s11600-013-0140-2.

Wright, T. J., Parsons, B., England, P. C., & Fielding, E. J. (2004). InSAR observations of low slip rates on the major faults of western Tibet. *Science*, *305*(5681), 236–239. https://doi.org/10.1126/science.1096388.

Xia, K., Rosakis, A. J., & Kanamori, H. (2005). Supershear and subrayleigh to supershear transition observed in laboratory earthquake experiments. *Experimental Techniques*, *29*, 63–66.

Yoon, C. E., O'Reilly, O., Bergen, K. J., Beroza, G. C., O'Reilly, O., Bergen, K. J., & Beroza, G. C. (2015). Earthquake detection through computationally efficient similarity search. *Science Advances*, *1*(11), e1501057. https://doi.org/10.1126/sciadv.1501057.

Zhang, X., Zhang, J., Yuan, C., Liu, S., Chen, Z., & Li, W. (2020). Locating induced earthquakes with a network of seismic stations in Oklahoma via a deep learning method. *Scientific Reports*, *10*(1), 1–12. https://doi.org/10.1038/s41598-020-58908-5.

Zhao, Y., & Takano, K. (1999). An artificial neural network approach for broadband seismic phase picking. *Bulletin of the Seismological Society of America*, *89*(3), 670–680.

Zhu, L., Peng, Z., McClellan, J., Li, C., Yao, D., Li, Z., & Fang, L. (2019). Deep learning for seismic phase detection and picking in the aftershock zone of 2008 Mw7.9 Wenchuan Earthquake. *Physics of the Earth and Planetary Interiors*, *293*, 106261. https://doi.org/10.1016/j.pepi.2019.05.004.

Zhu, W., & Beroza, G. C. (2019). {PhaseNet}: A deep-neural-network-based seismic arrival-time picking method. *Geophysical Journal International*, *216*(1), 261–273. https://doi.org/10.1093/gji/ggy423.

CHAPTER THREE

Machine learning techniques for fractured media

Shriram Srinivasan[a,*], Jeffrey D. Hyman[b], Daniel O'Malley[b], Satish Karra[b], Hari S. Viswanathan[b], and Gowri Srinivasan[c]

[a]Center for Nonlinear Studies (CNLS) & Computational Earth Science (EES-16), Earth and Environmental Sciences Division, Los Alamos National Laboratory, Los Alamos, NM, United States
[b]Computational Earth Science (EES-16), Earth and Environmental Sciences Division, Los Alamos National Laboratory, Los Alamos, NM, United States
[c]Computational Physics Division (XCP-8), Los Alamos National Laboratory, Los Alamos, NM, United States
*Corresponding author: e-mail address: shrirams@lanl.gov

Contents

1. Introduction	109
2. Preliminaries	113
2.1 Governing equations	113
2.2 DFN to graph mappings	116
3. Graph as a DFN reduced-order model	122
4. Pruned DFN as a reduced-order model	122
4.1 Existence of backbones	122
5. Machine learning methods for backbone identification	124
5.1 Fracture-classification: Labeling by FTG membership	125
5.2 Fracture-classification: Labeling by mass flux	130
5.3 Path-classification: Labeling by FTG membership	136
6. Further scope for ML in fractured media	144
References	144
Further reading	150

1. Introduction

Every aspect of the problem of flow through porous media, be it modeling, computation or experiment, has been plagued with stiff challenges that stem from the fundamental nature of the problem, to wit the interaction between the porous medium and the other flowing constituents, and the question of the appropriate length scale at which to describe the heterogeneous structure of the porous medium itself (Bear, 2013). Accordingly, there have been efforts that try to model phenomena at the pore-scale while

Advances in Geophysics, Volume 61
ISSN 0065-2687
https://doi.org/10.1016/bs.agph.2020.08.001

© 2020 Elsevier Inc.
All rights reserved.

109

others approximate the gross behavior at the macroscale through continuum models. For the majority of applications, however, the pore-scale description is untenable due to the large size of the domains involved and there is no alternative but to utilize appropriate continuum models such as those due to Srinivasan and Rajagopal (2014), Srinivasan (2016), or Biot (1956 a, 1956 b). In these models, the effect of local heterogeneity and porosity are all lumped into a single spatial parameter called the permeability that can be determined experimentally. Indeed, one may think of these models as "averaged" versions of the pore-scale equations in the sense of homogenization (Hornung, 1997).

These continuum descriptions work well for classical porous media (such as oil sands, clay, gravel) but there exists a subclass within porous materials whose fundamental character is defined not by permeability but by the material topology. These are fractured media, a feature of geological strata composed of shale, granite, or other crystalline rock. In these materials, flow and transport primarily occurs through the network of interconnected fractures embedded within a rock matrix which may itself be considered impermeable in comparison. The simulation of fluid flow and transport through such fractured media (Kang, Dentz, Le Borgne, & Juanes, 2015; Sherman, Hyman, Bolster, Makedonska, & Srinivasan, 2019) is necessary for many subsurface applications. These include hydrocarbon extraction (Hyman et al., 2016; Karra, Makedonska, Viswanathan, Painter, & Hyman, 2015; Middleton et al., 2015), environmental restoration of contaminated fractured media (National Research Council, 1996; Neuman, 2005; VanderKwaak & Sudicky, 1996), CO_2 sequestration (Jenkins, Chadwick, & Hovorka, 2015), and aquifer storage and management (Kueper & McWhorter, 1991). Classical continuum representations of such media fail to incorporate the flow and transport characteristics that are dictated by network topology (Botros, Hassan, Reeves, & Pohll, 2008; Jackson, Hoch, & Todman, 2000; Karimi-Fard, Gong, & Durlofsky, 2006; Painter & Cvetkovic, 2005). Thus, explicit representation and resolution of network topology in the paradigm of Discrete Fracture Networks (DFN) models has emerged as a tool to model flow and transport through fractured media. Note that this is not the only way to model fractured media; the review by Berre, Doster, and Keilegavlen (2019) summarizes other paradigms such as dual porosity and discrete fracture matrix models. However, in this article, we shall be concerned with the use of machine learning within the context of the DFN approach alone.

In this approach (Berrone, Pieraccini, & Scialo, 2013; Berrone, Pieraccini, Scialò, & Vicini, 2015; Erhel, de Dreuzy, & Poirriez, 2009;

Hyman, Karra, et al., 2015; Mustapha & Mustapha, 2007; Pichot, Erhel, & de Dreuzy, 2012; Pichot, Erhel, & de Dreuzy, 2010), the fracture network in three-dimensions is represented by intersections of two-dimensional planar surfaces, such as circles, ellipses or polygons. To represent a particular geological site, fracture data from the site is used to model physical and topological attributes (fracture lengths/radii, location, orientation, and aperture/thickness) as probability distributions from which realizations of said network may be constructed by repeated sampling. Once the modeled network is equipped with a computational grid, one can solve the governing flow and transport equations to predict the quantities of interest.

However, due to the explicit representation of fractures and network structure, DFN models are computationally expensive. This is true even for networks of modest size. Due to this reason, the application of DFN models has been limited to one-dimensional pipe-network approximations (Cacas et al., 1990; Dershowitz & Fidelibus, 1999), two-dimensional systems (de Dreuzy, Darcel, Davy, & Bour, 2004; de Dreuzy, Davy, & Bour, 2001, 2002), or relatively small three-dimensional systems (Bogdanov, Mourzenko, Thovert, & Adler, 2007). However, while recent advances in the areas of mesh generation (Hyman, Gable, Painter, & Makedonska, 2014), discretization techniques (Erhel et al., 2009; Pichot et al., 2012, 2010), computational frameworks (Berrone et al., 2013, 2015), and high-performance computing have made computing in DFN models feasible for larger systems, they are not sufficient for the wide-spread usage of DFN simulations.

The increase in model fidelity with usage of DFNs comes at a large computational cost because of the large number of mesh elements required to represent thousands of fractures across a range of sizes for proper resolution of pressure at fracture intersections with large gradients. Fig. 1 provides two examples showcasing the cost of DFN simulations. In Fig. 1A, the network of a field site with layered stratigraphy is made up of around 100,000 fractures drawn from fifteen families. It is meshed with over 40 millions nodes in a conforming Delaunay triangulation. A magnified view of a smaller network is shown in Fig. 1B where an 8 fracture DFN is shown along with the mesh that contains close to 90,000 nodes. This mesh must be further refined if in-fracture plane variability is considered because in that case the mesh resolution needs to be small enough to resolve correlations in the aperture field (de Dreuzy, Méheust, & Pichot, 2012; Makedonska et al., 2016).

As fracture locations and attributes are modeled stochastically, a large number of network realizations are required to bound system uncertainty in an uncertainty quantification (UQ) framework, which exacerbates the problem of computational expense. In a basic UQ framework, an ensemble

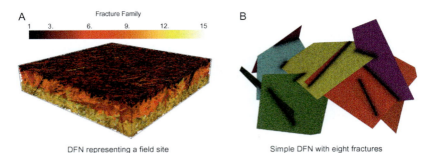

Fig. 1 (A) A DFN composed of around 100,000 fractures with 15 fracture families in three geological layers (B) DFN composed of 8 fractures meshed with close to 90,000 nodes. *From Viswanathan, H. S., Hyman, J. D., Karra, S., OMalley, D., Srinivasan, S., Hagberg, A., & Srinivasan, G. (2018). Advancing graph-based algorithms for predicting flow and transport in fractured rock. Water Resources Research, 54(9), 6085–6099. https://doi.org/10.1029/2017WR022368.*

is generated using computational models and the Monte Carlo method (MC) is used to estimate the likelihood of a scenario. Since the standard MC method suffers from slow convergence, variants of it, such as multifidelity Monte Carlo (MFMC) (Ng & Willcox, 2014) and multilevel Monte Carlo (MLMC) (Giles, 2008), have been developed to approximate the ensemble mean of the quantity of interest with small variance and faster convergence rates. However, in case of fracture networks, MLMC (Berrone, Canuto, Pieraccini, & Scialò, 2018) relies on several levels of accuracy in the space discretization to utilize more samples at the cheaper (less accurate) levels while improving the estimates with few samples at the finer levels. The method presumes the existence of a solver and a meshing strategy that can use very coarse meshes, but meshing remains a major challenge in fracture networks. MFMC (O'Malley, Karra, Hyman, Viswanathan, & Srinivasan, 2018), on the other hand, utilizes two or more models with varying levels of fidelity and computational efficiency, and requires lower-fidelity models that are highly correlated but significantly increase computational performance. Lower fidelity or reduced-order models (ROMs) for DFN are thus useful as forward models in a UQ framework (MC or MFMC) but they also serve as forward models for forecasting in unconventionals (Mudunuru et al., 2020), and as an efficient way to generate synthetic data for machine learning (ML). Established ROMs for classical porous media such as those based on multiscale finite elements (Efendiev, Galvis, & Wu, 2011; Hou & Wu, 1997) and multiscale finite volume discretizations (Jenny, Lee, & Tchelepi, 2003; Minev, Srinivasan, & Vabishchevich, 2018;

Srinivasan, Lazarov, & Minev, 2016) are inapplicable for DFN due to the same reason that continuum upscaling models failed as described earlier. Thus, development of low-fidelity models or ROMs is an important and challenging issue in fracture networks, and this observation provides context for the subject of this article—an account of the role of machine learning allied with graph theory in their development.

After describing the construction of a DFN, the necessary governing equations for flow and transport through DFN and various graph representations in Section 2, we shall organize the development of ROMs in DFN using graphs and ML into different unifying themes in Sections 3,4. Then in Section 5, we place into an abstract perspective, three different approaches that utilize machine learning to develop ROMS for fractured media and explain the fundamental ideas of each approach. As this is an overview/tutorial article, our intent is to explain broad themes and generalizations, not the intricate details of the approaches. Finally, in Section 6, we list some future directions where machine learning in fracture media may be employed gainfully.

2. Preliminaries
2.1 Governing equations

The transport of a solute through a fracture network is represented in a Lagrangian approach as a collection of passive, conservative particles moving through the network. The alternative, an Eulerian approach, would solve an advection-diffusion equation but the inherent numerical dissipation gives the advantage to the Lagrangian approach.

We use the DFNWORKS (Hyman, Karra, et al., 2015) suite to generate each DFN, solve the steady-state flow equations and determine transport properties of the network. DFNWORKS combines the feature rejection algorithm for meshing (FRAM) (Hyman et al., 2014), the LaGriT meshing toolbox (LaGriT, 2013), the parallelized subsurface flow and reactive transport code PFLOTRAN (Lichtner et al., 2015), and an extension of the WALKABOUT particle-tracking method (Makedonska, Painter, Bui, Gable, & Karra, 2015; Painter, Gable, & Kelkar, 2012). FRAM is used to generate the three-dimensional fracture networks, while LaGriT creates a computational mesh representation of the DFN in parallel. Finally, PFLOTRAN numerically integrates the governing flow equations while WALKABOUT is used to determine path-lines through the DFN and simulate solute transport. Details of the suite, its abilities, applications, and references for detailed implementation are provided in Hyman, Karra, et al. (2015).

In a DFN approach, the network of fractures in three-dimensions are represented explicitly as intersections of objects with dimension 2, i.e., intersections of planar shapes such as rectangles or ellipses. Moreover, each representative fracture is endowed with attributes such as size, orientation, aperture, and permeability. Such attributes are drawn from a probability distribution that best fits data sampled from a geological site (Alemanni et al., 2011; SKB, 2011).

The equations governing steady flow in a fracture with uniform aperture are the aperture-averaged approximations due to Boussinesq (1868) in conjunction with the assumption of incompressibility

$$\mathbf{Q} = -\frac{b^3}{12\mu}\nabla p \tag{1}$$

$$\mathrm{div}(\mathbf{Q}) = 0 \tag{2}$$

where $\mathbf{Q}$ is the volumetric flow rate per unit fracture width normal to the direction of flow, ∇p the pressure gradient, μ the viscosity of the fluid, and b the aperture of the fracture. Defining $\mathbf{q} := \frac{\mathbf{Q}}{b}$ as the Darcy flux per unit cross-sectional area normal to the flow gives

$$\mathbf{q} = -\frac{b^2}{12\mu}\nabla p \tag{3}$$

$$\mathrm{div}(\mathbf{q}) = 0 \tag{4}$$

The factor $\frac{b^2}{12}$ plays a role analogous to permeability which is why the governing equation may be thought of as Darcy's law, expressing the proportionality between the volumetric flux and the pressure gradient. These two equations when combined yield

$$\mathrm{div}\left(\frac{b^2}{12\mu}\nabla p\right) = 0 \ . \tag{5}$$

In the context of DFN models, the viscosity of the fluid μ is typically assumed to be a constant, so that Eq. (5) is a linear, elliptic, partial differential equation. However, it is well known that the viscosity of fluids depends on pressure (Bridgman, 1931), and in subsurface applications where flow is driven by large pressure gradients, recognition of this effect leads to interesting nonlinear behavior (Rajagopal & Srinivasan, 2014; Srinivasan, Bonito, & Rajagopal, 2013; Srinivasan & Rajagopal, 2016).

The domain is assumed to be a cube of side L meters, and that flow is driven by the pressure gradient resulting from prescribed pressures at the two ends (faces) of the domain, with no-flow boundary conditions at the other boundaries. Aligning a Cartesian coordinate axis with the flow direction, without loss of generality we assume that flow is along the x-axis so that the inlet and outlet planes have the representations $x = 0$ and $x = L$ respectively.

In the next step, an Eulerian velocity field $\mathbf{u}(\mathbf{x})$ consistent with the Darcy fluxes is reconstructed by using the technique outlined in Painter et al. (2012) and Makedonska et al. (2015). The trajectory of a particle starting at location $\mathbf{a}$ on the inlet plane with a mass $m(\mathbf{a})$ through $\mathbf{u}(\mathbf{x})$ is given by the kinematic relationship

$$\frac{d\mathbf{x}(t;\mathbf{a})}{dt} = \mathbf{v}(t;\mathbf{a}), \ \mathbf{x}(0;\mathbf{a}) = \mathbf{a}, \tag{6}$$

where $\mathbf{v}(t;\mathbf{a}) := \mathbf{u}(\mathbf{x}(t;\mathbf{a}))$ is the Lagrangian description of the known velocity field. The initial distribution of particles on the inlet plane is assigned by flux-weighting, i.e., the number of particles at a location is proportional to the inflow flux at that location (Hyman, Painter, Viswanathan, Makedonska, & Karra, 2015). We take every particle to have the same mass $m(\mathbf{a})$ and also assume complete mixing at fracture intersections so that the probability to enter an outgoing fracture is weighted by the outgoing flux therein (Sherman et al., 2019).

Once (6) is integrated to obtain $\mathbf{x}(t;\mathbf{a})$, we can gather the first passage time of a particle to cross the outlet plane, $\tau(\mathbf{a})$. The distribution of $\tau(\mathbf{a})$ across the ensemble represents the probability of solute mass breaking through at a time t, given by the Lebesgue integral

$$\Phi(t) = \frac{1}{M_0} \int_{\Omega_a} dm(\mathbf{a})\delta[t - \tau(\mathbf{a})], \tag{7}$$

where Ω_a is the set of all particles, $\delta[\cdot]$ is the Dirac-delta measure, and M_0 is the total mass of the particles. The function $\Phi(\cdot)$ is referred to as the breakthrough curve, and it is the main quantity of interest in applications related to fractured media.

In addition to the time taken for a particle to pass through the system, it is possible to track the sequence of fractures through which the particle travels as it moves through the fracture network. As the location and dimensions of all fractures in the network are known, the instantaneous value $\mathbf{x}(t;\mathbf{a})$ permits us to determine the particular fracture within which the particle is instantaneously located. If we represent a DFN composed of n fractures as a set $F = \{f_i\}$ for $i = 1, \ldots, n$, the sequence of fractures traversed by the k-th particle is the finite set $S^{(k)} = \{f_i, f_j, \ldots\}$, the particle-trace. Once constructed, the collection $S^{(1)}$, $S^{(2)}$, ... across Ω_a allows us to determine the number of particles (and thus proportion of the fluid mass) that passed through a particular fracture. Intuitively, it should be clear that this construction can be used to identify sequences of fractures through which a lion's share of fluid mass passes through.

Fig. 2 illustrates how particle information can be included into the graph of a DFN. Subfigure (A) shows a DFN composed of 45 fractures colored by the number of particles that pass through each fracture and (B) is the corresponding topology graph representation (Section 2.2.1) of the DFN shown in (A). Here, the vertex size is proportional to number of particles that pass through the corresponding fracture. The source (inflow) is colored blue and the target (outflow) is red.

2.2 DFN to graph mappings

The mathematical construct of a graph consists of a set of vertices connected by a set of edges, and it has been found to be a well-suited tool to represent key network structure information in a DFN. The mapping of a fracture network to a graph has been done in different, but related ways, and here we define and present them so that we may refer to them in context later.

Viewing a DFN as a collection of intersecting fractures leads to a natural association with the mathematical construct of a graph. A graph is a structure used to depict a set of objects along with the relationship between them (taken in pairs). Specifically, a graph is an ordered tuple $G(V, E, \mathcal{F})$ consisting of a set of vertices $V = \{v_1, v_2, \ldots\}$, a set of edges $E = \{e_{ij} | \mathcal{F}(v_i, v_j) = +1\}$ for "related" vertices v_i, v_j, and a function $\mathcal{F} : V \times V \mapsto \{+1, -1\}$ that defines whether or not two vertices are "related." For brevity, we suppress the third argument and denote a graph simply by $G(V, E)$. Thus, one can construct different graphs for a given problem by changing some or all of the objects that make up the vertex set V, the edge-set E, or the definition of when elements in V are said to be related. Both the vertices and edges can also have weights assigned to them to further encode information about the vertices themselves or the connection between vertices.

Machine learning techniques for fractured media 117

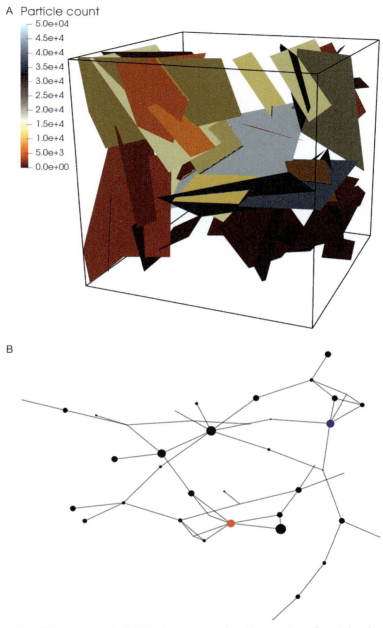

Fig. 2 (A) A DFN composed of 45 fractures colored by the number of particles that pass through each fracture. (B) A graph representation of the DFN shown in (A). Vertex size is proportional to number of particles that pass through the corresponding fracture. The source (inflow) is *colored blue* and the target (outflow) is *red. From Srinivasan, S., Cawi, E., Hyman, J. D., Osthus, D., Hagberg, A., Viswanathan, H., & Srinivasan, G. (2020). Physics-informed machine-learning for backbone identification in discrete fracture networks. Computational Geo- sciences, 24, 1429–1444. https://doi.org/10.1007/s10596-020-09962-5.*

2.2.1 Topology map

One mathematical representation of a mapping between a fracture network F and a graph G is the following (Andresen, Hansen, Le Goc, Davy, & Hope, 2013; Hope, Davy, Maillot, Le Goc, & Hansen, 2015; Huseby, Thovert, & Adler, 1997).

Let $F = \{f_i\}$ for $i = 1, \ldots, n$ denote a fracture network composed of n fractures. We define a mapping, ϕ, that transforms F into a graph $G(V, E)$ composed of $n = |V|$ vertices, and $m = |E|$ edges. For every $f_i \in F$, there is a unique vertex $u_i \in V$,

$$\phi : f_i \rightarrow u_i . \tag{8}$$

If two fractures f_i and f_j intersect, $f_i \cap f_j \neq \emptyset$, then there is an edge in E connecting the corresponding vertices,

$$\phi : f_i \cap f_j \neq \emptyset \rightarrow e_{ij} = (u_i, u_j) , \tag{9}$$

where $(u, v) \in E$ denotes an edge between vertices u and v. One can also include source s and target t vertices into G to incorporate flow direction. Every fracture that intersects the inlet plane $\mathbf{x_0}$ is connected to the source vertex,

$$\phi : f_i \cap \mathbf{x_0} \neq \emptyset \rightarrow e_{si} = (s, u_i) , \tag{10}$$

and every fracture that intersects the outlet plane $\mathbf{x_L}$ is connected to the target vertex t,

$$\phi : f_i \cap \mathbf{x_L} \neq \emptyset \rightarrow e_{it} = (u_i, t) . \tag{11}$$

This choice to include boundary conditions via source and target nodes is crucial if flow-based quantities of interest (QoI) are considered instead of geometric or topological properties (Neuman, 2005).

In this regard, many research efforts have used this mapping for various purposes, such as to study the topology of fracture networks (Andresen et al., 2013; Hope et al., 2015), to investigate how topology influences upscaled properties (Ghaffari, Nasseri, & Young, 2011; Sævik & Nixon, 2017) and flow behavior (Santiago, Romero-Salcedo, Velasco-Hernández, Velasquillo, & Hernández, 2013; Santiago, Velasco-Hernández, & Romero-Salcedo, 2014; Santiago, Velasco-Hernández, & Romero-Salcedo, 2016), or to design a domain-decomposition for high-performance computing for DFN (Berrone et al., 2015).

A *path* in a graph $G(V, E)$ is a finite sequence $\{v_i, v_j, \ldots v_n\}$ consisting of elements in the vertex set V. In this instance, the path is said to be from vertex v_i to vertex v_n. At this juncture, it should be clear that each of the particle-traces $S^{(1)}$, $S^{(2)}$, ... that represents the sequence of fractures traversed by a particle in the DFN is equivalent to a sequence of vertices corresponding to a path through the graph, from the source vertex to the target vertex. Thus the graph representation has naturally inherited the concepts of Lagrangian transport, and serves as a surrogate to the DFN.

2.2.2 Pipe-network map

A different mapping that considers the set of fracture intersections as the vertex set can be defined as follows (Cacas et al., 1990; Dershowitz & Fidelibus, 1999; Nœtinger & Jarrige, 2012). Let $F = \{f_i\}$ for $i = 1, ..., n$ denote a DFN composed of n fractures (Fig. 3). Define a mapping, ψ, that transforms F into a graph $G(V, E)$ composed of $N = |V|$ vertices, and $M = |E|$ edges. If two fractures f_i and f_j intersect, $f_i \cap f_j \neq \emptyset$, then there is a node $u \in V$, that represents the line of intersection

$$\psi : f_i \cap f_j \to u . \tag{12}$$

If $f_i \cap f_j \neq \emptyset$ and $f_i \cap f_k \neq \emptyset$, then there is an edge in E connecting the corresponding vertices,

$$\psi : f_i \cap f_j \neq \emptyset \text{ and } f_i \cap f_k \neq \emptyset \to e(u, v), \tag{13}$$

where $e(u, v) \in E$ denotes an edge between vertices u and v. Under this mapping, each fracture in the network is represented by a clique in G. Representative source s and target t vertices in G incorporate flow direction, so that if f_i intersects the inflow or outflow plane, then a node is added to V that represents the line of intersection.

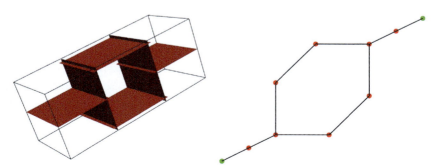

Fig. 3 (A) A simple DFN and (B) its pipe-network graph representation. The *red vertices* correspond to fracture intersections, while the *green vertices* represent source and target vertices where the fracture intersects the inlet or outlet boundary.

Under this mapping ψ, each fracture is represented by a k-clique where k is the number of intersections on the fracture. Thus, each edge can be thought of as residing on a single fracture and edge weights can represent hydrological and geometric properties of that fracture. Unlike ϕ, ψ is not a bijective mapping between the fractures in F and $G(V, E)$, because it is not surjective.

Note that among the multiple graph representations possible for a DFN, the most general is a bipartite graph where fractures and intersections are two disjoint vertex sets of the graph (Hyman et al., 2018) from which the two representations discussed can be obtained via graph projections.

2.2.3 Flow topology graph (FTG) construction

Based on the topology map representation and results from transport simulations, it is possible to construct a flow topology graph (FTG) that encodes information about mass transport as edge weights of the graph using the method provided by Aldrich et al. (2017). Since the construction of FTG is crucial for later development, we briefly recount it here for completeness.

Given a DFN with n fractures and k particle-traces, create the vertex set $V = \{v_1, \ldots, v_n, v_s, v_t\}$ consisting of the fractures and source and target vertices as before. Now, corresponding to a given particle-trace, there is a path on the graph that starts at the source vertex and ends at the target vertex.

1. In the particle-trace $S^{(k)}$, each time a particle transitions from vertex v_i to v_j, create an edge directed from v_i to v_j or add 1 to the edge-weight $w_{i,j}$ if it already exists.
2. Once this process is followed for all $\{S^{(i)} : i = 1, 2, \ldots, k\}$, remove cycles by eliminating the edge with lesser edge-weight.
3. Finally, reset the edge-weight to be the reciprocal, i.e., $1/w_{ab}$. This ensures that edges with high proportion of particles flowing through have low weights.
4. Thus computing the $i = 1, 2, \ldots, i_{max}$ shortest paths from source vertex (s) to target vertex (t), obtain the backbone.

The value of i_{max} is found by plotting the percent of total mass moving through the paths as a function of i until there is no more increase. Thus, the derivative is computed to estimate the change in flow volume as the number of paths in the graph is increased, and $i = i_{max}$ is chosen such that the derivative is close to 0.

Fig. 4A shows the backbone defined for the network provided in Fig. 2, and also demonstrates how fractures with very few particles on them have been removed. The corresponding graph shown in Fig. 4B highlights how there are no dead-end fractures on the backbone and no isolated fractures.

Machine learning techniques for fractured media 121

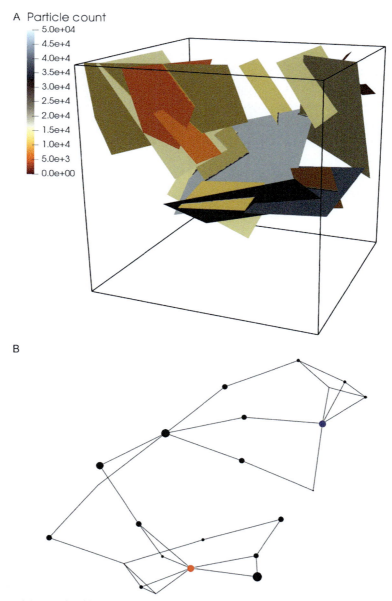

Fig. 4 (A) DFN backbone identified by the FTG for the network provided in Fig. 2. (B) Corresponding FTG. *From Srinivasan, S., Cawi, E., Hyman, J. D., Osthus, D., Hagberg, A., Viswanathan, H., & Srinivasan, G. (2020). Physics-informed machine-learning for backbone identification in discrete fracture networks. Computational Geo- sciences, 24, 1429–1444. https://doi.org/10.1007/s10596-020-09962-5.*

3. Graph as a DFN reduced-order model

In the quest to create a ROM for a DFN, one direct way is to use its graph representation as a surrogate for the DFN, and perform flow and transport calculations on the graph representation itself. This represents a drastic simplification, for it eliminates spatial complexity and the need for a mesh by reducing the DFN to a lumped-parameter system. The pipe-network model described in Section 2.2.2 is a natural choice, and indeed, Karra, O'Malley, Hyman, Viswanathan, and Srinivasan (2018) derived the governing equations for single-phase, steady flow on the graph and investigated the performance of the model. The governing equations have the familiar form $L\hat{p} = c$, where L is the Laplacian matrix of the graph, $\hat{p}$ is the vector of unknown vertex pressures and c is a vector with contributions from the known pressures at the boundary vertices. Karra et al. found a systematic error in the transport results obtained from the ROM due to the fact that fracture intersections (which are regions of sharp gradients) had now been upscaled to graph vertices. The systematic error manifested itself in slower flow, and consequently slower transport on the graph, and grew in significance as network size increased. Based on data of actual time of transport along a fracture obtained from a single high-fidelity DFN simulation, an a posteriori power-law was fitted to correct the time of transport for particles along a graph edge. Over all, the authors found a computational speed up of the order of 10^4 over the high-fidelity DFN simulations, and when coupled with the accuracy obtained by the a posteriori correction, it represented a remarkable result. The applicability of this ROM is limited only by the assumptions that underpin it, namely, steady-state and single-phase flow. While the model has been extended to include transient flow (Srinivasan, O'Malley, et al., 2020), for multiphase flow or reactive transport, it is not clear if a graph ROM can inherit, or correlate closely with the behavior of the high-fidelity model. This provides a segue into a second avenue to create a ROM for DFN networks.

4. Pruned DFN as a reduced-order model
4.1 Existence of backbones

If a graph ROM be inapplicable, one may imagine discarding a subset of the DFN to reduce the system, or alternatively, use a subset of the DFN as a ROM. However, there is a fundamental difference with the philosophy of the graph ROM, where, at the least, the topology of the entire network was being incorporated into the graph representation. Thus, given the primacy of network topology in determining transport characteristics, there is a physical rationale for the ROM. To justify the idea that subsets of a DFN may be used in place of the whole, one needs some evidence based on physics. Fortunately, there is copious evidence that supports the idea that not all fractures of a DFN are equally important, and that one may replace a DFN with a suitable subset or "backbone" and still approximate upscaled transport observables and quantities of interest with desired fidelity.

The idea to do so emerged from the observation of preferential flow-paths and channeling effects (Abelin et al., 1991; Abelin, Neretnieks, Tunbrant, & Moreno, 1985; Hyman, Painter, et al., 2015; Tsang & Neretnieks, 1998). This implied that not all fractures in a DFN are crucial to the flow and transport through it, and one could get reasonably similar transport characteristics by using a subnetwork of the DFN, often called a "backbone" (Maillot, Davy, Goc, Darcel, & De Dreuzy, 2016). Studies in sparse fracture networks with fracture sizes spanning a range of length scales indicate that network structure (de Dreuzy et al., 2001, 2012; Frampton & Cvetkovic, 2011; Hyman, Aldrich, Viswanathan, Makedonska, & Karra, 2016) and flow direction (Neuman, 2005) are more dominant in determining regions of flow channeling than geometric or hydraulic properties of the fractures.

In the context of system reduction Hyman, Hagberg, Srinivasan, Mohd-Yusof, and Viswanathan (2017) obtained a subgraph corresponding to the union of k-shortest paths, by utilizing the topology map graph representation described in Section 2.2.1. Srinivasan et al. (2018), starting from the full graph, obtained a subgraph corresponding to the two-core subgraph of the network.

A different method to identify subnetworks was proposed by Aldrich et al. (2017) based on the FTG (Section 2.2.3) but the subnetworks could be identified only after running simulations on the full network. These methods accurately predict the first passage times which are important for applications such as detection of gas seepage. However, these are not useful for applications such as hydrocarbon extraction or contaminant remediation, where the entire distribution of passage times is of interest.

Srinivasan et al. (2018) used the graph mapping (Section 2.2.2) to solve for the flow-physics on the graph and then used the flow field to extract a subgraph by thresholding on the Darcy flux. The algorithm allowed control over the size of the backbone, so it was claimed to be useful for predicting both first passage times as well as the full breakthrough.

5. Machine learning methods for backbone identification

There has been a surge in the use of machine learning (ML) in the geosciences (Bergen, Johnson, de Hoop, & Beroza, 2019). This surge is fueled, among other reasons, by better sensors, improvements in large scale simulations, and the appearance of large, publicly available datasets stored over the cloud (Karpatne, Ebert-Uphoff, Ravela, Babaie, & Kumar, 2018). These data come from multiple sources such as remote sensing (Cracknell & Reading, 2014; Zhang, Zhang, & Du, 2016), in situ sensors such as weather stations or instruments on ocean buoys (National Oceanic and Atmospheric Administration, 2018), and ensembles of climate models (World Climate Research Program, 2018). Aside from applications in the geosciences, there are success stories of machine learning in modeling disease spread (Scavuzzo et al., 2018), soil health (Hengl et al., 2017), land cover classification (Dev, Wen, Lee, & Winkler, 2016), and clinical medicine (Zhang, 2016). The challenging features of these data include dimensionality, high spatio-temporal correlation, dynamic domains, small sample sizes, missing data, and rare events of interest (Griewank, Reich, Roulstone, & Stuart, 2017; Karpatne et al., 2018; Lary, Alavi, Gandomi, & Walker, 2016).

There is an abundance of existing simulation data from high-fidelity simulations on fracture networks. This knowledge, coupled with the fact that large number of fractures are contained in each network suggests that a data-driven (machine learning) approach could be useful to address the problem of backbone identification. In this context, machine learning plays a role in DFN modeling that is not limited solely to reduction of computational expense through backbone identification. Rather, machine learning also advances our understanding of how the network determines the structure of the flow field within the DFN.

Machine learning techniques can be grouped into supervised, unsupervised, and reinforcement learning algorithms (Abu-Mostafa, Magdon-Ismail, & Lin, 2012). Supervised learning algorithms require input data along with labels or values while unsupervised learning algorithms do not require labels or values associated with the input data. Reinforcement learning is the process of formulating a policy of action based on feedback in the form of rewards and penalties for prior actions. The problem of system reduction of DFN fits naturally into a supervised learning paradigm. Under the umbrella of supervised learning algorithms, one can look to either predict a quantity of interest directly (a regression problem) or assign it to one of the labeled classes or groups (classification problem). System reduction is a binary classification problem, where each fracture is either selected or rejected.

Before moving on, let us digress to discuss an abstract formulation of the binary classification problem. Given a training set S_{train}, consisting of labeled objects (labels $\{+1, -1\}$) $\{F_1, F_2, \ldots F_n\}$, each of which is characterized by a set of m attributes (features) $\{x_j^i \,|\, j = 1, 2, \ldots m\}$. Further assume that the set $\mathfrak{F}$ consists of those objects that have label $+ 1$. One can then pose a learning problem based on the object features as follows:

$$\text{Find } g : \mathbb{R}^m \mapsto \{+1, -1\} \ni \tag{14a}$$

$$\forall\, i = 1, 2, \ldots n, \tag{14b}$$

$$g(x_1^i, x_2^i, \ldots, x_m^i) = \begin{cases} +1 \ \forall\, F_i \in \mathfrak{F}, \\ -1 \ \text{otherwise.} \end{cases} \tag{14c}$$

In words, (14) expresses the idea that given a set of features and class-assignments for some observations, the algorithm attempts to approximate the underlying function that maps features to classes. For instance, the objects could be images of cats (labeled $+ 1$) and dogs (label $- 1$), and the features the pixels of the image.

We point out that implicit in the description of the binary classification problem above are three assumptions that have nothing to do with the mathematical theory but instead are rooted in domain knowledge of the problem:

1. The identity of the objects being classified
2. The features used for a given object
3. The criteria used to label a given object

In this context, we can now place the works of Valera et al. (2018), Srinivasan, Karra, Hyman, Viswanathan, and Srinivasan (2019), and Srinivasan, Cawi, et al. (2020).

5.1 Fracture-classification: Labeling by FTG membership

Like typical problems of machine learning classification, we start with data (a set of DFN realizations generated from the same underlying assumptions) partitioned into two disjoint proper subsets: a training set S_{train} consisting of DFNs for which the backbones have already been identified from high-fidelity DFN particle-tracking simulations, and a test set S_{test} (a set of DFN realizations) for which the backbones will be predicted by the algorithm.

Earlier, we stated the point of view that the most important fractures in the network, the ones through which *most* of the flow takes place, comprise the backbone of the DFN. If we look to quantify the phrase "most of the flow," we are led naturally to the idea of the FTG described in Section 2.2.3. Valera et al. (2018) chose to use fractures as the element of classification, and their labeling for fractures in a given DFN was based on the criteria that if a fracture appeared as a vertex in the FTG of the network, then it was labeled + 1 as a backbone fracture.

Regarding the features, they used both topological and physical/hydrological attributes of a fracture in a DFN. For instance, each fracture has a size, aperture, orientation and permeability. Moreover, its global and local topological attributes, namely betweenness centrality, degree centrality, source-to-target simple paths and the projected volume, are also easily evaluated. These quantities are now defined below.

Degree centrality: The *degree centrality* of a vertex is a normalized value representing the number of edges touching a vertex. For a fracture in a network, it is an indicator of the number of fractures that intersect it. Vertices with low degree centrality will usually be on the periphery of the network or the low flow branches. For a vertex i, its degree centrality is given by:

$$D(i) = \frac{1}{n-1} \sum_{j=1}^{n} A_{ij}, \tag{15}$$

where A_{ij} is the ij-th element of the adjacency matrix A of the graph and n is the number of vertices in the graph.

Betweenness centrality: Betweenness centrality of a vertex is a global topological measure. It quantifies the extent to which a vertex controls communication in a network by estimating how frequently paths through the network include the vertex. We define a geodesic path as a path (sequence of vertices connected by edges) between vertices u and v with the fewest possible edges, and denote the number of geodesics as σ_{uv}. We also denote as $\sigma_{uv}(i)$ the number of geodesic paths through u and v that pass through vertex i. *Betweenness centrality* is a normalized metric that indicates the fraction of geodesic paths in G that pass through vertex i via:

$$B(i) = \frac{1}{(n-1)(n-2)} \sum_{u,v=1, u \neq i \neq v}^{n} \frac{\sigma_{uv}(i)}{\sigma_{uv}} \tag{16}$$

Vertices with high betweenness centrality thus represent hubs that many paths pass through, and represent either highways or bottlenecks for the flow.

Current flow betweenness centrality: Source to target current flow is a centrality measure based on analogy with a resistive circuit where every edge has unit resistance, and a unit current is injected at a "source" node, with flow measured from source to target. Define the *Graph Laplacian* as $L = D - A$, where D is a diagonal matrix specifying the degree of each node and the pseudo-inverse of L as L^{+}. Current flow centrality of a vertex is defined as the total amount of current that flows through the vertex. The potential of the i-th vertex is $L_{is}^{+} - L_{it}^{+}$, so the magnitude of the current through the edge that connects vertices i, j is $|(L_{is}^{+} - L_{it}^{+}) - (L_{js}^{+} - L_{jt}^{+})|$. To get the total current, we use the i-th row of the adjacency matrix A to sum over all the edges that connect to vertex i, and thus the current flow centrality

$$C(i) = \sum_{j=1}^{n} A_{ij} |(L_{is}^{+} - L_{it}^{+}) - (L_{js}^{+} - L_{jt}^{+})| \qquad (17)$$

This centrality measure can also be modified by using an adjacency matrix that is weighted by the resistance of each edge.

Permeability: Another fracture feature is permeability, which is related to the idea of "conductivity," i.e., how much fluid can flow through the material. For a fracture with aperture b, its permeability is $k = b^2/12$.

Projected volume: This is a measure of that component of a fracture's volume that is oriented along the direction of flow (from inlet to outlet plane). If a fracture has volume V and orientation vector (n_x, n_y, n_z) (unit vector normal to the fracture plane), then taking flow to be oriented along the x-direction, the projected volume is expressed by the projection of the volume onto the yz plane, i.e., $V\sqrt{n_y^2 + n_z^2}$

Since the topological features are defined in a normalized fashion, they are independent of the number of fractures (vertices) in the network. All the topological features can be calculated using the Python package NetworkX (Hagberg, Schult, & Swart, 2008).

Three popular classification algorithms are logistic regression (LR), support vector machines (SVM), and random forest classifier (RF) (Friedman, Hastie, & Tibshirani, 2001). Based on decision trees that determine the classification of each element based on feature values, RF yields a predicted classification as well as an estimate of importance of each feature for the classification. SVM constructs a separating hyperplane and partitions the data into classified sets, while LR estimates and returns a probability of membership to each class for each data-point.

Valera et al. (2018) used both RF and SVM and trained the models with a set of 100 network realizations. Eighty were used for training while twenty were set aside for testing. Note that the model has no tunable parameters other than the hyperparameters of the algorithms which were trained by `gridsearchcv in scikit-learn` (Pedregosa et al., 2011).

An important result that emerged from the RF was the estimate of feature importance (see Fig. 5). Global topological features are seen to have the greatest importance, local topological features have significant but lower importance, and physical features play a very small role in classification. This is in line with the observations that flow and transport in fracture networks has topology rather than permeability as the dominant influence.

We have adhered to the convention that a positive classification (+1) of a fracture corresponds to assignment to the backbone and a negative classification (−1) is an assignment to its complement. True positives (TP) and true negatives (TN) correspond to correct classification. Classification errors can thus be due to either a false positive (FP) where a fracture that ought to belong to the complement of the backbone is misclassified as belonging to the backbone, or a false negative (FN) where a fracture that ought to belong to the backbone is grouped in the complement instead.

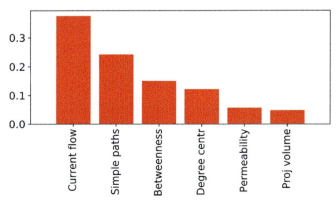

Fig. 5 Relative importance of features based on training data. *From Valera, M., Guo, Z., Kelly, P., Matz, S., Cantu, V. A., Percus, A. G., … Viswanathan, H. S. (2018). Machine learning for graph-based representations of three-dimensional discrete fracture networks. Computational Geosciences, 22, 695–710. https://doi.org/10.1007/s10596-018-9720-1.*

This leads to the metrics of precision and recall, which are nonnegative numbers between 0 and 1 that measure the performance of the classifying algorithm used. Mathematically,

$$\text{Precision} = \frac{TP}{TP + FP}, \ \text{Recall} = \frac{TP}{TP + FN}. \tag{18}$$

Ideally, one would like high precision and high recall (close to 100%), but all classification algorithms have a trade-off between the two (Abu-Mostafa et al., 2012) and in most ML classification problems, calculation of these metrics and calibration of parameters to attain the optimal metric is the end goal. However, in the context of applications in geosciences, these metrics alone are inconsequential. Since our purpose was to find a DFN backbone or subnetwork that would have similar transport properties as the full network, we need a subset of fractures that conduct a significant share of the flow and hence provide transport properties that are reasonably close to the full network. In this context, recall is more important than precision because low precision simply adds more fractures to the backbone (see Table 1, but lower recall implies that some fractures which are primarily responsible for flow and transport are not being selected. Thus we aim for high recall. Perfect recall, however, is not necessary, since one can obtain significant system reduction in terms of number of fractures even with low precision.

Indeed, Valera et al. (2018) observe that with very low recall values, disconnected backbones or backbones with a single connected path from the inlet to outlet are obtained.

An instance showing comparison of the breakthrough curves obtained from the subnetworks is shown in Fig. 6. The breakthrough curve is shown as a CDF. As expected, larger backbone subnetworks are closer to the breakthrough curve of the full DFN.

Table 1 Random forest classifiers labeled by `min_samples_leaf` parameter value that controls branching.

Classifier	Precision %	Recall %	Fractures remaining %
RF(1400)	18	90	36
RF(30)	26	75	21
RF(15)	30	65	15
RF(1)	58	20	2.5

Source: From Valera, M., Guo, Z., Kelly, P., Matz, S., Cantu, V. A., Percus, A. G., ... Viswanathan, H. S. (2018). Machine learning for graph-based representations of three-dimensional discrete fracture networks. Computational Geosciences, 22, 695–710. https://doi.org/10.1007/s10596-018-9720-1.

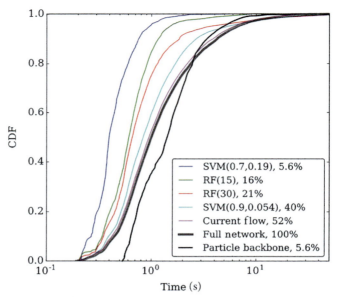

Fig. 6 BTC predictions for a DFN visualized as a CDF. Representative results for four models are given along with parameters and size of reduced network as a percentage of the full network. Particle backbone corresponds to the FTG, while current flow thresholding eliminates fractures with zero current flow. *From Valera, M., Guo, Z., Kelly, P., Matz, S., Cantu, V. A., Percus, A. G., ... Viswanathan, H. S. (2018). Machine learning for graph-based representations of three-dimensional discrete fracture networks. Computational Geosciences, 22, 695–710. https://doi.org/10.1007/s10596-018-9720-1.*

This algorithm, the first to use ML to select DFN backbones, suffers from two drawbacks: (1) there is no parameter that controls the extent of system reduction and (2) there is no constraint to ensure that the reduced sub-network is connected, so there is no guarantee that a physically meaningful subnetwork will be generated. The work by Srinivasan et al. (2019) which we describe next, tries to remedy these shortcomings.

5.2 Fracture-classification: Labeling by mass flux

In the previous subsection, fracture labels were assigned based on whether or not they are included in the FTG. Srinivasan et al. (2019) modified the criteria to instead label fractures based on a user-defined threshold of mass flux.

The mass flux through the fractures is obtained from the particle-tracking simulations of a nonreactive tracer advecting with the fluid, and is motivated by the steps used to construct the FTG. If one starts with N particles to simulate transport through particle-tracking, and the number of particles that pass through the i-th fracture is n_i, then the dimensionless ratio $\eta_i = n_i/N$, $0 \leq \eta_i \leq 1$ is a measure of mass flux through the i-th fracture. The quantity η_i is valid for resident injection, but if one were to consider flux-weighted injection, then the mass of the particles would have to be considered too. This dimensionless measure of mass flux through the fracture is different from the volumetric flux obtained from the flow equations on the graph in Srinivasan et al. (2018) which corresponded to flow intersecting fractures.

Given $0 \leq \epsilon \leq 1$, a DFN network with n fractures $\mathfrak{F} = \{F_i | i = 1, 2, \ldots n\}$, one can define the ϵ-backbone as the subnetwork $\mathfrak{F}_\epsilon := \{F_i | \eta_i \geq \epsilon\}$. Clearly, $\mathfrak{F}_0 = \mathfrak{F}$ and $\mathfrak{F}_1 = \{\}$ so that the only cases of interest are when $0 < \epsilon < 1$.

Now, if each fracture F_i in the network has a set of m attributes (features) $\{x^i{}_j, | j = 1, 2, \ldots m\}$, then one can pose a learning problem of backbone identification based on the features of the fractures as follows:

$$\text{Find } g_\epsilon : \mathbb{R}^m \mapsto \{+1, -1\} \ni \tag{19a}$$

$$\forall \ i = 1, 2, \ldots n, \tag{19b}$$

$$g_\epsilon(x_1^i, x_2^i, \ldots, x_m^i) = \begin{cases} +1 \ \forall \ F_i \in \mathfrak{F}_\epsilon, \\ -1 \ \text{ otherwise.} \end{cases} \tag{19c}$$

As stated, (19) is for a single DFN network. For the case where we are given a training set, an $\epsilon > 0$, $S_{train} = \{\mathfrak{F}^k | k = 1, 2, \ldots K\}$, we shall have individual backbones $\mathfrak{F}_\epsilon^k$, $k = 1, 2, \ldots K$ and network sizes n_k corresponding to each of the K networks in S_{train}, and F_i^k will denote the i-th fracture in the k-th network while x_j^{ik} represents the j-th feature of F_i^k.

The general learning problem, given S_{train}, then is

$$\text{Find } g_\epsilon : \mathbb{R}^m \mapsto \{+1, -1\} \ni \tag{20a}$$

$$\forall \ k = 1, 2, \ldots K, \tag{20b}$$

$$\forall \ i = 1, 2, \ldots n_k, \tag{20c}$$

$$g_\epsilon(x_1^{ik}, x_2^{ik}, \ldots, x_m^{ik}) = \begin{cases} +1 \ \forall \ F_i^k \in \mathfrak{F}_\epsilon^k, \\ -1 \ \text{ otherwise.} \end{cases} \tag{20d}$$

The set of features used by Valera et al. (2018) is reused with one exception—current flow centrality is replaced by fracture radii.

As before, flow and transport simulations are performed for 100 networks, and the mass flux through all fractures in the networks are recorded, so that they are ready to be used as training data in the machine learning algorithm.

Since the least nonzero value and the mean of the mass flux for the fractures in the dataset were found to be $O(10^{-6})$ and $O(10^{-2})$, respectively, values of $\epsilon = 10^{-6}, 10^{-5}, 10^{-4}, 10^{-3}, 10^{-2}, 5 \times 10^{-2}, 10^{-1}, 2 \times 10^{-1}, 3 \times 10^{-1}$ were considered. For a given ϵ, we are interested in knowing the average number of fractures making up the subnetworks as per the definition, as well as whether the subnetworks are disconnected. It was found that the network size decreases steadily as expected for increasing values of ϵ, and the size reduces to approximately 7% for $\epsilon = 1 \times 10^{-1}$ while staying connected. However, network disconnection sets in for $\epsilon > 1 \times 10^{-1}$.

Using RF once again, the machine learning algorithm is trained via (20), and we have an approximation $g_\epsilon \approx \hat{g}_\epsilon$. From $\hat{g}_\epsilon$, the backbone corresponding to each network $f \in S_{test}$ is obtained as:

$$f_\epsilon = \{ f_i | \hat{g}_\epsilon(x_1^i, x_2^i, ... x_m^i) = 1 \}. \tag{21}$$

The transport characteristics of f and f_ϵ can then be compared to assess the quality of the predicted backbone.

The 80-20 random split is repeated 6 times so that we have 6 different realizations of the training and test sets. We set aside one of these training-test splits to tune parameters.

We consider values of $\epsilon = 10^{-6}, 10^{-5}, 10^{-4}, 10^{-3}, 10^{-2}, 5 \times 10^{-2}, 10^{-1}, 2 \times 10^{-1}$.

Despite knowing that the backbone fractures that comprise the training set in the case $\epsilon = 2 \times 10^{-1}$ form disconnected networks, the case $\epsilon = 2 \times 10^{-1}$ is included. This is because the unit of data in our method, as in Valera et al. (2018), is a single fracture, and the ML classifier makes decisions based solely on features of an individual fracture. Indeed, there is no way to incorporate or consider the connectivity of the fractures within the current algorithm. Due to the same reason, Valera et al. (2018) obtained disconnected subnetworks despite ensuring connected backbones in the training set through construction of an FTG. Here the authors take the point of view that the best chance to obtain connected backbones from the ML classifier is by having a good definition of backbone that offers a selection of fractures to choose from, and the connectivity of the subnetworks obtained must be checked a posteriori.

Corresponding to each value of ϵ, a parameter search was performed to determine optimal parameters for maximum recall by using the scikit-learn function gridsearchcv. These determine the number of trees (*n_estimators*), the number of features to consider for the best split (*max_features*), and the minimal number of training samples in a leaf node (*min_samples_leaf*) as indicated in Srinivasan et al. (2019). Apart from these, *criterion*= '*entropy*' was set to measure the value of a split and weights were set to be inversely proportional to class frequency by *class_weight*= '*balanced_subsample*' to alleviate the class imbalance in the problem.

We have already seen how an RF can give insight into a problem through the values of feature importance. In the previous method, we saw that betweenness centrality, a global topological feature, is the most crucial in characterizing a fracture that belongs to a backbone. However, Fig. 7 for this method gives an even more nuanced picture. In fact, we see that local topological features (simple paths and degree centrality) are the most important for small values of ϵ, and their importance then declines as the backbone class becomes sparser for larger ϵ. The trends for betweenness centrality are the reverse, i.e., the importance of betweenness centrality grows as the backbone becomes smaller. The radius and aperture are insignificant in comparison. Srinivasan et al. (2019) explain the variation of importance between the global and local topological features in the following way: for large backbones (small ϵ), the fracture will readily meet the threshold flux if it intersects more neighbors, leading to more pathways of fluid conductance through it. However, for small backbones, the flux through each fracture needs to be

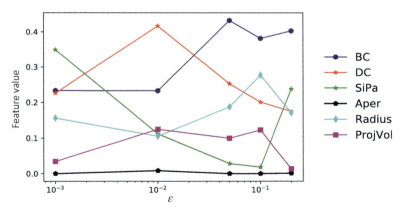

Fig. 7 Relative importance of different features in classification of a fracture, shown as a function of the threshold ϵ. The abbreviations BC, DC, SiPa, Aper, and ProjVol in the legend stand for betweenness centrality, degree centrality, (source-to-target) simple paths, aperture, and projected volume, respectively. *From Srinivasan, S., Karra, S., Hyman, J. D., Viswanathan, H. S., & Srinivasan, G. (2019). Model reduction for fractured porous media: A machine-learning approach for identifying main flow pathways. Computational Geosciences, 23(3), 617–629. https://doi.org/10.1007/s10596-019-9811-7.*

much larger, locally important fractures may not be globally important enough to conduct a large proportion of the flux, and thus betweenness centrality as a global attribute is most important.

When the performance of the RF in the context of the abstract machine learning problem was examined, the training error (in-sample error) and the test error (out-of-sample error) were found to be approximately equal, showing that the classifier generalizes well to the test set.

Finally, we are ready to evaluate the performance of the classifier in terms of the eventual quantity of interest—computational efficiency and accuracy. As a general rule, the smaller the backbone, greater the computational efficiency while accuracy is the degree to which the breakthrough curve of the subnetwork matches that of the full network. However, these two notions are on two ends of the spectrum, since smaller backbones will necessarily have fewer fractures and deviate more from the full breakthrough curve. One difference from the previous section is that one can vary ϵ to control backbone size, and indirectly, the accuracy.

From the results presented in Table 2, it is clear that the subnetworks can be as small as 12% of the full network by increasing by increasing ϵ. The reduction to small subnetworks leads to expected overall computational

Table 2 Statistics on the size of the backbones obtained from the RF for various values of ϵ.

| ϵ | Number of fractures in subnetwork | | Number of dofs in mesh of subnetwork | | Percentage of disconnected subnetworks | Average connectivity index |
| | In % of full DFN | | In % of full DFN | | | |
	Mean	SD	Mean	SD		
1×10^{-3}	38	5	50	8	0	7
1×10^{-2}	35	5	47	7	0	7
5×10^{-2}	21	3	32	5	0	6
1×10^{-1}	17	2	27	4	0	6
2×10^{-1}	12	2	19	3	0	5

As ϵ increases, the size of the subnetwork decreases. Most importantly, the subnetworks are never disconnected; even when subnetwork is less than 15% of the full network, there are no disconnected networks. The connectivity index gives the average number of fracture–fracture intersections that must be removed before the subnetwork is disconnected.
Source: From Srinivasan, S., Karra, S., Hyman, J. D., Viswanathan, H. S., & Srinivasan, G. (2019). Model reduction for fractured porous media: A machine-learning approach for identifying main flow pathways. Computational Geosciences, 23(3), 617–629. https://doi.org/10.1007/s10596-019-9811-7.

savings in terms of simulation time (total time for meshing, flow, and transport). It is known, from the documented results in Srinivasan et al. (2018) that simulation time scales roughly with network size. Thus, large reductions in computational time can be expected for small subnetworks, yielding as much as 90% computational savings for subnetworks that are 10%–12% of the full network. However, obtaining connected subnetworks is more important than computational efficiency since this was a primary shortcoming in the technique outlined in Valera et al. (2018). As shown by Table 2, this algorithm does address that. *Disconnected networks were never produced for any value of ϵ considered.* The possibility of disconnection is greater for smaller backbones, but Srinivasan et al. report that even for $\epsilon = 3 \times 10^{-1}$ (average network size 10%, not shown in table), connected subnetworks were obtained.

The connectivity index, defined as the minimum number of edges that must be removed before the network becomes disconnected, supports this expectation. For as ϵ increases, fewer edges need to be removed to disconnect the network. Compared to the algorithms in both Srinivasan et al. (2018) and Valera et al. (2018), this method produced connected backbones due to two reasons: (1) physics that utilized particle-tracking data from non-reactive tracer transport simulation instead of the graph flow solution as in Srinivasan et al. (2018), and (2) the precision–recall trade-off where the low values of precision explain how we still obtained connected subnetworks. Due to the low precision, more fractures are selected, increasing the size, but contributing to make the subnetwork connected. A perfect classifier would have led to disconnected networks, but the algorithm uses the trade-off between precision and recall to advantage.

An instance of comparison of the breakthrough curves obtained from the subnetworks is shown in Fig. 8. The breakthrough curve is shown both as a PDF and a CDF. The smallest backbone (largest ϵ) deviates most from the full breakthrough, while the largest subnetwork (smallest ϵ) displays the best agreement. Moreover, most of the subnetworks capture both the peak and the decay of the breakthrough curve.

This second system-reduction technique for DFNs using supervised machine learning via RF improves and complements the method of Valera et al. (2018) described in the previous section. It overcomes the disadvantages of the previous method proposed by allowing the size to be controlled by a single dimensionless parameter. Moreover, the parameter is physically motivated and subnetworks produced by the method remain connected.

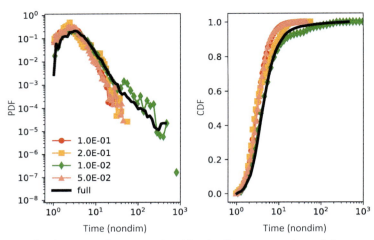

Fig. 8 Time has been nondimensionalized by the first passage time of the network so that all the breakthrough curves have a common starting point. The larger backbones capture the behavior of the full breakthrough curve. From Srinivasan, S., Karra, S., Hyman, J. D., Viswanathan, H. S., & Srinivasan, G. (2019). Model reduction for fractured porous media: A machine-learning approach for identifying main flow pathways. *Computational Geosciences, 23(3), 617–629.* https://doi.org/10.1007/s10596-019-9811-7.

It is important to emphasize that the technique we proposed here, just like Srinivasan et al. (2018) and Valera et al. (2018), *still does not* explicitly build in the feature of connected subnetworks. As subnetworks decrease in size, at some stage ($\epsilon = 1$ offers a trivial example), one will invariably get disconnected backbones. The next section will describe a system-reduction technique which *guarantees* that no disconnected networks would be generated.

5.3 Path-classification: Labeling by FTG membership

As we have seen, machine learning methods are mathematical constructs that are agnostic to the source of data or the physics models that may have produced the data. Physics-informed machine learning (PIML) methods (Karpatne et al., 2017; Pan & Duraisamy, 2018) attempt to correct this deficiency. They combine knowledge of the physics of the phenomena into the machine learning framework by viewing the governing equations or physical principles as constraints on the system or search space of solutions.

In the construction of a backbone through the removal of fractures, the constraint is the requirement of a connected path through the backbone to mimic the flow and transport through the full network. The constraint on our physical system limits the space of solutions, which is the power set (set of all subsets) of the set of fractures in the full network.

Both the previous methods are unable to theoretically guarantee connected networks because the fundamental unit of selection in the classification algorithm is a fracture, where it is impossible to limit the selection to connected subnetworks.

Srinivasan, Cawi, et al. (2020) address this issue by viewing a DFN as a union of sequences of fractures, i.e., connected paths through the network, rather than a union of individual fractures. The motivation for this choice emerged from the recognition that a particle travels through a sequence of fractures. Thus the backbone itself is a sequence of fractures along which particles pass through the system most often. Within the framework of our abstract classification problem, the fundamental unit of selection is now a sequence of fractures, rather than individual fractures. As a by product, features for classification are also based on the sequence of fractures. This effectively restricts the sample space to ensure that every selected sample subnetwork (which is the union of all selected sequences of fractures) is connected.

The criteria to label paths is similar to that of Section 5.1. The 40 shortest paths in the FTG of a DFN are chosen to be its backbone. Hence, if a simple path in the graph of the DFN coincides with one of the 40, it is labeled a backbone path. In order to realize the graph of the DFN as a union of paths, the complication is that computing all possible simple paths through the graph representation of a DFN is combinatorially impractical because the number of paths grows exponentially. However, computing the first k unweighted shortest paths is feasible. Srinivasan, Cawi, et al. (2020) confirmed that 97% of vertices/fractures in the graph/DFN are contained in the first 1000 shortest paths through the graphs. However, they found that to include the remaining 3% of the vertices/fractures would require roughly double the number of shortest paths and hence elected to use only the 1000 shortest paths for each DFN as elements for classification.

We defined a path in a graph earlier as a sequence of vertices so that the edges when taken in order connect the first and last vertex in the sequence. A simple path is a path with no repeating vertices (i.e., a sequence with distinct elements) and the set of all simple paths from source to target is denoted as G_P. The concept of paths on graphs is inherited from that of particle-trace on the corresponding DFN as was described in Section 2.

Denoting the set of paths used to construct the backbone as G_{BP}, assign the label "1" to paths in G_{BP}, and "0" to those in $G_P - G_{BP}$.

The learning problem for the mapping for paths proceeds in steps. First, each path $p \in G_P$ is transformed into a feature vector $x \in \mathbb{R}^{N_P}$ (with length N_P) by the action of a function $\phi(\bullet)$.

$$\phi_P : G_P \to \mathbb{R}^{N_P}, \tag{22}$$

The features represent physical and graph-based properties of the paths, and the ones used for this method are discussed subsequently.

Next, a model function $M(\bullet)$ is constructed, which assigns probability of backbone membership corresponding to a feature vector of a path.

$$M_P : \mathbb{R}^{N_P} \to \mathbb{R}, \tag{23}$$

$$\forall \, x \in \mathbb{R}^{N_P}, \, M_P(x) = \text{Prob}(\phi_P^{-1}(x) \in G_{BP}) \tag{24}$$

Finally, a threshold $T > 0$ is defined, and it assigns backbone membership to paths with probability higher than T through a function $f_P(\bullet)$ as follows:

$$f_P : G_P \to \{0, 1\}, \tag{25}$$

$$\forall \, p \in G_P, \, f_P(p) = \begin{cases} 1 & M_P(\phi_P(p)) > T \\ 0 & \text{else} \end{cases} \tag{26}$$

Therefore, one has $G_{BP} = f_P^{-1}(\{1\})$. Once all the paths (sequences of vertices/fractures) that comprise the set G_{BP} are known, so too are all the fractures that comprise the set G_B, thus completing the description of the backbone.

For every path $p \in G_p$, Srinivasan, Cawi, et al. (2020) used five features based on individual vertex properties along that path to classify backbone membership: degree centrality, betweenness centrality, current flow betweenness centrality, and both the arithmetic and harmonic mean of permeability.

The values of degree centrality, betweenness centrality, and current flow betweenness centrality are the mean of these features computed at every vertex on the path, i.e., for a vertex feature $f(i)$, we calculate the path-feature to be $\Sigma_{i=1}^{N} f(i)/N$ for a path with N vertices. The mean permeability along the path is defined as the mean of the edge permeabilities along the path, while for the harmonic permeability, it is the harmonic average.

A measure of classification accuracy that combines both precision and recall is the "F1-score," being the harmonic average of the two. The sensitivity of the harmonic average to low values ensures that high scores can result only from classifiers with high values of both precision and recall.

The classification problem in this method is solved for paths, but we note that our eventual goal is a set of fractures. We realize that the aggregate of the selected paths is the set of fractures, and thus the backbone realized as a union of paths preserves the network connectivity.

The RF and LR models were used for the model function $M(\cdot)$ in (26) since these methods are representative of two different families of algorithms. We recall that the RF algorithm is based on decision trees, while LR is based on gradient descent.

The same procedure was then used to split the data, with 80% used for training and 20% used for testing. Specifically, out of a total of 100 networks, 80 were randomly selected for training and the remaining 20 were set aside for testing. Then, for each of the 1000 simple paths for a network, 5 features and a label were generated. The training set was then further split by 10-fold cross-validation using the subroutines available in `scikit-learn` to train the hyperparameters of the model for the best "F1-score".

5.3.1 Logistic regression

Logistic regression is a generalized linear model used for binary classification. If $x_i \in R^{N_p}$ is the feature vector for path i, and $Y_i \in \{0, 1\}$ is the binary variable representing backbone membership, then the probability of backbone membership is expressed as:

$$P(Y_i = 1 | x_i) = \frac{1}{1 + \exp(\beta_0 + \boldsymbol{\beta}^T x_i)} \tag{27}$$

where $\boldsymbol{\beta}$ is a vector of regression coefficients. $\boldsymbol{\beta}$ and β_0 are typically found by maximum likelihood estimation given the training data. Logistic regression is implemented by using maximum likelihood estimation (MLE) which is performed using a gradient descent algorithm. Maximizing the likelihood function determines the parameters that are most likely to produce the observed data. The coefficient β_k can be interpreted as the influence of the k-th feature on the log-odds of backbone membership. For example if $\beta_k = 5$ then if the k-th feature is increased by one the log-odds of backbone membership are increased fivefold. Typically, a regularization parameter C is also added to penalize misclassification, with smaller C corresponding to a stronger regularization (Hosmer, Lemeshow, & Sturdivant, 2013). In cross-validation using `scikit-learn`, Srinivasan et al. found $C = 1.291$.

5.3.1.1 Random forest

The total number of features in the data is denoted by N_p, and in this case, $N_p = 5$. The parameters of interest are the number of trees in the ensemble, `n_estimators`, and the number of features to use at each split in each tree, `max_features`. Srinivasan et al. performed cross-validation on a parameter range of `n_estimators` in [10, 50, 100, 200, 400, 500] and `max_features` one of $\{N_p, \sqrt{N_p}, \log_2(N_p)\}$. The best parameters were found to be `n_estimators`=100 and `max_features` $= \log_2 N_p$ features used in each split.

The immediate result of the algorithm is an assignment of probability of class-membership for each item being classified. The hyperparameters of the RF and LR model were tuned by cross-validation to obtain the best "F1-score," but it is more informative to list the precision and recall scores separately and construct the PR curve (precision–recall curve). The PR curves were computed based on the realizations designated for testing alone. To be clear, for each of the 100 realizations, 1000 labeled shortest paths were considered. The testing data comprised 20 realizations chosen out of the 100, and PR curves were computed by comparing the predicted labels with the true labels for each path. The threshold probability T that finally decides membership in (26) is a user-defined parameter. Its modification allows control of precision and recall, and thus indirectly, the extent of system reduction. An approximate one-to-one correspondence between the threshold probability T and the resultant precision p and recall r may be set up by sampling values of the probability $T \in [0, 1]$, with precision and recall computed for each value.

Since the classification problem is for the identification of the set G_{BP}, these measures for paths are immediately relevant, hence one can measure the precision and recall in terms of the fractures too, with the knowledge that the set of selected fractures G_B, realized as a union of paths, preserves the network connectivity. The two PR curves obtained for these cases are shown in Fig. 9. It is worth emphasizing that no fracture-classification problem has been solved—the results of the path-classification problem have been cast in terms of both paths G_{BP} and fractures G_B. If it were a perfect classifier, it would have a precision and recall of hundred percent, and the curve would be horizontal at precision $= 1$ across every recall value. Since that is impossible, PR curves close to the top right corner are deemed desirable. Note that the two algorithms perform very similarly, with LR having higher precision at almost all recall values.

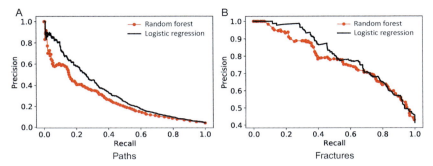

Fig. 9 The precision–recall (PR) curves have been constructed for the original path-classification problem in (A). The path-classification problem results in identification of the set G_{BP} that comprises backbone paths. The aggregate of the selected paths in G_{BP} yields a set of selected fractures G_B. Measures of precision–recall for the derived set of fractures are shown in (B). (A) Paths; (B) Fractures.

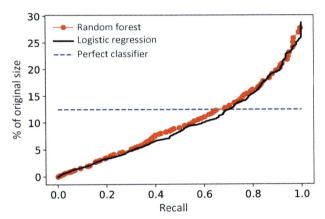

Fig. 10 Size reduction (the number of fractures in the backbone as a percentage of the full network) as a function of recall for both RF and LR. As recall increases, precision decreases and hence the backbone includes a larger number of fractures. The perfect classifier is one that would produce neither false positives nor false negatives, and this corresponds to the true backbone of the networks in the test dataset.

The resultant system reduction is shown in Fig. 10, and one may conclude that for a given recall, both methods yield backbones of approximately similar size. Minimizing the loss of important fractures due to false negative classification is of most interest, and on this parameter both algorithms yield small backbones of approximately 30% size even at very high recall. The

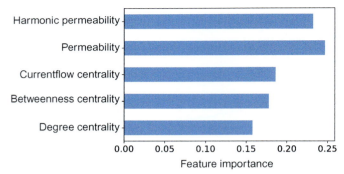

Fig. 11 An estimate of feature importance for the path-classification problem obtained from the random forest classifier.

figure also shows the size of the backbone that could be obtained by a perfect classifier. Both RF and LR yield backbones of comparable size for values of precision and recall around 70%.

As before, the values of feature importance provided by the RF are also examined. These are shown in Fig. 11, where it becomes clear that no features are superfluous. In this context, the feature importance values also shed light on an important distinction between fracture-classification and path-classification. Whereas fracture-classification (Srinivasan et al., 2019; Valera et al., 2018) was dominated by global topological features, permeability regains importance in path-classification. The fundamental change to the classification problem (path classification rather than fracture-classification) has led to this insight into flow channeling.

For recall values of 50%, 80%, and 90%, respectively, let us examine the performance of the algorithms more closely. In Table 3, system reduction of the DFN is quantified. We note that for each recall value, LR has greater network size, and the explanation for these numbers lies in the threshold probability T that corresponds to the recall. Thus to get a recall of 50%, $T = 0.48$ for RF but 0.30 for LR and this leads to more paths (and fractures) selected by LR.

Moving on from the metrics of the classification problem to the transport characteristics of the backbone, it is expected that adding more fractures will capture more of the particle flow than the true backbone. Comparing the breakthrough curves of RF and LR for 50% and 90% recall in Fig. 12, it is observed that the breakthrough curve of LR matches that of the full network better, again due to the lower threshold probability and larger backbone.

Table 3 Reduction measures for random forests and logistic regression at 50%, 80%, and 90% recall thresholds.

Model	Recall	Network size (fracture)	Network size (mesh)
RF	50	8%	16%
LOG	50	12%	22%
RF	80	17%	30%
LOG	80	20%	34%
RF	90	20%	34%
LOG	90	23%	38%

The precision–recall curves for random forest and logistic regression were almost indistinguishable from each other, and these numbers further emphasize the trend.

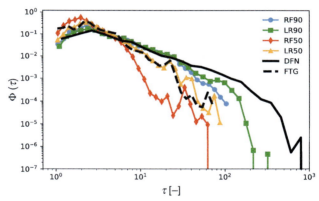

Fig. 12 An instance showing the breakthrough curves obtained from the classifiers for 50% and 90% recall for both random forest (RF) and logistic regression (LR). The numbers in the legend indicate the percentage value of recall. The breakthrough curves obtained from the full network (legend DFN) and the true backbone which is the flow-topology-graph (FTG) is also shown. *From Srinivasan, S., Cawi, E., Hyman, J. D., Osthus, D., Hagberg, A., Viswanathan, H., & Srinivasan, G. (2020). Physics-informed machine-learning for backbone identification in discrete fracture networks. Computational Geosciences, 24, 1429–1444. https://doi.org/10.1007/s10596-020-09962-5.*

In contrast to the methods in Section 5.1 and 5.2 (Srinivasan et al., 2019; Valera et al., 2018) where the fundamental unit of selection was a fracture, the novel perspective that views a network as a union of connected paths ensures backbone connectivity. Unlike fracture-classification, where the global topological features were overwhelmingly dominant, in path-classification, we saw that permeability does play an important role, although as a global property associated with a path. In this light, the proposed method is the first physics-informed machine learning algorithm for backbone identification within discrete fracture networks simulations.

6. Further scope for ML in fractured media

One disadvantage of all the ML methods presented so far is the computational expense of the high-fidelity simulations to generate the training data. If low-fidelity data from reduced-order models could be used for training, it will further increase computational efficiency.

At this juncture, we also note that all the ROMs alluded to thus far correspond to *steady flow*. However, steady flow is an idealization that is not always appropriate. For the observation of any physical phenomena of interest, the notion of an initial time and initial condition exist naturally and all phenomena may be considered transient problems with an initial condition. For a large class of problems of interest, (a) the forcing function of the problem may be assumed to be time-independent and (b) it is not the initial transients that are of interest but the long term, steady-state behavior.

However, either (a) or (b) or both could be invalid, in which case the transient response of a system needs to be considered. It was mentioned earlier that the steady-state problem on DFNs is computationally expensive, but the situation in case of the transient problem is even worse, because of the need for repeated solution in a time marching loop. In that case, even the reduced system produced here by machine learning is unviable as a ROM for transient problems. This is especially relevant for fractured hydrocarbon reservoirs where the quantity of interest is the time varying production rate or an optimal well pressure management strategy to maximize the production rate. Recently, Mudunuru et al. (2020) proposed a framework where machine learning as an inverse model is suggested in conjunction with the ROM proposed by Patzek, Male, and Marder (2013) as the forward model. However, research is still ongoing and more results and evidence are required before there can be clarity on a clear strategy.

References

Abelin, H., Birgersson, L., Moreno, L., Widén, H., Ågren, T., & Neretnieks, I. (1991). A large-scale flow and tracer experiment in granite: 2. results and interpretation. *Water Resources Research, 27*(12), 3119–3135.

Abelin, H., Neretnieks, I., Tunbrant, S., & Moreno, L. (1985). *Final report of the migration in a single fracture: Experimental results and evaluation.* (Tech. Rep. Nos. SKB-SP-TR–85-03).

Abu-Mostafa, Y. S., Magdon-Ismail, M., & Lin, H.-T. (2012). Learning from data. In (Vol. 4). New York, NY, USA: AMLBook.

Aldrich, G., Hyman, J. D., Karra, S., Gable, C. W., Makedonska, N., Viswanathan, H., ... Hamann, B. (2017). Analysis and visualization of discrete fracture networks using a flow topology graph. *IEEE Transactions on Visualization and Computer Graphics*, *23*(8), 1896–1909. https://doi.org/10.1109/tvcg.2016.2582174.

Alemanni, A., Battaglia, M., Bigi, S., Borisova, E., Campana, A., Loizzo, M., & Lombardi, S. (2011). A three dimensional representation of the fracture network of a CO2 reservoir analogue (Latera Caldera, Central Italy). *Energy Procedia*, *4*, 3582–3587. https://doi.org/10.1016/j.egypro.2011.02.287.

Andresen, C. A., Hansen, A., Le Goc, R., Davy, P., & Hope, S. M. (2013). Topology of fracture networks. *Frontiers in Physics*, *1*, Art 7.

Bear, J. (2013). Dynamics of fluids in porous media. Courier Dover Publications.

Bergen, K. J., Johnson, P. A., de Hoop, M. V., & Beroza, G. C. (2019). Machine learning for data-driven discovery in solid earth geoscience. *Science*, *363*(6433). https://doi.org/10.1126/science.aau0323.

Berre, I., Doster, F., & Keilegavlen, E. (2019). Flow in fractured porous media: A review of conceptual models and discretization approaches. *Transport in Porous Media*, *130*(1), 215–236.

Berrone, S., Canuto, C., Pieraccini, S., & Scialò, S. (2018). Uncertainty quantification in discrete fracture network models: Stochastic geometry. *Water Resources Research*, *54*, 1338–1352. https://doi.org/10.1002/2017wr021163.

Berrone, S., Pieraccini, S., & Scialo, S. (2013). A PDE-constrained optimization formulation for discrete fracture network flows. *SIAM Journal on Scientific Computing*, *35*(2), B487–B510.

Berrone, S., Pieraccini, S., Scialò, S., & Vicini, F. (2015). A parallel solver for large scale DFN flow simulations. *SIAM Journal on Scientific Computing*, *37*(3), C285–C306.

Biot, M. A. (1956a). Theory of elastic waves in a fluid-saturated porous solid. I. High frequency range. *Journal of the Acoustical Society of America*, *28*, 179–191.

Biot, M. A. (1956b). Theory of elastic waves in a fluid-saturated porous solid. I. Low frequency range. *Journal of the Acoustical Society of America*, *28*, 168–178.

Bogdanov, I. I., Mourzenko, V. V., Thovert, J. F., & Adler, P. M. (2007). Effective permeability of fractured porous media with power-law distribution of fracture sizes. *Physical Review E*, *76*(3), 036309.

Botros, F. E., Hassan, A. E., Reeves, D. M., & Pohll, G. (2008). On mapping fracture networks onto continuum. *Water Resources Research*, *44*(8). https://doi.org/10.1029/2007wr006092.

Boussinesq, J. (1868). Mémoire sur l'influence des frottements dans les mouvements réguliers des fluids. *J Math Pures Appl*, *13*(377–424), 21.

Bridgman, P. W. (1931). The physics of high pressure. New York, USA: MacMillan Company.

Cacas, M.-C., Ledoux, E., Marsily, G. d., Tillie, B., Barbreau, A., Durand, E., ... Peaudecerf, P. (1990). Modeling fracture flow with a stochastic discrete fracture network: Calibration and validation: 1. The flow model. *Water Resources Research*, *26*(3), 479–489.

Cracknell, M. J., & Reading, A. M. (2014). Geological mapping using remote sensing data: A comparison of five machine learning algorithms, their response to variations in the spatial distribution of training data and the use of explicit spatial information. *Computers & Geosciences*, *63*, 22–33.

de Dreuzy, J. R., Darcel, C., Davy, P., & Bour, O. (2004). Influence of spatial correlation of fracture centers on the permeability of two-dimensional fracture networks following a power law length distribution. *Water Resources Research*, *40*(1).

de Dreuzy, J. R., Davy, P., & Bour, O. (2001). Hydraulic properties of two-dimensional random fracture networks following a power law length distribution 2. Permeability of networks based on lognormal distribution of apertures. *Water Resources Research, 37*(8), 2079–2095.

de Dreuzy, J. R., Davy, P., & Bour, O. (2002). Hydraulic properties of two-dimensional random fracture networks following power law distributions of length and aperture. *Water Resources Research, 38*(12).

de Dreuzy, J. R., Méheust, Y., & Pichot, G. (2012). Influence of fracture scale heterogeneity on the flow properties of three-dimensional discrete fracture networks. *Journal of Geophysical Research-Solid Earth, 117*(B11).

Dershowitz, W. S., & Fidelibus, C. (1999). Derivation of equivalent pipe network analogues for three-dimensional discrete fracture networks by the boundary element method. *Water Resources Research, 35*(9), 2685–2691.

Dev, S., Wen, B., Lee, Y. H., & Winkler, S. (2016). Ground-based image analysis: A tutorial on machine-learning techniques and applications. *IEEE Geoscience and Remote Sensing Magazine, 4*(2), 79–93.

Efendiev, Y., Galvis, J., & Wu, X. H. (2011). Multiscale finite element methods for high-contrast problems using local spectral basis functions. *Journal of Computational Physics, 230*, 937–955.

Erhel, J., de Dreuzy, J. R., & Poirriez, B. (2009). Flow simulation in three-dimensional discrete fracture networks. *SIAM Journal on Scientific Computing, 31*(4), 2688–2705.

Frampton, A., & Cvetkovic, V. (2011). Numerical and analytical modeling of advective travel times in realistic three-dimensional fracture networks. *Water Resources Research, 47*(2).

Friedman, J., Hastie, T., & Tibshirani, R. (2001). The elements of statistical learning *(Vol. 1) (No. 10)*. New York, NY, USA: Springer.

Ghaffari, H. O., Nasseri, M. H. B., & Young, R. P. (2011). Fluid flow complexity in fracture networks: Analysis with graph theory and LBM. *arXiv preprint arXiv:11074918.*

Giles, M. B. (2008). Multilevel Monte Carlo path simulation. *Operations Research, 56*(3), 607–617. https://doi.org/10.1287/opre.1070.0496.

Griewank, A., Reich, S., Roulstone, I., & Stuart, A. M. (2017). Mathematical and algorithmic aspects of data assimilation in the geosciences. *Oberwolfach Reports, 13*(4), 2705–2748.

Hagberg, A. A., Schult, D. A., & Swart, P. (2008). Exploring network structure, dynamics, and function using NetworkX. In *Proceedings of the 7th python in science conferences (SciPy 2008)* (Vol. 2008. pp. 11–16).

Hengl, T., de Jesus, J. M., Heuvelink, G. B. M., Gonzalez, M. R., Kilibarda, M., Blagotić, A., … Kempen, B. (2017). SoilGrids250m: Global gridded soil information based on machine learning. *PLoS ONE, 12*(2), e0169748.

Hope, S. M., Davy, P., Maillot, J., Le Goc, R., & Hansen, A. (2015). Topological impact of constrained fracture growth. *Frontiers in Physics, 3*, 75.

Hornung, U. (Ed.). (1997). In (Vol. 6.), New York: Springer. https://doi.org/10.1007/978-1-4612-1920-0.

Hosmer, D. W., Jr, Lemeshow, S., & Sturdivant, R. X. (2013). Applied logistic regression. In (Vol. 398)John Wiley & Sons.

Hou, T.-Y., & Wu, X. (1997). A multiscale finite element method for elliptic problems in composite materials and porous media. *Journal of Computational Physics, 134*, 169–189.

Huseby, O., Thovert, J. F., & Adler, P. M. (1997). Geometry and topology of fracture systems. *Journal of Physics A: Mathematical and General, 30*, 1415–1444. https://doi.org/10.1088/0305-4470/30/5/012.

Hyman, J., Aldrich, G., Viswanathan, H., Makedonska, N., & Karra, S. (2016). Fracture size and transmissivity correlations: Implications for transport simulations in sparse three-dimensional discrete fracture networks following a truncated power law distribution of fracture size. *Water Resources Research*, *52*(8), 6472–6489.

Hyman, J. D., Gable, C. W., Painter, S. L., & Makedonska, N. (2014). Conforming Delaunay triangulation of stochastically generated three dimensional discrete fracture networks: A feature rejection algorithm for meshing strategy. *SIAM Journal on Scientific Computing*, *36*(4), A1871–A1894.

Hyman, J. D., Hagberg, A., Osthus, D., Srinivasan, S., Srinivasan, G., & Viswanathan, H. S. (2018). Identifying backbones in three-dimensional discrete fracture networks: A bipartite graph-based approach. *Multiscale Modeling & Simulation*, *16*(4), 1948–1968. https://doi.org/10.1137/18M1180207.

Hyman, J. D., Hagberg, A., Srinivasan, G., Mohd-Yusof, J., & Viswanathan, H. (2017). Predictions of first passage times in sparse discrete fracture networks using graph-based reductions. *Physical Review E*, *96*(1), 013304.

Hyman, J. D., Jiménez-Martínez, J., Viswanathan, H. S., Carey, J. W., Porter, M. L., Rougier, E., … Makedonska, N. (2016). Understanding hydraulic fracturing: A multi-scale problem. *Philosophical Transactions of the Royal Society of London A: Mathematical, Physical and Engineering Sciences*, *374*(2078), 20150426.

Hyman, J. D., Karra, S., Makedonska, N., Gable, C. W., Painter, S. L., & Viswanathan, H. S. (2015). dfnWorks: A discrete fracture network framework for modeling subsurface flow and transport. *Computers & Geosciences*, *84*, 10–19.

Hyman, J. D., Painter, S. L., Viswanathan, H., Makedonska, N., & Karra, S. (2015). Influence of injection mode on transport properties in kilometer-scale three-dimensional discrete fracture networks. *Water Resources Research*, *51*(9), 7289–7308.

Jackson, C. P., Hoch, A. R., & Todman, S. (2000). Self-consistency of a heterogeneous continuum porous medium representation of a fractured medium. *Water Resources Research*, *36*, 189–202. https://doi.org/10.1029/1999wr900249.

Jenkins, C., Chadwick, A., & Hovorka, S. D. (2015). The state of the art in monitoring and verification—Ten years on. *International Journal for Greenhouse Gas Control*, *40*, 312–349.

Jenny, P., Lee, S. H., & Tchelepi, H. A. (2003). Multi-scale finite-volume method for elliptic problems in subsurface flow simulation. *Journal of Computational Physics*, *187*(1), 47–67.

Kang, P. K., Dentz, M., Le Borgne, T., & Juanes, R. (2015). Anomalous transport on regular fracture networks: Impact of conductivity heterogeneity and mixing at fracture intersections. *Physical Review E*, *92*(2), 022148. https://doi.org/10.1103/PhysRevE.92.022148.

Karimi-Fard, M., Gong, B., & Durlofsky, L. J. (2006). Generation of coarse-scale continuum flow models from detailed fracture characterizations. *Water Resources Research*, *42*. https://doi.org/10.1029/2006wr005015.

Karpatne, A., Atluri, G., Faghmous, J. H., Steinbach, M., Banerjee, A., Ganguly, A., … Kumar, V. (2017). Theory-guided data science: A new paradigm for scientific discovery from data. *IEEE Transactions on Knowledge and Data Engineering*, *29*(10), 2318–2331. https://doi.org/10.1109/TKDE.2017.2720168.

Karpatne, A., Ebert-Uphoff, I., Ravela, S., Babaie, H. A., & Kumar, V. (2018). Machine learning for the geosciences: Challenges and opportunities. *IEEE Transactions on Knowledge and Data Engineering*, 1. https://doi.org/10.1109/TKDE.2018.2861006.

Karra, S., Makedonska, N., Viswanathan, H. S., Painter, S. L., & Hyman, J. D. (2015). Effect of advective flow in fractures and matrix diffusion on natural gas production. *Water Resources Research*, *51*(10), 8646–8657.

Karra, S., O'Malley, D., Hyman, J. D., Viswanathan, H. S., & Srinivasan, G. (2018). Modeling flow and transport in fracture networks using graphs. *Physical Review E*, *97*(3), 033304.

Kueper, B. H., & McWhorter, D. B. (1991). The behavior of dense, nonaqueous phase liquids in fractured clay and rock. *Ground Water*, *29*(5), 716–728.

LaGriT. (2013). In *Los Alamos Grid Toolbox, (LaGriT)*. http://lagrit.lanl.gov. (Last Checked : May 20, 2019).

Lary, D. J., Alavi, A. H., Gandomi, A. H., & Walker, A. L. (2016). Machine learning in geosciences and remote sensing. *Geoscience Frontiers, 7*(1), 3–10.

Lichtner, P. C., Hammond, G. E., Lu, C., Karra, S., Bisht, G., Andre, B., ... Kumar, J. (2015). *PFLOTRAN user manual: A massively parallel reactive flow and transport model for describing surface and subsurface processes* (Tech. Rep.). (Report No.: LA-UR-15-20403) Los Alamos National Laboratory. https://doi.org/10.2172/1168703.

Maillot, J., Davy, P., Goc, R. L., Darcel, C., & De Dreuzy, J. R. (2016). Connectivity, permeability, and channeling in randomly distributed and kinematically defined discrete fracture network models. *Water Resources Research, 52*(11), 8526–8545.

Makedonska, N., Hyman, J. D., Karra, S., Painter, S. L., Gable, C. W., & Viswanathan, H. S. (2016). Evaluating the effect of internal aperture variability on transport in kilometer scale discrete fracture networks. *Advanced Water Resources, 94*, 486–497.

Makedonska, N., Painter, S. L., Bui, Q. M., Gable, C. W., & Karra, S. (2015). Particle tracking approach for transport in three-dimensional discrete fracture networks. *Computational Geosciences, 19*(5), 1123–1137.

Middleton, R. S., Carey, J. W., Currier, R. P., Hyman, J. D., Kang, Q., Karra, S., ... Viswanathan, H. S. (2015). Shale gas and non-aqueous fracturing fluids: Opportunities and challenges for supercritical CO_2. *Applied Energy, 147*, 500–509.

Minev, P., Srinivasan, S., & Vabishchevich, P. (2018). Flux formulation of parabolic equations with highly heterogeneous coefficients. *Journal of Computational and Applied Mathematics, 340*, 582–601. https://doi.org/10.1016/j.cam.2017.12.003.

Mudunuru, M. K., O'Malley, D., Srinivasan, S., Hyman, J. D., Sweeney, M. R., Frash, L., ... Viswanathan, H. S. (2020). Physics-informed machine learning for real-time reservoir management. In J. Lee, E. F. Darve, P. K. Kitanidis, M. W. Farthing, & S. C. Tyler Hesser 03 (Eds.), *Proceedings of the AAAI 2020 spring symposium on combining artificial intelligence and machine learning with physical sciences (AAAI-MLPS 2020)*. http://ceur-ws.org/Vol-2587/.

Mustapha, H., & Mustapha, K. (2007). A new approach to simulating flow in discrete fracture networks with an optimized mesh. *SIAM Journal on Scientific Computing, 29*, 1439.

National Oceanic and Atmospheric Administration. In *National Centers for Environmental Information*. www.ncdc.noaa.gov.

National Research Council. (1996). Rock fractures and fluid flow: Contemporary understanding and applications. National Academy Press.

Neuman, S. P. (2005). Trends, prospects and challenges in quantifying flow and transport through fractured rocks. *Hydrogeology Journal, 13*(1), 124–147.

Ng, L. W. T., & Willcox, K. E. (2014). Multifidelity approaches for optimization under uncertainty. *International Journal for Numerical Methods in Engineering, 100*(10), 746–772. https://doi.org/10.1002/nme.4761.

Nœtinger, B., & Jarrige, N. (2012). A quasi steady state method for solving transient Darcy flow in complex 3D fractured networks. *Journal of Computational Physics, 231*(1), 23–38.

O'Malley, D., Karra, S., Hyman, J. D., Viswanathan, H. S., & Srinivasan, G. (2018). Efficient Monte Carlo with graph-based subsurface flow and transport models. *Water Resources Research, 54*, 3758–3766. https://doi.org/10.1029/2017wr022073.

Painter, S., & Cvetkovic, V. (2005). Upscaling discrete fracture network simulations: An alternative to continuum transport models. *Water Resources Research, 41*(2).

Painter, S. L., Gable, C. W., & Kelkar, S. (2012). Pathline tracing on fully unstructured control-volume grids. *Computational Geosciences, 16*(4), 1125–1134.

Pan, S., & Duraisamy, K. (2018). Data-driven discovery of closure models. *SIAM Journal on Applied Dynamical Systems, 17*(4), 2381–2413.

Patzek, T. W., Male, F., & Marder, M. (2013). Gas production in the Barnett shale obeys a simple scaling theory. *Proceedings of the National Academy of Sciences, 110*(49), 19731–19736.

Pedregosa, F., Varoquaux, G., Gramfort, A., Michel, V., Thirion, B., Grisel, O., ...Duchesnay, E. (2011). Scikit-learn: Machine Learning in Python. *Journal of Machine Learning Research, 12*, 2825–2830.

Pichot, G., Erhel, J., & de Dreuzy, J. R. (2012). A generalized mixed hybrid mortar method for solving flow in stochastic discrete fracture networks. *SIAM Journal on Scientific Computing, 34*(1), B86–B105.

Pichot, G., Erhel, J., & de Dreuzy, J. R. (2010). A mixed hybrid mortar method for solving flow in discrete fracture networks. *Applicable Analysis, 89*(10), 1629–1643.

Rajagopal, K. R., & Srinivasan, S. (2014). Flow of fluids through porous media due to high pressure gradients: Part 2—Unsteady flows. *Journal of Porous Media, 17*(9), 751–762. https://doi.org/10.1615/JPorMedia.v17.i9.10.

Sævik, P. N., & Nixon, C. W. (2017). Inclusion of topological measurements into analytic estimates of effective permeability in fractured media. *Water Resources Research, 53*(11), 9424–9443. https://doi.org/10.1002/2017WR020943.

Santiago, E., Romero-Salcedo, M., Velasco-Hernández, J. X., Velasquillo, L. G., & Hernández, J. A. (2013). An integrated strategy for analyzing flow conductivity of fractures in a naturally fractured reservoir using a complex network metric. In I. Batyrshin & M. G. Mendoza (Eds.), *Advances in computational intelligence: 11th Mexican international conference on artificial intelligence, MICAI 2012, San Luis Potosí, Mexico, October 27–November 4, 2012. Revised Selected Papers, Part II, Berlin, Heidelberg* (pp. 350–361). Springer.

Santiago, E., Velasco-Hernández, J. X., & Romero-Salcedo, M. (2014). A methodology for the characterization of flow conductivity through the identification of communities in samples of fractured rocks. *Expert Systems With Applications, 41*(3), 811–820. https://doi.org/10.1016/j.eswa.2013.08.011.

Santiago, E., Velasco-Hernández, J. X., & Romero-Salcedo, M. (2016). A descriptive study of fracture networks in rocks using complex network metrics. *Computers & Geosciences, 88*, 97–114.

Scavuzzo, J. M., Trucco, F., Espinosa, M., Tauro, C. B., Abril, M., Scavuzzo, C. M., & Frery, A. C. (2018). Modeling dengue vector population using remotely sensed data and machine learning. *Acta Tropica, 185*, 167–175.

Sherman, T., Hyman, J. D., Bolster, D., Makedonska, N., & Srinivasan, G. (2019). Characterizing the impact of particle behavior at fracture intersections in three-dimensional discrete fracture networks. *Physical Review E, 99*(1), 013110. https://doi.org/10.1103/PhysRevE.99.013110.

SKB. (2011) *Long-term safety for the final repository for spent nuclear fuel at Forsmark* (Tech. Rep. No. SKB TR-11-01). Swedish Nuclear Fuel and Waste Management Co., Stockholm.

Srinivasan, G., Hyman, J. D., Osthus, D., Moore, B., O'Malley, D., Karra, S., ... Viswanathan, H. (2018). Quantifying topological uncertainty in fractured systems using graph theory and machine learning *Nature Scientific Reports, 8*(11665).

Srinivasan, S. (2016). A generalized Darcy-Dupuit-Forchheimer model with pressure-dependent drag coefficient for flow through porous media under large pressure gradients. *Transport in Porous Media, 111*(3), 741–750. https://doi.org/10.1007/s11242-016-0625-y.

Srinivasan, S., Bonito, A., & Rajagopal, K. R. (2013). Flow of a fluid through a porous solid due to high pressure gradients. *Journal of Porous Media, 16*(3), 193–203. https://doi.org/10.1615/JPorMedia.v16.i3.20.

Srinivasan, S., Cawi, E., Hyman, J. D., Osthus, D., Hagberg, A., Viswanathan, H., & Srinivasan, G. (2020). Physics-informed machine-learning for backbone identification in discrete fracture networks. *Computational Geosciences, 24*, 1429–1444. https://doi.org/10.1007/s10596-020-09962-5.

Srinivasan, S., Hyman, J. D., Karra, S., O'Malley, D., Viswanathan, H. S., & Srinivasan, G. (2018). Robust system size reduction of discrete fracture networks: A multi-fidelity method that preserves transport characteristics. *Computational Geosciences*, *22*(6), 1515–1526. https://doi.org/10.1007/s10596-018-9770-4.

Srinivasan, S., Karra, S., Hyman, J. D., Viswanathan, H. S., & Srinivasan, G. (2019). Model reduction for fractured porous media: A machine-learning approach for identifying main flow pathways. *Computational Geosciences*, *23*(3), 617–629. https://doi.org/10.1007/s10596-019-9811-7.

Srinivasan, S., Lazarov, R., & Minev, P. (2016). Multiscale direction-splitting algorithms for parabolic equations with highly heterogeneous coefficients. *Computers and Mathematics with Applications*, *72*(6), 1641–1654. https://doi.org/10.1016/j.camwa.2016.07.032.

Srinivasan, S., O'Malley, D., Hyman, J. D., Karra, S., Viswanathan, H. S., & Srinivasan, G. (2020). Transient flow modelling in fractured media using graphs. *Physical Review E: Statistical Physics, Plasmas, Fluids, and Related Interdisciplinary Topics.* (under review).

Srinivasan, S., & Rajagopal, K. R. (2014). A thermodynamic basis for the derivation of the Darcy, Forchheimer and Brinkman models for flows through porous media and their generalizations. *International Journal of Non-Linear Mechanics*, *58*, 162–166. https://doi.org/10.1016/j.ijnonlinmec.2013.09.004.

Srinivasan, S., & Rajagopal, K. R. (2016). On the flow of fluids through inhomogeneous porous media due to high pressure gradients. *International Journal of Non-Linear Mechanics*, *78*, 112–120. https://doi.org/10.1016/j.ijnonlinmec.2015.09.003.

Tsang, C. F., & Neretnieks, I. (1998). Flow channeling in heterogeneous fractured rocks. *Reviews of Geophysics*, *36*(2), 275–298.

Valera, M., Guo, Z., Kelly, P., Matz, S., Cantu, V. A., Percus, A. G., ... Viswanathan, H. S. (2018). Machine learning for graph-based representations of three-dimensional discrete fracture networks. *Computational Geosciences*, *22*, 695–710. https://doi.org/10.1007/s10596-018-9720-1.

VanderKwaak, J. E., & Sudicky, E. A. (1996). Dissolution of non-aqueous-phase liquids and aqueous-phase contaminant transport in discretely-fractured porous media. *Journal of Contaminant Hydrology*, *23*(1–2), 45–68.

World Climate Research Program. (2018). In *Coupled Model Intercomparison Project*. cmip-pcmdi.llnl.gov.

Zhang, L., Zhang, L., & Du, B. (2016). Deep learning for remote sensing data: A technical tutorial on the state of the art. *IEEE Geoscience and Remote Sensing Magazine*, *4*(2), 22–40.

Zhang, Z. (2016). When doctors meet with AlphaGo: Potential application of machine learning to clinical medicine. *Annals of Translational Medicine*, *4*(6), 125. https://doi.org/10.21037/atm.2016.03.25125.

Further reading

Viswanathan, H. S., Hyman, J. D., Karra, S., O'Malley, D., Srinivasan, S., Hagberg, A., & Srinivasan, G. (2018). Advancing graph-based algorithms for predicting flow and transport in fractured rock. *Water Resources Research*, *54*(9), 6085–6099. https://doi.org/10.1029/2017WR022368.

CHAPTER FOUR

Seismic signal augmentation to improve generalization of deep neural networks

Weiqiang Zhu*, S. Mostafa Mousavi, and Gregory C. Beroza
Department of Geophysics, Stanford University, Stanford, CA, United States
*Corresponding author: e-mail address: zhuwq@stanford.edu

Contents

1. Introduction 151
2. Benchmark data and training procedure 154
3. Augmentations 156
   3.1 Random shift 156
   3.2 Superimposing events 159
   3.3 Superposing noise 161
   3.4 False positive noise 163
   3.5 Channel dropout 164
   3.6 Resampling 166
   3.7 Augmentation for synthetic data generation 167
4. Discussion 170
5. Conclusions 173
Acknowledgments 174
References 174

1. Introduction

Deep learning has been successfully applied to a wide range of seismic problems, such as earthquake detection (Dokht, Kao, Visser, & Smith, 2019; Mousavi, Zhu, Sheng, & Beroza, 2019; Perol, Gharbi, & Denolle, 2018; Wu et al., 2018; Zhou, Yue, Kong, & Zhou, 2019; Zhu et al., 2019), clustering (Mousavi, Zhu, Ellsworth, & Beroza, 2019), phase detection and picking (Ross, Meier, Hauksson, & Heaton, 2018; Wang, Xiao, Liu, Zhao, & Yao, 2019; Zheng, Lu, Peng, & Jiang, 2018; Zhu & Beroza, 2018), polarity determination (Ross, Meier, & Hauksson, 2018), earthquake location (Mousavi & Beroza, 2019; Zhang et al., 2020), magnitude estimation (Mousavi & Beroza, 2020), denoising (Si & Yuan, 2018; Zhu, Mousavi, & Beroza, 2019), phase association (McBrearty, Delorey, & Johnson, 2019; Ross, Yue, Meier, Hauksson, & Heaton, 2019), among others. Based on deep neural network layers, deep learning algorithms exploit structures in seismic data, extract useful features and representations of seismic waveforms, and learn a map to a target distribution of interest, such as probabilities to separate earthquake from non-earthquake signals. Because the performance of a deep neural network improves with the number of diverse training samples, large seismic datasets like STEAD (Mousavi, Sheng, Zhu, & Beroza, 2019) have been created for deep-learning-based research. However, large-scale training datasets do not exist for every problem, e.g., there is a limited number of very large earthquakes, which due to the power-law distribution of earthquake magnitudes are (fortunately) rare. Moreover, building a high-quality large training dataset, with sufficient labeling and quality control, requires significant effort and time. One way to circumvent these problems is through data augmentation, which consists of various techniques to generate new training samples based on collected datasets to expand the size and variety of training samples. Data augmentation has proven successful in avoiding overfitting and improving the generalization of deep learning models trained on small training datasets and thus shows potential for applications on seismic datasets.

"Generalization" commonly is used to refer to the process of recognizing that a specific feature belongs to a larger category. In deep learning, "generalization" denotes the ability of a trained neural network to perform well on data that were not used in its training or validation (e.g., a unique seismic source or seismograms from a region not used in training). Among the factors that affect generalization are the network architectures, optimization techniques, and training datasets. The size, accuracy (of labels), and completeness of the training datasets are key elements for developing a well-performing model. A training dataset may lack any or all of these properties; for this situation, data augmentation can provide an effective option to improve a model's performance. In high-dimension data space (Fig. 1), the limited training samples, including signals and noise, may only span a small subspace and provide weak constraints on the possible decision boundaries learned by the neural network. This can result in poor performance either in the form of low true positives or high false positives. Data augmentation is designed to increase the training sample size and complexity, thus expanding the sampled feature space such that the neural network

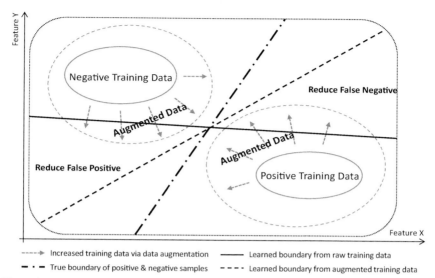

Fig. 1 Data augmentation expands the data space (small dashed arrows) spanned by collected seismic signals and noise, and results in improved constraints on the decision boundary (the shift from solid to dashed lines). With a broader data space exploited by data augmentation, the trained deep neural networks can generalize better, with reduced false negative and reduced false positive rates, to unseen signals and noise.

can learn an improved decision boundary to reduce both false positives and false negatives and to improve generalization on unseen samples.

Data augmentation is commonly used in training deep neural networks to boost performance in classification and recognition problems for computer vision (Krizhevsky, Sutskever, & Hinton, 2012; Mikołajczyk & Grochowski, 2018; Perez & Wang, 2017; Simard, Steinkraus, Platt, et al., 2003), audio processing (Cui, Goel, & Kingsbury, 2015; Salamon & Bello, 2017) and other areas (Fadaee, Bisazza, & Monz, 2017; Frid-Adar et al., 2018; Um et al., 2017). When implemented correctly, data augmentation reduces the risk of overfitting by increasing the number and variability of training data. It can be especially effective for applications with scarce labeled data. Most data augmentation methods are designed for images, for which semantic meaning can be easily preserved. However, some augmentations appropriate for vision, e.g., horizontal flipping, rotating, and shearing, are inappropriate for seismic data because these processes violate the physical properties of the waveform data of interest. Very few studies exist on augmentation techniques for seismic data. Because seismic data are usually collected and organized in a standard way, i.e., using a fixed time window around the P-wave arrival, obvious augmentations are needed to prevent the neural networks from learning and memorizing any artifacts in how the training data are organized. Less obvious augmentations, such as random shifting, recombining events and noise, and channel (and station) dropout, are consistent with the character of seismic signals and can mitigate bias in training data and increase model performance. In this paper, we demonstrate these techniques and explain how they help to improve generalization of the model to cases of interest including low quality data, complex noise, and closely recorded earthquakes in earthquake swarms. Our examples demonstrate that with appropriate augmentation, we can expand the application of deep learning to smaller datasets than would otherwise be possible for purposes of earthquake monitoring.

2. Benchmark data and training procedure

We collected a small, high signal-to-noise ratio (SNR) training dataset with accurate manual labels of 500 earthquake waveforms from the Northern California Earthquake Data Center (NCEDC) (NCEDC, 2014) recorded before 2018 to demonstrate the application of augmentation for training deep learning models on seismic data. In the same way, we created a validation dataset with another 500 high SNR earthquake waveforms before 2018, used to choose the best model during training. We evaluated the augmentation methods with a much larger test dataset of 10,000 earthquake waveforms recorded in 2018. This choice of ratios between training, validation, and test datasets, is purposely designed to evaluate the effect of augmentation on small datasets. For real applications, we would need to choose a larger sample size for the training dataset than the validation and test datasets. The SNR distribution of the data is shown in Fig. 2. Here the SNR is calculated based on the standard deviations of waveforms before and after the first manually picked P-wave arrival. The distribution of epicentral distances of the test dataset is shown in Fig. 2D, with source-station paths mainly ranging from 0 to 120 km. The training and validation sets (not illustrated) have similar distributions of epicentral distances.

We used the now well-studied phase picking problem as an exemplar for the effects of data augmentation for training deep neural networks. We used the same neural network architecture as PhaseNet (Zhu & Beroza, 2018), i.e., a fully convolutional neural network designed for seismic phase picking problems. To avoid the complex effects of hyperparameter tuning, we removed dropout, learning rate decay, and weight decay (regularization) and keep only batch normalization in the architecture.

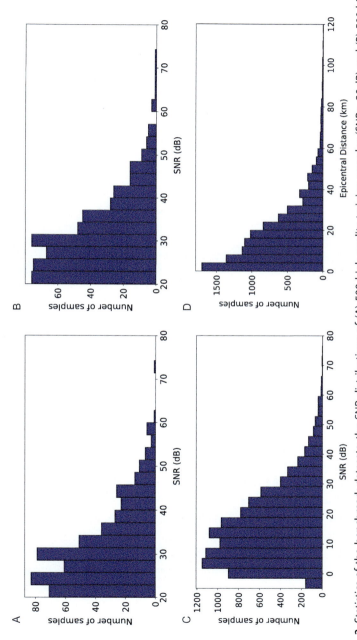

Fig. 2 Statistics of the benchmark datasets: the SNR distributions of (A) 500 high quality training samples (SNR > 20dB) and (B) 500 high quality validation samples (SNR > 20dB), both sampled from earthquakes that occurred prior to 2018; (C) the SNR distribution of 10,000 test samples from earthquakes that occurred in 2018; (D) the epicentral distance distribution of the test samples. The training and validation datasets have similar epicentral distance distributions, which are not shown here.

We used the Adam optimization with a learning rate of 0.01, a batch size of 20, and 100 total epochs. Augmentation increases the size of the training dataset by generating a large number of new training samples. Here, we kept the total number of training samples the same in each epoch (500 samples) and apply data augmentation on the fly, so that augmentation increases the variety of training samples in each epoch. The neural network architecture, training procedure, and optimization method will also affect the generalization for different problems and data formats (He et al., 2019). In this paper, we keep the training procedure simple and focus on the effect of data augmentation for seismic data. With suitable modification, our results should apply to related problems like earthquake detection, phase detection, as well as other deep learning applications to seismic data.

3. Augmentations

Many data augmentation techniques have been proposed and applied in image and acoustic signal processing. Here we examine those augmentations of particular relevance to processing seismic data. Note that due to the randomness in initialization and optimization of neural networks and data augmentation, the reported performance will have a certain randomness.

3.1 Random shift

Seismic training data are usually collected based on the phase arrival information from earthquake catalogs. Therefore, it is tempting to define a cutout window based on the time relative to a particular seismic phase such as the first P-wave arrival. Training neural networks using data with a fixed reference time or with a limited time shift of a few seconds around the reference time point can introduce positional bias into the model. Such a neural network model is prone to memorizing the location of the anchor time point rather than learning more general functions from the feature space.

Here, we train three models on data with a 30-s time window using: no random shift, limited random shift from 10 to 15 s, and full random shift from 0 to 30 s. The random shift is applied on the fly so that each training waveform is shifted differently in each epoch. We examine the P-wave arrival predictions by sliding the test waveform from the left side to the right side of the window and record the predicted activation scores at the true P-arrival locations (Fig. 3). For the model trained without random shift, regardless of where the true P-wave arrival is, the neural network continues to predict a high probability score at 10 s, which is the fixed P-wave arrival time during

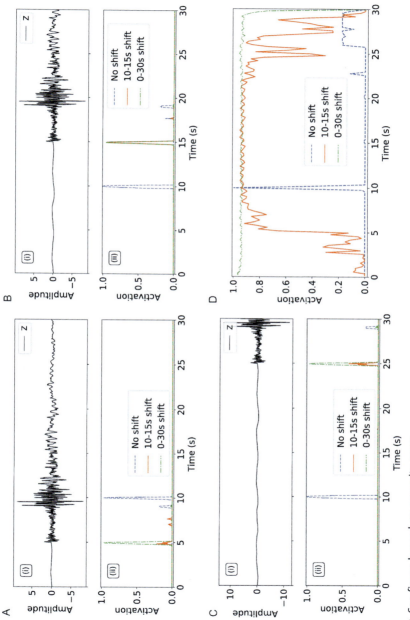

Fig. 3 See figure legend on opposite page.

training (Fig. 3A–C). For the model trained with limited shift within 10–15 s, the result shows a statistical bias that phases are expected to exist within a limited window. The predicted activations fall off at the edges of the time window (Fig. 3D). This bias can be neglected in evaluation if the test dataset is shifted within the same window from 10 to 15 s; however, the bias will result in deteriorated performance when applied to continuous data where the signal arrival time is not known a priori. A very narrow shift range can make a model ineffective for detecting earthquakes outside the shift window used in training. In contrast, the case with random shift from 0 to 30 s has high activation scores across the whole window. This potential bias from random shift could also occur in other deep learning applications, such as earthquake detection, based on a specific window size.

In addition to mitigating the potential bias across the prediction window, random shift can also increase sample variety and improve detection performance. Table 1 compares two implementations of random shift: applying a fixed set of shifting times to the 500 training samples, and applying random shift on the fly so that each sample is shifted differently in each epoch. Random shift on the fly allows a greater variety of time shifting on different training epochs and leads to better performance. In contrast, a fixed set of time shifting for all epochs may only sample limited time points, especially for a small number of training data. In order to effectively apply random shift on the fly within the 30 s window in this case, we cut the training sample to a larger window of 90 s to avoid the need for zero padding. Zero padding may itself introduce a subtle bias by having the neural network learn the transition from zeros to signals and use this artifact as the basis for its predictions. Investigating the distribution of arrival times in a training dataset prior to and after the augmentation is a good practice for deep-learning-based seismic phase detecting/picking models.

Fig. 3 Activation scores from three models trained with different random shift augmentations. (A)–(C) show the predicted probability sequences of neural networks trained with no random shift (blue short-dashed line), random shift within 10–15 s (orange solid line), and random shift within 0–30 s (green alternate short-long dashed line). The test waveform slides from the left to the right edge of the window, and the cases with P-wave arrivals at 5, 15, 25 s are shown in (A)–(C). (D) shows the P-wave arrival predictions at every time point when sliding waveform across the window. Training without random shift leads the neural network to learn a fixed time regardless of the waveform content. Training using incomplete shift between 10 and 15 s leads to activations decay at the edges of the considered window, causing missed detections when applied to continuous waveforms.

Table 1 Comparison between two implementations of random shift: (a) precomputing a fixed random shift for training samples; (b) calculating random shifts on the fly so that each training sample has a different shift in each epoch.

Random shifting type		Precomputed	On the fly
P-wave	Precision	**0.908**	0.902
	Recall	0.550	**0.588**
	F1 score	0.685	**0.712**
S-wave	Precision	0.738	**0.748**
	Recall	0.529	**0.571**
	F1 score	0.616	**0.647**

The higher scores are marked in bold.

3.2 Superimposing events

With the exception of earthquake swarms and aftershock sequences, cataloged earthquakes tend to occur in isolation; thus, training samples contain only one earthquake, commonly referred to as an "event" in the training window. However, training only on single event data can lead neural networks to learn a subtle bias of expecting only one event within the duration of the window and suppressing the detection of smaller events that are also present in the time window. This bias can result in missed events for semantic-segmentation-based methods (Mousavi, Zhu, Sheng, & Beroza, 2019; Zhu & Beroza, 2018), which are designed to detect every event in a time window. We would like a well-trained neural network to perform appropriately on normal earthquakes, but also to generalize to extreme cases, like earthquake swarms and induced earthquakes, when information is dense and earthquakes occur at much more frequent intervals than is usually the case.

An effective augmentation to address the case of multiple events in a short window is to artificially superimpose events in a way that mimics such cases and removes the bias of only one event existing in each window. Superimposition simply means adding two or more time series together, often referred to as "stacking" in seismological parlance. During superimposition, we also apply a random ratio between event amplitudes, which further enhances the neural network's ability to detect smaller earthquakes that occur close in time to larger ones. Fig. 4 shows one example with two earthquakes occurring close to one another in time. The neural network

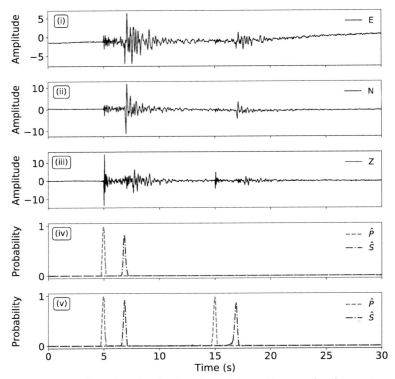

Fig. 4 Comparison of predicted activations for training without and with superimposed events: (i)–(iii) waveforms of ENZ channels; (iv) training without superimposed events; (v) training with superimposed events.

model trained without superimposed events only detects the first large event and neglects the second smaller one (Fig. 4(iv)); however, the model trained with superimposed events predicts the second smaller event with high probability scores for both P and S waves (Fig. 4(v)).

Although event waveforms in real data may completely overlap each other when a station simultaneously records two earthquakes, we avoid synthesizing these cases. These are event waveforms that usually left unused during the manual processing. Because neural networks' performance is highly dependent on training data, these cases have the potential to increase false positives on common seismic waveforms. Thus, we avoid implicitly introducing fully masked events to the training through augmentation. Our observations suggest that a well-trained model should be capable of generalizing to the events with overlapping waveforms. In special applications to realistically replicate an earthquake swarm, stacking much more overlapping events could be useful. Since most of the datasets provide no information regarding the end of waveform (or end of the earthquake coda), we can roughly estimate the end of the earthquake using the time between the P and S arrival times. Another option is to use measurements based on envelope functions similar to those used in coda magnitude estimation.

3.3 Superposing noise

Although most manual labels are selected from high-quality seismic data, a robust neural network should also be able to work on low-quality or otherwise complex data. Superposing noise is a straightforward way to increase the performance of neural networks applied to low signal-to-noise ratio (SNR) data. A distinct advantage of this augmentation is that the high reliability of labels from high SNR data can be retained when the waveforms are superposed with strong noise. Because the augmented weak signals are de-amplified versions of known high SNR signals, their labels are more accurate than those on low SNR signals. By controlling the ratio between the signal and the superposed noise, we can influence the detection limits of the neural network. In particular, by superposing strong noise, we can push the neural network to detect weak signals hidden inside the background noise; however, it should be noted that the potential for false positives may increase as well. Nevertheless, superposing noise is also an effective way to mitigate overfitting on a small training dataset because noise samples are easily obtained from continuous seismic recordings or from synthetically generated random noise.

Fig. 5 compares the neural network's performance with and without superposing noise. It is clear that high scores of precision, recall, and F1 score can be obtained on the high SNR test samples ($>20\,dB$) training with only a small training dataset. However, recall is much lower for the low SNR test samples ($\leq20\,dB$) when only high-quality samples are used for training. After training with superposing noise as augmentation, the recall and F1 score are significantly improved for the low SNR data; meanwhile, the performance for high SNR data is maintained. Note that the increase of recall is more significant than precision; this is because the neural network trained with augmented noisy data becomes more sensitive to weak signals buried inside noise and recovers more events, but this may also increase the potential of false positives, thus limits the improvement in precision. Here we have fixed the activation score threshold for comparison. In practice, we can tune a sequence of activation thresholds and generate a

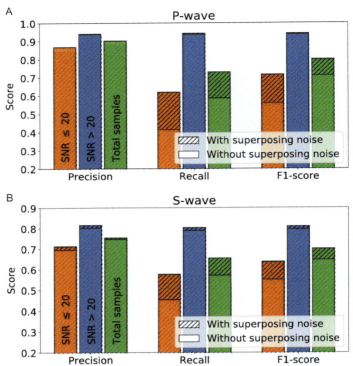

Fig. 5 Comparison of detection performance with and without noise superposition: (A) P-wave arrival; (B) S-wave arrival. For high quality test data (SNR > 20 dB), both models achieve a high performance; however, the model with augmentation significantly increases the recall and F1 score for low quality test data (SNR ≤ 20 dB).

receiver operating characteristic (ROC) curve to balance between precision and recall and to determine a better activation threshold with both improved precision and recall.

The choice of superposed noise depends on the specific environment, such as applications to borehole data, urban data, ocean bottom seismometer (OBS) data. Using real seismic noise recorded by seismic instruments in situ should result in a more realistic augmentation and better performance; however, care must be taken to avoid including undetected events within the noise windows, which would inadvertently increase the mis-labeling rate in the dataset and deteriorate the performance. There is also a risk of biasing the performance for a specific type of instrument if the noise waveforms are not sampled appropriately. Although the use of Gaussian noise would be safe in terms of avoiding mislabeling, the result would be a less realistic representation of real seismic noise, which is usually strongly coherent and non-stationary. Thus, real seismic noise is preferred for augmentation provided that a reliable set of noise samples that have been checked for the existence of non-cataloged earthquakes is available.

3.4 False positive noise

As shown above, superposing noise on earthquake signals helps improve the performance of neural networks for low SNR data. Adding false positive signals (non-earthquake signals) is another way to deal with complex noise effects, such as shaped pulses from urban vibration. This is particularly effective for training neural networks to recognize negative samples (i.e., reducing the false positive rate). Due to the complex noise and non-stationary sources of noise in continuous seismic data, a small training dataset can only cover a limited range of noise. As a result, the neural network trained on limited noise samples may result in many false positives on unseen noise with features similar to seismic signals. To tackle this problem, we can add these false positive noise examples or synthesize similar non-earthquake signals into the training dataset to retrain or fine-tune the neural network to learn features to recognize these false positives and correct their predictions.

Fig. 6 presents a common case in seismic data acquisition when part of the data is missing due to instrumental or telemetry errors. This introduces abrupt changes in continuous data that may result in false positives. These types of false positives are common in traditional methods, like STA/LTA, and they can confuse deep learning methods if this kind of noise is absent from the training dataset. The model trained without appropriate augmentation can produce a false prediction of P and S waves at the abrupt waveform change, as shown in Fig. 6(iv). Adding a small number of similar kinds of noise samples into the training data, however, can effectively suppress false-positive predictions of this type. The same logic applies to other types of common false positive noise, such as impulsive signals from human activities. In practice, identifying different classes of common false positives is challenging due to the need for comprehensive test data and manual examination. To improve the efficiency of identifying false positive samples, active learning (Bergen & Beroza, 2017; Cohn, Atlas, & Ladner, 1994; Kirsch, van Amersfoort, & Gal, 2019) could be one option. Because manually recognizing false positives from a large number of unlabeled samples during application is often difficult, active learning aims to design a strategy to rank these unlabeled samples and annotate the most informative samples

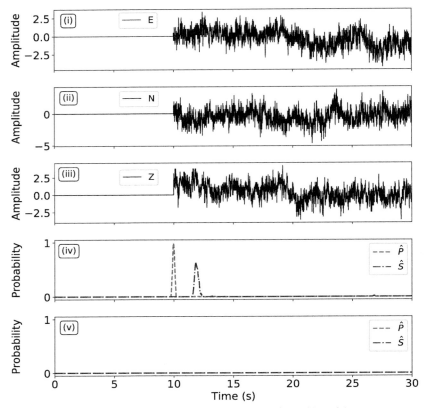

Fig. 6 Comparison of predicted activations before and after adding false positive noise: (i–iii) waveforms of ENZ channels; (iv) training without false positive noise; (v) training with false positive noise.

first, such as samples with the largest uncertainty. These samples are more likely to be recognized as false positives or false negatives, and adding them into training can improve learning efficiency.

3.5 Channel dropout

Three-component seismic data are the most common data form in modern earthquake seismology; however, single channel recordings dominate many historical archives and are still used in some deployments. Moreover, with three-component recordings, it is not uncommon for one of the channels to fail due to either instrument malfunctions or errors in telemetry. Augmentation is an effective strategy to improve the performance of models trained using three-component data on single-channel data. A suitable approach is to use a technique similar to dropout (Gal & Ghahramani, 2016; Srivastava, Hinton, Krizhevsky, Sutskever, & Salakhutdinov, 2014). In the input layer, we randomly drop one or two channels from the EZN input channels. This channel dropout trains the neural network to also predict on data with missing channels. For applications like phase association, where the training is done based on data from multiple stations, we can apply a similar approach to randomly dropping data from part of the stations during training (Zhu, Tai, Mousavi, & Beroza, 2019). This can prevent the network from overfitting on dominant stations and increases the neural network's robustness in the inevitable cases where data from some stations are missing or corrupted.

Table 2 compares the performance between training with and without channel dropout. We apply the trained model on high quality test data (SNR $> 20\,$dB) and examine the performance on different components.

Table 2 Comparison of detection performances on different channels.

		Z	E	N	EN	ENZ
P-wave performance						
With channel dropout	Precision	**0.952**	**0.869**	**0.886**	**0.893**	**0.957**
	Recall	**0.944**	**0.825**	**0.856**	**0.869**	**0.943**
	F1 score	**0.948**	**0.846**	**0.870**	**0.881**	**0.950**
Without channel dropout	Precision	0.944	0.718	0.712	0.795	0.938
	Recall	0.928	0.684	0.705	0.777	0.937
	F1 score	0.936	0.700	0.709	0.786	0.938
S-wave performance						
With channel dropout	Precision	0.523	0.748	0.777	**0.824**	**0.827**
	Recall	**0.436**	**0.732**	**0.767**	**0.808**	**0.810**
	F1 score	**0.476**	**0.740**	**0.772**	**0.816**	**0.818**
Without channel dropout	Precision	**0.630**	**0.771**	**0.793**	0.806	0.803
	Recall	0.019	0.683	0.740	0.787	0.789
	F1 score	0.036	0.724	0.766	0.796	0.796

The higher scores are marked in bold.

Both models show performance similar to that of three-component data; however, the model trained with channel dropout works better on single-component data. The performance on single E-, N-, Z-, and EN-component combinations is instructive and reflects the information learned by the neural network to distinguish P and S waves. For picking the P-wave arrival, performance scores using only the Z-component are similar to those using all ENZ-components, reflecting that the Z-component provides most of the information used for picking P-wave arrivals. In contrast, the horizontal EN-components contain the essential information for picking S-wave arrivals. This is in agreement with the polarization of P and S waves–the P waves appear stronger on the vertical component, while the S waves show up more strongly on the horizontal components.

3.6 Resampling

In deep learning, effective training using imbalanced datasets can be challenging (He & Garcia, 2009; Kotsiantis, Kanellopoulos, Pintelas, et al., 2006). This issue may be especially significant when training a neural network using earthquake signals due to the imbalance of earthquake magnitude distributions. A power law relationship (Gutenberg & Richter, 1944) exists between earthquake magnitude and the number of earthquakes, meaning that the number of large magnitude earthquakes for the training is much more limited compared to the number of small magnitude earthquakes. This imbalance directly impacts applications like magnitude estimation using neural networks (Mousavi & Beroza, 2020). Similar issues can exist for distance, depth, geographic location, tectonic setting, source mechanism, magnitude type, instrument type, and SNR in a specific training set. Station coverage and configuration can also vary significantly among seismic monitoring networks. These imbalances can reduce the generalizability of a model trained on a specific dataset to a broader range of earthquakes. For this reason, it is necessary to investigate data properties during the construction of a training dataset. Based on such preliminary investigations, an appropriate resampling approach can be developed to address possible imbalance problems within a dataset.

Random resampling is a technique to deal with the imbalance issue by oversampling the minority class or undersampling the majority class during training, so that the class distribution does not become biased toward a few specific classes, and better generalization can be achieved by training on a more balanced sample distribution. Resampling can, however, bring about undesirable side effects. Undersampling the majority classes comes with the cost of losing part of the training data and reducing the training size. Extreme oversampling, by repeating a few minority samples of similar magnitudes or from a same region, can also bias the neural network to simply memorizing these samples, which clearly works against generalization. Moreover, oversampling may have limited applications to large earthquakes. Not only are large events rarer, but they are also are more complex compared with small ones. Large earthquakes usually exhibit complex spatial and temporal rupture patterns involving multiple faults. Thus, oversampling may be insufficient to capture the full variety of large magnitude earthquakes. Combining oversampling with the augmentation methods discussed above could be a more effective way to increase both the ratio and variety of the minority samples. Another alternative would be synthesizing training samples from existing instances using more advanced approaches, such as SMOTE (Chawla, Bowyer, Hall, & Kegelmeyer, 2002), ADASYN (He, Bai, Garcia, & Li, 2008), and GAN (Goodfellow et al., 2014).

3.7 Augmentation for synthetic data generation

In some applications, augmentation techniques can be used to generate semi-synthetic data for training. Two such examples are the seismic denoising problem (Zhu, Mousavi, & Beroza, 2019) and earthquake detection on scanned analog-seismograms (Wang, Zhu, Ellsworth, & Beroza, 2019), for which the ground truth (the training target) is unknown and infeasible to obtain from manual labeling. To tackle this problem, we can use augmentation to synthesize input and target pairs from the abundant seismic waveforms. For example, Zhu, Mousavi, and Beroza (2019) generated an accurate denoising mask as the training target for neural networks based on high SNR earthquake signals and a group of noise waveforms. As a result, this augmentation provided a sufficiently large number of training samples by randomly combining signal and noise with a random ratio on the fly during training. In this way, the neural network is trained to learn a challenging inverse process to separate signal and noise in opposite to the forward synthesizing process.

Here we show another example of clipped seismic waveform recovery. Clipped waveforms commonly occur for moderate to large earthquakes recorded on nearby weak motion instruments (Yang & Ben-Zion,

2010; Zhang et al., 2016). Because the true unclipped waveform cannot be observed at the station, we cannot directly get training data from historical waveforms. We can, however, synthesize training data from unclipped waveforms by manually clipping these waveforms. In this way, the input data for the neural network is the synthetically clipped waveforms and the training target the true unclipped waveforms, so that we can easily collect a large number of training data through augmentation. As in denoising, this augmentation has the advantage of being derived from a signal where the (unclipped) ground truth is known and provides an accurate training label. In this case, we use the same network architecture as the other cases but use a mean squared error (MSE) to measure the waveform difference between the recovered and the true unclipped waveforms. Fig. 7 shows the recovered waveforms using the neural network model trained on the synthetic clipped waveforms.

Applying augmentation to synthesizing training data solves the problem of unknown ground truth for some applications. The idea is similar to generating training data using numerical simulations; however, the augmentation method generates the training data based on real seismic waveforms, which is efficient and results in samples that are ipso facto realistic. The trained model can generalize better from the semi-synthetic data to real seismic recordings. If we think of the data generation process as a forward operation, the neural network essentially learns an inverse modeling from the synthesized training data to the true signal of interest that underlies the synthetic data. On the other hand, for cases where not only the label is missing but the real data is also scarce, numerical simulations could become a source for training data, such as finite fault modeling of large complex earthquakes. In this case, we could combine the synthetic earthquake waveforms with real noise to generate training data to improve detection on large magnitude earthquakes. However, the model trained with simulation data may have a generalization issue when applied to real seismic data. Model fine-tuning or transfer learning on a few real seismic waveforms would be needed to narrow the generalization gap. Many other algorithms in computer vision can also be used to bridge the domain gap between simulation and real word, such as adversarial discriminative domain adaptation (Tzeng, Hoffman, Saenko, & Darrell, 2017). The importance of large earthquakes provides strong motivation for future research in this direction.

Seismic signal augmentation

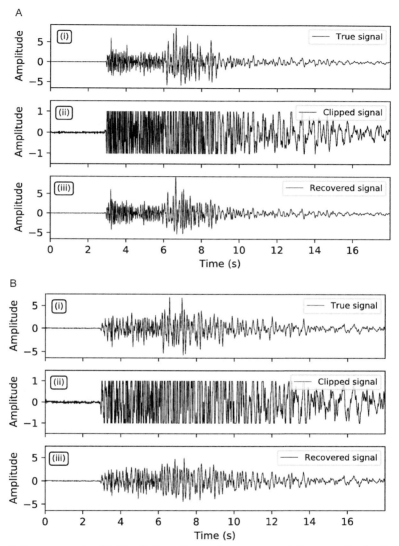

Fig. 7 Two examples (A, B) of clipped waveform recovery: (i) true seismic signals; (ii) manually clipped signals; (iii) recovered signals based on neural networks trained with input-target pairs between (ii) and (i).

4. Discussion

In this paper, we have introduced and discussed several augmentation techniques that can improve the performance of deep learning methods for seismological applications. Combining these augmentations can expediently increase the possible training samples and improve the model generalization even on small training datasets. In addition to the augmentations discussed above, other augmentation methods used in image and speech processing could also be applicable to seismic data, such as (1) filtering seismograms in different narrow frequency bands; (2) time or frequency stretching (Salamon & Bello, 2017); (3) masking part of signal by zeros (DeVries & Taylor, 2017b; Zhong, Zheng, Kang, Li, & Yang, 2017); (4) vertically or possibly horizontally flipping signal; (5) rotating the horizontal components to account station orientation issues and create novel source-station paths; (6) scaling between three components using PCA augmentation (Krizhevsky et al., 2012); and (7) feature space augmentation (DeVries & Taylor, 2017a). Some augmentation, such as time stretching or vertically flipping, can result in potential side effects like changing the phase or polarization information. Hence, some caution is advised in selecting augmentation techniques. Beyond signal-processing-based augmentation, the Generative Adversarial Network (GAN) approach can be used as a synthetic signal generator to make new training samples (Frid-Adar et al., 2018; Li, Meier, Hauksson, Zhan, & Andrews, 2018; Shin et al., 2018; Yi, Walia, & Babyn, 2019). AutoML-based methods, such as AutoAugment, can be used to automatically search for appropriate data augmentations for different problems and datasets (Cubuk, Zoph, Mane, Vasudevan, & Le, 2019). The effectiveness of these methods for seismic data remains as an important area for future research.

The augmentations discussed above were designed during training neural networks, while test time augmentation (He, Zhang, Ren, & Sun, 2016; Krizhevsky et al., 2012) can also help to improve the prediction performance after training. In image classification, several fixed crops and scales are applied to the test image. Similar to ensemble learning, the final prediction score is averaged over these augmentations to improve the overall prediction. The training time and test time augmentations serve different purposes. Training time augmentations, like superposing noise, aim to increase the variety and complexity of training samples, which renders the recognition task much more challenging and pushes the neural networks to learn more accurate decision boundaries between signal and noise (Fig. 1). The test time augmentations aim to make the recognition task easier by sampling certain transforms that properly represent the data features and make the prediction more robust by aggregating different augmentations. For seismic data, data pre-processing methods, such as filtering, can be used as test time augmentation. Filtering can transform the signal into certain high SNR frequency bands covered by the training dataset, thus improving the prediction accuracy on noisy data. Table 3 shows the improved prediction performance by applying 1 Hz high-pass filtering on the test dataset. Another potential method for test time augmentation is to compress the large epicentral distance waveform into a shorter time window in order to mitigate the learning bias due to the imbalanced distribution of epicentral distances (91% samples <40 km) in the training dataset. We applied a 50% down-sampling to compress the waveforms of earthquakes with epicentral distance larger than 40 km (Fig. 2D) into half of the original time window. Table 4 shows the improved prediction performance after time compression. Note that we used prior information that these observations are from 40 km away before applying the compression, which may not be available in applications.

In addition to using data augmentation methods to improve neural network performance for small datasets, generalization of deep learning models is also determined by other factors, such as the neural network architecture, loss functions, optimization methods, and various training techniques, including batch normalization (Ioffe & Szegedy, 2015), layer normalization (Ba, Kiros, & Hinton, 2016), dropout (Srivastava et al., 2014), early stopping (Prechelt, 1998), learning rate decay, weight decay (weight regularization),

Table 3 Detection performance using a 1 Hz high-pass filtering as test time augmentation.

Filtering as test time augmentation		Without filtering	With filtering
P-wave	Precision	0.902	0.902
	Recall	0.588	**0.626**
	F1 score	0.712	**0.739**
S-wave	Precision	**0.748**	0.725
	Recall	0.571	**0.650**
	F1 score	0.647	**0.685**

The higher scores are marked in bold.

Table 4 Detection performance using time compressing as test time augmentation.

Time compressing as test time augmentation		Without compressing (>40 km)	With compressing (>40 km)
P-wave	Precision	0.785	**0.889**
	Recall	0.374	**0.393**
	F1 score	0.507	**0.545**
S-wave	Precision	0.545	**0.708**
	Recall	0.342	**0.424**
	F1 score	0.421	**0.531**

The higher scores are marked in bold.

label smoothing (Müller, Kornblith, & Hinton, 2019), and model ensemble (Deng & Platt, 2014; He et al., 2016). Most of these training techniques aim to stabilize training and prevent overfitting. Note that overfitting and poor generalization do not necessarily go hand-in-hand. A model without overfitting can still suffer from poor generalization to unseen datasets with significantly different characteristics from the training dataset, while an over-fitted model can always have generalization problems. With the development of deep learning, a variety of neural network architectures have been designed, modified and tested on seismic signals, and this trend will continue. The data augmentation techniques discussed here can generally apply to these deep learning models to improve their performance and generalization.

For seismic data, choosing the data processing domains, such as the time domain, time-frequency domain, or wavelet domain, is also important for model performance depending on specific applications. Neural networks can flexibly process multi-dimensional data, like time sequence, image, and videos, which allows presenting seismic signals in either time or frequency domains when designing deep learning models for earthquake detection, denoising, and other problems. For convolutional neural networks, the convolutional kernels have a certain degree of similarity to sine/cosine or wavelet kernels used in Fourier or wavelet transform. The convolutional kernels after training show a variety of frequency and orientation features in image recognition (Krizhevsky et al., 2012). Although a neural network can learn frequency information directly from time domain waveforms, transforming seismic data into time-frequency domain explicitly presents the change of frequency distribution with time, making it easier to capture the frequency information during training. But training and testing in the time-frequency also comes with an expensive computational cost and a loss of high time resolution.

Another approach to addressing the issue of insufficient labeled training data is transfer learning and domain adaptation methods, which are developed to adapt the features and knowledge learned from a large training dataset to improve the training and generalization on a new dataset or task of interest, which often has much less training data (Bengio, 2012; LeCun, Bengio, & Hinton, 2015; Pan & Yang, 2009). For example, pre-trained models on ImageNet dataset (Deng et al., 2009; Oquab, Bottou, Laptev, & Sivic, 2014) has been used for a wide range of problems, such as object detection (Girshick, Donahue, Darrell, & Malik, 2014; Ren, He, Girshick, & Sun, 2015), image segmentation (He, Gkioxari, Dollár, & Girshick, 2017; Long, Shelhamer, & Darrell, 2015), medical image recognition (Tajbakhsh et al., 2016) and remote sensing (Marmanis, Datcu, Esch, & Stilla, 2015). Commonly, transfer learning means transferring low-level features and representations trained on large datasets from the designed task to a different task on a new dataset, while domain adaptation refers to the case of the same task on two different datasets. Because of the similarity among earthquake signals, pre-trained deep neural networks on large datasets like STEAD (Mousavi, Sheng, Zhu, & Beroza, 2019) extract common low-level features for seismic waves, which could be used for applications without enough training data by transfer learning or domain adaptation. Unsupervised pretraining, such as auto-encoding (Vincent, Larochelle, Bengio, & Manzagol, 2008), can also be used to extract good data representation for transfer learning. In contrast to pre-training, self-training is another way to use the extensive unlabeled data (Zoph et al., 2020). Self-training first trains a model on the labeled dataset, then generates pseudo labels on a much large unlabeled dataset. The new pseudo-labeled data is combined with the labeled dataset to train a new model. This same process can be iterated a few times to advance model performance.

5. Conclusions

Data augmentation is an efficient way to improve the performance and generalization of deep neural networks for common cases where labeled data is scarce. We have presented and analyzed data augmentation methods that are well-suited for seismic data. Although we used only a small training dataset, our results show data augmentation can mitigate the bias in training data and improve the performance on a dataset with different statistics. Because data augmentation is independent of the particular neural network model and the computational cost is negligible compared with training, augmentation methods can widely benefit the training of deep learning models on seismic data.

Acknowledgments

We thank Dr. William L. Yeck and Dr. Zachary E. Ross for their constructive comments and suggestions. This work was supported by the Department of Energy Basic Energy Sciences; Award DE-SC0020445.

References

Ba, J. L., Kiros, J. R., & Hinton, G. E. (2016). *Layer normalization*. arXiv preprint arXiv:1607.06450.

Bengio, Y. (2012). Deep learning of representations for unsupervised and transfer learning. In *Proceedings of ICML workshop on unsupervised and transfer learning* (pp. 17–36).

Bergen, K., & Beroza, G. C. (2017). In *Automatic earthquake detection by active learning. AGU Fall Meeting 2017.*

Chawla, N. V., Bowyer, K. W., Hall, L. O., & Kegelmeyer, W. P. (2002). SMOTE: Synthetic minority over-sampling technique. *Journal of Artificial Intelligence Research*, *16*, 321–357.

Cohn, D., Atlas, L., & Ladner, R. (1994). Improving generalization with active learning. *Machine Learning*, *15*(2), 201–221.

Cubuk, E. D., Zoph, B., Mane, D., Vasudevan, V., & Le, Q. V. (2019). Autoaugment: Learning augmentation strategies from data. In *Proceedings of the IEEE conference on computer vision and pattern recognition* (pp. 113–123).

Cui, X., Goel, V., & Kingsbury, B. (2015). Data augmentation for deep neural network acoustic modeling. *IEEE/ACM Transactions on Audio, Speech, and Language Processing*, *23*(9), 1469–1477.

Deng, J., Dong, W., Socher, R., Li, L.-J., Li, K., & Fei-Fei, L. (2009). ImageNet: A large-scale hierarchical image database. In *2009 IEEE conference on computer vision and pattern recognition* (pp. 248–255).

Deng, L., & Platt, J. C. (2014). Ensemble deep learning for speech recognition. In *Fifteenth annual conference of the international speech communication association.*

DeVries, T., & Taylor, G. W. (2017a). *Dataset augmentation in feature space*. arXiv preprint arXiv:1702.05538.

DeVries, T., & Taylor, G. W. (2017b). *Improved regularization of convolutional neural networks with cutout*. arXiv preprint arXiv:1708.04552.

Dokht, R. M., Kao, H., Visser, R., & Smith, B. (2019). Seismic event and phase detection using time–frequency representation and convolutional neural networks. *Seismological Research Letters*, *90*(2A), 481–490.

Fadaee, M., Bisazza, A., & Monz, C. (2017). *Data augmentation for low-resource neural machine translation*. arXiv preprint arXiv:1705.00440.

Frid-Adar, M., Diamant, I., Klang, E., Amitai, M., Goldberger, J., & Greenspan, H. (2018). GAN-based synthetic medical image augmentation for increased CNN performance in liver lesion classification. *Neurocomputing*, *321*, 321–331.

Gal, Y., & Ghahramani, Z. (2016). Dropout as a bayesian approximation: Representing model uncertainty in deep learning. In *International conference on machine learning* (pp. 1050–1059).

Girshick, R., Donahue, J., Darrell, T., & Malik, J. (2014). Rich feature hierarchies for accurate object detection and semantic segmentation. In *Proceedings of the IEEE conference on computer vision and pattern recognition* (pp. 580–587).

Goodfellow, I., Pouget-Abadie, J., Mirza, M., Xu, B., Warde-Farley, D., Ozair, S., et al. (2014). Generative adversarial nets. In *Advances in neural information processing systems* (pp. 2672–2680).

Gutenberg, B., & Richter, C. F. (1944). Frequency of earthquakes in California. *Bulletin of the Seismological Society of America, 34*(4), 185–188.

He, H., Bai, Y., Garcia, E. A., & Li, S. (2008). ADASYN: Adaptive synthetic sampling approach for imbalanced learning. In *2008 IEEE international joint conference on neural networks (IEEE world congress on computational intelligence)* (pp. 1322–1328).

He, H., & Garcia, E. A. (2009). Learning from imbalanced data. *IEEE Transactions on Knowledge and Data Engineering, 21*(9), 1263–1284.

He, K., Gkioxari, G., Dollár, P., & Girshick, R. (2017). Mask R-CNN. In *Proceedings of the IEEE international conference on computer vision* (pp. 2961–2969).

He, K., Zhang, X., Ren, S., & Sun, J. (2016). Deep residual learning for image recognition. In *Proceedings of the IEEE conference on computer vision and pattern recognition* (pp. 770–778).

He, T., Zhang, Z., Zhang, H., Zhang, Z., Xie, J., & Li, M. (2019). Bag of tricks for image classification with convolutional neural networks. In *Proceedings of the IEEE conference on computer vision and pattern recognition* (pp. 558–567).

Ioffe, S., & Szegedy, C. (2015). *Batch normalization: Accelerating deep network training by reducing internal covariate shift.* arXiv preprint arXiv:1502.03167.

Kirsch, A., van Amersfoort, J., & Gal, Y. (2019). Batchbald: Efficient and diverse batch acquisition for deep bayesian active learning. In *Advances in neural information processing systems* (pp. 7024–7035).

Kotsiantis, S., Kanellopoulos, D., Pintelas, P., et al. (2006). Handling imbalanced datasets: A review. *GESTS International Transactions on Computer Science and Engineering, 30*(1), 25–36.

Krizhevsky, A., Sutskever, I., & Hinton, G. E. (2012). ImageNet classification with deep convolutional neural networks. In *Advances in neural information processing systems* (pp. 1097–1105).

LeCun, Y., Bengio, Y., & Hinton, G. (2015). Deep learning. *Nature, 521*(7553), 436–444.

Li, Z., Meier, M.-A., Hauksson, E., Zhan, Z., & Andrews, J. (2018). Machine learning seismic wave discrimination: Application to earthquake early warning. *Geophysical Research Letters, 45*(10), 4773–4779.

Long, J., Shelhamer, E., & Darrell, T. (2015). Fully convolutional networks for semantic segmentation. In *Proceedings of the IEEE conference on computer vision and pattern recognition* (pp. 3431–3440).

Müller, R., Kornblith, S., & Hinton, G. E. (2019). When does label smoothing help? In *Advances in neural information processing systems* (pp. 4696–4705).

Marmanis, D., Datcu, M., Esch, T., & Stilla, U. (2015). Deep learning earth observation classification using ImageNet pretrained networks. *IEEE Geoscience and Remote Sensing Letters, 13*(1), 105–109.

McBrearty, I. W., Delorey, A. A., & Johnson, P. A. (2019). Pairwise association of seismic arrivals with convolutional neural networks. *Seismological Research Letters, 90*(2A), 503–509.

Mikołajczyk, A., & Grochowski, M. (2018). Data augmentation for improving deep learning in image classification problem. In *2018 International interdisciplinary PhD workshop (IIPhDW)* (pp. 117–122).

Mousavi, S. M., & Beroza, G. C. (2019). *Bayesian-deep-learning estimation of earthquake location from single-station observations.* arXiv preprint arXiv:1912.01144.

Mousavi, S. M., & Beroza, G. C. (2020). A machine-learning approach for earthquake magnitude estimation. *Geophysical Research Letters, 47*(1).

Mousavi, S. M., Sheng, Y., Zhu, W., & Beroza, G. C. (2019). STanford EArthquake Dataset (STEAD): A global data set of seismic signals for AI. *IEEE Access, 7.*

Mousavi, S. M., Zhu, W., Ellsworth, W., & Beroza, G. (2019). Unsupervised clustering of seismic signals using deep convolutional autoencoders. *IEEE Geoscience and Remote Sensing Letters, 16*(11), 1693–1697.

Mousavi, S. M., Zhu, W., Sheng, Y., & Beroza, G. C. (2019). CRED: A deep residual network of convolutional and recurrent units for earthquake signal detection. *Scientific Reports, 9*(1), 1–14.

NCEDC. (2014). *Northern California earthquake data center.* UC Berkeley Seismological Laboratory.

Oquab, M., Bottou, L., Laptev, I., & Sivic, J. (2014). Learning and transferring mid-level image representations using convolutional neural networks. In *Proceedings of the IEEE conference on computer vision and pattern recognition* (pp. 1717–1724).

Pan, S. J., & Yang, Q. (2009). A survey on transfer learning. *IEEE Transactions on Knowledge and Data Engineering, 22*(10), 1345–1359.

Perez, L., & Wang, J. (2017). *The effectiveness of data augmentation in image classification using deep learning.* arXiv preprint arXiv:1712.04621.

Perol, T., Gharbi, M., & Denolle, M. (2018). Convolutional neural network for earthquake detection and location. *Science Advances, 4*(2).

Prechelt, L. (1998). Early stopping-but when? In *Neural networks: Tricks of the trade* (pp. 55–69), Springer.

Ren, S., He, K., Girshick, R., & Sun, J. (2015). Faster R-CNN: Towards real-time object detection with region proposal networks. In *Advances in neural information processing systems* (pp. 91–99).

Ross, Z. E., Meier, M.-A., & Hauksson, E. (2018). P wave arrival picking and first-motion polarity determination with deep learning. *Journal of Geophysical Research: Solid Earth, 123*(6), 5120–5129.

Ross, Z. E., Meier, M.-A., Hauksson, E., & Heaton, T. H. (2018). Generalized seismic phase detection with deep learning. *Bulletin of the Seismological Society of America, 108*(5A), 2894–2901.

Ross, Z. E., Yue, Y., Meier, M.-A., Hauksson, E., & Heaton, T. H. (2019). PhaseLink: A deep learning approach to seismic phase association. *Journal of Geophysical Research: Solid Earth, 124*(1), 856–869.

Salamon, J., & Bello, J. P. (2017). Deep convolutional neural networks and data augmentation for environmental sound classification. *IEEE Signal Processing Letters, 24*(3), 279–283.

Shin, H.-C., Tenenholtz, N. A., Rogers, J. K., Schwarz, C. G., Senjem, M. L., Gunter, J. L., et al. (2018). Medical image synthesis for data augmentation and anonymization using generative adversarial networks. In *International workshop on simulation and synthesis in medical imaging* (pp. 1–11).

Si, X., & Yuan, Y. (2018). Random noise attenuation based on residual learning of deep convolutional neural network. In *SEG technical program expanded abstracts 2018* (pp. 1986–1990), Society of Exploration Geophysicists.

Simard, P. Y., Steinkraus, D., Platt, J. C., et al. (2003). Best practices for convolutional neural networks applied to visual document analysis. In *3. Icdar.*

Srivastava, N., Hinton, G., Krizhevsky, A., Sutskever, I., & Salakhutdinov, R. (2014). Dropout: A simple way to prevent neural networks from overfitting. *The Journal of Machine Learning Research, 15*(1), 1929–1958.

Tajbakhsh, N., Shin, J. Y., Gurudu, S. R., Hurst, R. T., Kendall, C. B., Gotway, M. B., et al. (2016). Convolutional neural networks for medical image analysis: Full training or fine tuning? *IEEE Transactions on Medical Imaging, 35*(5), 1299–1312.

Tzeng, E., Hoffman, J., Saenko, K., & Darrell, T. (2017). Adversarial discriminative domain adaptation. In *Proceedings of the IEEE conference on computer vision and pattern recognition* (pp. 7167–7176).

Um, T. T., Pfister, F. M., Pichler, D., Endo, S., Lang, M., Hirche, S., et al. (2017). Data augmentation of wearable sensor data for Parkinson's disease monitoring using convolutional neural networks. In *Proceedings of the 19th ACM international conference on multimodal interaction* (pp. 216–220).

Vincent, P., Larochelle, H., Bengio, Y., & Manzagol, P.-A. (2008). Extracting and composing robust features with denoising autoencoders. In *Proceedings of the 25th international conference on machine learning* (pp. 1096–1103).

Wang, J., Xiao, Z., Liu, C., Zhao, D., & Yao, Z. (2019). Deep learning for picking seismic arrival times. *Journal of Geophysical Research: Solid Earth, 124*(7), 6612–6624.

Wang, K., Zhu, W., Ellsworth, W. L., & Beroza, G. C. (2019). Earthquake detection in develocorder films: An image-based detection neural network for analog seismograms. In *AGU Fall Meeting 2019.*

Wu, Y., Lin, Y., Zhou, Z., Bolton, D. C., Liu, J., & Johnson, P. (2018). DeepDetect: A cascaded regionbased densely connected network for seismic event detection. *IEEE Transactions on Geoscience and Remote Sensing, 57*(1), 62–75.

Yang, W., & Ben-Zion, Y. (2010). An algorithm for detecting clipped waveforms and suggested correction procedures. *Seismological Research Letters, 81*(1), 53–62.

Yi, X., Walia, E., & Babyn, P. (2019). Generative adversarial network in medical imaging: A review. *Medical Image Analysis, 101552.*

Zhang, J., Hao, J., Zhao, X., Wang, S., Zhao, L., Wang, W., et al. (2016). Restoration of clipped seismic waveforms using projection onto convex sets method. *Scientific Reports, 6,* 39056.

Zhang, X., Zhang, J., Yuan, C., Liu, S., Chen, Z., & Li, W. (2020). Locating induced earthquakes with a network of seismic stations in Oklahoma via a deep learning method. *Scientific Reports, 10*(1), 1–12.

Zheng, J., Lu, J., Peng, S., & Jiang, T. (2018). An automatic microseismic or acoustic emission arrival identification scheme with deep recurrent neural networks. *Geophysical Journal International, 212*(2), 1389–1397.

Zhong, Z., Zheng, L., Kang, G., Li, S., & Yang, Y. (2017). *Random erasing data augmentation.* arXiv preprint arXiv:1708.04896.

Zhou, Y., Yue, H., Kong, Q., & Zhou, S. (2019). Hybrid event detection and phase-picking algorithm using convolutional and recurrent neural networks. *Seismological Research Letters, 90*(3), 1079–1087.

Zhu, W., & Beroza, G. C. (2018). *PhaseNet: A deep-neural-network-based seismic arrival time picking method.* arXiv preprint arXiv:1803.03211.

Zhu, W., Mousavi, S. M., & Beroza, G. C. (2019). Seismic signal denoising and decomposition using deep neural networks. *IEEE Transactions on Geoscience and Remote Sensing, 57*(11), 9476–9488.

Zhu, L., Peng, Z., McClellan, J., Li, C., Yao, D., Li, Z., et al. (2019). Deep learning for seismic phase detection and picking in the aftershock zone of 2008 Mw7. 9 Wenchuan Earthquake. *Physics of the Earth and Planetary Interiors, 293.*

Zhu, W., Tai, K. S., Mousavi, S. M., & Beroza, G. C. (2019). An end-to-end earthquake monitoring method for joint earthquake detection and association using deep learning. In *AGU Fall Meeting 2019.*

Zoph, B., Ghiasi, G., Lin, T.-Y., Cui, Y., Liu, H., Cubuk, E. D., et al. (2020). *Rethinking pre-training and self-training.* ArXiv:2006.06882 [Cs, Stat].

CHAPTER FIVE

Deep generator priors for Bayesian seismic inversion

Zhilong Fang[a,c,*], Hongjian Fang[b,c], and Laurent Demanet[a,b,c]

[a]Department of Mathematics, Massachusetts Institute of Technology, Cambridge, MA, United States
[b]Department of Earth, Atmospheric, and Planetary Sciences, Massachusetts Institute of Technology, Cambridge, MA, United States
[c]Earth Resource Laboratory, Massachusetts Institute of Technology, Cambridge, MA, United States
*Corresponding author: e-mail address: fangzl@mit.edu

Contents

1. Introduction		179
2. Methodology		183
	2.1 Bayesian inference	183
	2.2 Deep GAN prior model generator	184
	2.3 Sample the posterior PDF by pCN	186
3. Seismic inversion applications		187
	3.1 Traveltime tomography	187
	3.2 Full waveform inversion	188
4. Numerical examples		188
	4.1 Training the network architecture	188
	4.2 Quantitative performance evaluation of different generators	191
	4.3 Statistical inversion with Overthrust models	196
	4.4 Statistical inversion with Sigsbee models	203
5. Conclusions and discussion		212
Acknowledgments		214
References		214

1. Introduction

Earth scientists and exploration geophysicists study the Earth's interior to answer fundamental questions about the structure and dynamics of Earth. Since the current technology cannot offer researchers direct access to Earth's deep interior, researchers utilize cutting-edge seismic inversion techniques to infer unknown subsurface physical parameters (e.g., sound speed and density) from indirectly measured data (e.g., seismic traveltimes and seismic waveforms). With the developments in high-performance computing and

Advances in Geophysics, Volume 61
ISSN 0065-2687
https://doi.org/10.1016/bs.agph.2020.07.002

© 2020 Elsevier Inc.
All rights reserved.

179

advances in modern numerical and optimization methods in the past 30 years, seismic inversion techniques including traveltime tomography (TT) (Aki, Christoffersson, & Husebye, 1977; Dziewonski, 1984; Nolet, 2012; Romanowicz, 1979) and full waveform inversion (FWI) (Lailly et al., 1983; Pratt, 1999; Tarantola & Valette, 1982a; Virieux & Operto, 2009) enable researchers to generate high-resolution subsurface structures at different scales, from the inner core to the shallow subsurface within acceptable computational time (Tromp, 2019).

Observed seismic data contain noise from different sources including human activities and environmental movements. As the inversion proceeds, uncertainties in the observed data along with the nonuniqueness of the inverse problem would introduce uncertainties in final results. Because the inverted model parameters are inputs for a sequence of follow-up interpretations and processing, analyzing uncertainties in the inverted model parameters can lead to significant improvements in the understanding of Earth's internal structures (Osypov et al., 2013). The analysis of uncertainties requires a complete statistical description of the unknown model parameters. To that end, conventional deterministic approaches, which only find the best model parameters minimizing the misfit between the predicted and observed data, may not be appropriate candidates. On the other hand, statistical approaches, and in particular the Bayesian inference method (Kaipio & Somersalo, 2006), are desirable and necessary.

The Bayesian inference method is one of the most widely used approaches to solving statistical inverse problems and has been applied to many geophysical problems (Duijndam, 1988; Ely, Malcolm, & Poliannikov, 2017; Fang, Da Silva, Kuske, & Herrmann, 2018; Osypov, Nichols, Woodward, Zdraveva, & Yarman, 2008; Sambridge, Gallagher, Jackson, & Rickwood, 2006; Scales & Tenorio, 2001; Tarantola & Valette, 1982b). This approach formulates the inverse problem in the framework of statistical inference. It considers all unknown model parameters as random variables and incorporates statistical information in the observed data, the underlying forward modeling map, and researchers' prior knowledge about the unknown model parameters. The solution of the Bayesian inverse problem is a posterior probability density function (PDF) that incorporates all available statistical information from both the observations in a likelihood PDF and the prior knowledge in a prior PDF. Researchers can extract statistics of interests about the unknown parameters either by directly sampling the posterior PDF utilizing approaches like Markov chain Monte Carlo methods (McMC) (Kaipio & Somersalo, 2006; Matheron, 2012) or by locally approximating the posterior PDF with an easy-to-study Gaussian PDF (Bui-Thanh, Ghattas, Martin, & Stadler, 2013; Fang et al., 2018; Osypov et al., 2013; Zhu, Li, Fomel, Stadler, & Ghattas, 2016) as a compromise of a limited computational capacity, which in general will give reliable estimates of the moments of the full PDF. These statistics reflect the degree of confidence about the unknown model parameters and allow researchers to identify areas with high/low reliability in the model.

Prior knowledge plays a key role in the Bayesian inverse problem. Conventional prior knowledge is hand-crafted based on researchers' empirical observations about the subsurface structure. For instances, because subsurface structure images are generally piecewise smooth and sparse after wavelet or curvelet transformations, image priors that constrain the sparsity of wavelet or curvelet coefficients (Li, Aravkin, van Leeuwen, & Herrmann, 2012; Tu & Herrmann, 2015; Ying, Demanet, & Candes, 2005) or spatial gradients (Haber, Ascher, & Oldenburg, 2000) are widely utilized. While these hand-crafted priors can regularize the unknown parameters when solving deterministic inverse problems, they are usually too generic, in that prior models generated with these prior PDFs include but are far from limited to geophysically relevant structures. On the other hand, approaches that directly integrate geological information have shown the capability in providing results with high spatial resolution (Clapp, 2000; Li, Biondi, Clapp, & Nichols, 2016).

Recent developments in deep convolutional neural networks (DCNN) (Krizhevsky, Sutskever, & Hinton, 2012; LeCun, Bengio, & Hinton, 2015) provide researchers in the field of image/signal processing with a new way to design the prior knowledge, i.e., a learning-based prior generator. Instead of designing features by hands, learning-based approaches train DCNNs that straightforwardly learn features from existing training sets. The ultimate trained DCNN is designed to be able to create images sharing the same spatial distributions with the real/natural samples. Empirically, many researchers have shown the superior performance of the learning-based priors over the hand-crafted priors when dealing deterministic image related inverse problems such as denoising, super resolution, and inpainting (Arridge, Maass, Öktem, & Schönlieb, 2019; Ledig et al., 2017; Lunz, Öktem, & Schönlieb, 2018; Rick Chang, Li, Poczos, Vijaya Kumar, & Sankaranarayanan, 2017; Yeh et al., 2017).

The study of applying DCNNs to geophysical problems has been quite active during the last 5 years. Many geophysicists have shown potential applications to problems like data denoising (Yu, Ma, & Wang, 2019) and deblending (Sun et al., 2020), surface-related multiples removal

(Siahkoohi, Louboutin, & Herrmann, 2019), low-frequency data extrapolation (Sun & Demanet, 2018), and travel time tomography (Araya-Polo, Jennings, Adler, & Dahlke, 2018). Instead of training a generator as the prior information, most of the reported works aim at a network mapping input images/data to output images/data of interests. For example, a DCNN converts the data with noise to the data without noise for denoising problems and a DCNN converts travel time information to subsurface velocity images for traveltime tomography. Despite their potential superior performance, these specifically trained DCNN are designed to solve specific problems and usually require to retrain the DCNN for new problems.

Recent articles by Mosser, Dubrule, and Blunt (2020) and Herrmann, Siahkoohi, and Rizzuti (2019) study the application of the learning-based prior generator to statistical seismic inversion problems. In this work, we follow the recent works and study the Bayesian seismic inversion framework with prior information learned with a deep generative adversarial network (DGAN) (Goodfellow et al., 2014). One appealing property of DGAN for Bayesian inversion lies in its potential capability in generating perceptually appealing high-dimensional images from a low-dimensional latent parameter space, which has been empirically proven in many applications (Ledig et al., 2017). This property motivates us to train a DGAN that can map a low-dimensional normal distribution to the high-dimensional distribution of the training images. Employing such a DGAN to the Bayesian inversion framework has three main advantages. First, the trained DGAN enables us to conduct Bayesian inversions in the low-dimensional latent space instead of the original high-dimensional image space. As a result, we reduce the complexity of the problem and mitigate the curse of dimensionality that Bayesian inversions typically suffer from. Second, since the DGAN is directly learned from the existing training images, it can capture more detailed statistical information about the subsurface images compared to the conventional hand-crafted priors. This enables the DGAN prior generator to produce geologically plausible prior subsurface images that conventional hand-crafted priors cannot produce. Finally, according to the aforementioned setting of the training procedure, the prior PDF in the low-dimensional latent space is the simple normal distribution without complicated human-made design. This property motivates us to apply the dimension-free McMC method—preconditioner Cranks Nickos (pCN) method (Cotter, Roberts, Stuart, White, et al., 2013) to efficiently sample the posterior PDF and extract the statistics of interests.

The application of DGANs faces a key problem: the lack of quantitative criteria to evaluate trained generators. The problem is now the opposite of the generality alluded to earlier, i.e., the risk is now significant that the prior might be too restrictive, and will only generate models that are in some sense close to the training samples. To address this issue, we propose a quantitative evaluation criterion to help tune a DGAN, and establish its fitness as a Bayesian prior. Since the goal of DGANs is to map a low-dimensional latent vector to a high-dimensional subsurface image, a good generator should be able to generate artificial subsurface images that can match target images extracted from either the training or testing sets. Moreover, the corresponding latent vectors of the artificial subsurface images should follow the predefined prior distribution of the latent vectors and has a moderate magnitude to ensure that the generator is not overfitting. We propose to use the ℓ_2-norm relative model error between the generated and target images, and the logarithm ℓ_2-norm of the corresponding latent vector to evaluate the generality of the trained generators. Numerical examples illustrate that this criterion provides us with a direct representation of the generality of the trained generators.

The paper is organized as follows. First, we introduce the basic concepts about the Bayesian inference method. Following that, we introduce the proposed deep GAN prior model generator and the pCN algorithm. Then we present a short introduction about the traveltime tomography and full waveform inversion. After that, we introduce the proposed quantitative evaluation criterion and study the feasibility and generality of the proposed approach by two numerical examples. At last, we finalize the paper with a conclusion and discussion.

2. Methodology

2.1 Bayesian inference

The Bayesian inference method is a widely used approach to solving seismic statistical inverse problems. Unlike deterministic approaches that seek the best data-fit model, the Bayesian inference method aims at a comprehensive statistical description of the unknown parameters. For this purpose, the Bayesian inference method constructs a posterior PDF of the n_g-dimensional model parameters $\mathbf{m} \in \mathbb{R}^{n_g}$ that integrates statistical information from the forward map $F(\mathbf{m})$, n_d-dimensional observed data $\mathbf{d}_{obs} \in \mathbb{R}^{n_d}$, and researcher's prior knowledge. According to the Bayes' law, the posterior PDF $\rho_{post}(\mathbf{m}|\mathbf{d}_{obs})$ of $\mathbf{m}$ given $\mathbf{d}_{obs}$ is proportional to the product of a likelihood PDF $\rho_{\text{like}}(\mathbf{d}_{\text{obs}}|\mathbf{m})$ of $\mathbf{d}_{\text{obs}}$ given $\mathbf{m}$ and a prior PDF $\rho_{\text{prior}}(\mathbf{m})$ of $\mathbf{m}$ as follows:

$$\rho_{\text{post}}(\mathbf{m}|\mathbf{d}_{\text{obs}}) \propto \rho_{\text{like}}(\mathbf{d}_{\text{obs}}|\mathbf{m})\rho_{\text{prior}}(\mathbf{m}). \tag{1}$$

The prior PDF $\rho_{\text{prior}}(\mathbf{m})$ describes one's prior knowledge and beliefs in the unknown model parameters, and the likelihood PDF $\rho_{\text{like}}(\mathbf{d}_{\text{obs}}|\mathbf{m})$ describes the probability of observing data $\mathbf{d}_{\text{obs}}$ given the model $\mathbf{m}$. If we assume that the observed data contain additive Gaussian noise ϵ from the distribution $\mathcal{N}(0, \Sigma)$ with the covariance matrix Σ as follows:

$$\mathbf{d}_{\text{obs}} = F(\mathbf{m}) + \epsilon, \tag{2}$$

the likelihood PDF has the following expression:

$$
\begin{aligned}
\rho_{\text{like}}(\mathbf{d}_{\text{obs}}|\mathbf{m}) &\propto \exp\left(-\frac{1}{2}\|F(\mathbf{m}) - \mathbf{d}_{\text{obs}}\|_{\Sigma^{-1}}^2\right) \\
&= \exp\left(-\frac{1}{2}(F(\mathbf{m}) - \mathbf{d}_{\text{obs}})^\top \Sigma^{-1}(F(\mathbf{m}) - \mathbf{d}_{\text{obs}})\right).
\end{aligned}
\tag{3}
$$

With the posterior PDF in hands, we can extract statistical properties of interests including the maximum a posterior (MAP) estimate, the model covariance matrix, the model standard deviation (STD), and the confidence interval of $\mathbf{m}$. To conduct a successful Bayesian inversion, the primary issues are the construction of the posterior PDF and a computationally tractable method to extract statistical properties from the posterior PDF.

2.2 Deep GAN prior model generator

The selection of the prior PDF plays an important role in the statistical inversion. Conventional choices of the prior PDF are based on empirical observations of subsurface structures including the smoothness of the subsurface image, the sparsity of the coefficient of the image in certain transformed domains, or the total variation of the subsurface images. However, these kinds of prior information are too generic and cannot represent all the available prior information. Unlike the conventional prior choices, recent developments in learning-based methods in the field of imaging processing show a potential capability in estimating the prior distribution for images. In this work, we propose to use DGAN to directly extract prior information from existing subsurface images.

Given a training set $\mathcal{M}$, DGAN aims at generating artificial samples that share the same statistics as the training natural samples. A standard DGAN consists of two neural networks namely the generator $G(\mathbf{x}; \Theta_G)$ and the discriminator $D(\mathbf{m}; \Theta_D)$, which are parameterized by vectors Θ_G and Θ_D. The generator creates candidates of interests from vectors $\mathbf{x}$'s obeying a distribution $\mathcal{X}$ in a low-dimensional latent space, while the discriminator distinguishes these generated candidates from the true data distribution. The two networks contest with each other in a game. The discriminator tries to decrease the error rate of the discriminative network, while the generator tries to increase the error rate.

During the training step, DGAN solves the following *minmax* optimization problem:

$$\min_{\Theta_G} \max_{\Theta_D} \mathbb{E}_{\mathbf{m}\sim\mathcal{M}} \log\left(D(\mathbf{m}; \Theta_D)\right) + \mathbb{E}_{\mathbf{x}\sim\mathcal{X}}[\log\left(1 - D(G(\mathbf{x}; \Theta_G); \Theta_D)\right)],$$

$$(4)$$

where the expectation for $\mathbf{m}$ is over all the training samples $\mathcal{M}$, and the expectation for $\mathbf{x}$ is over the distribution $\mathcal{X}$. In general, $\mathcal{X}$ can be a set of images or data or any reasonable predefined distributions such as uniform and normal distributions. In this work, we select $\mathcal{X}$ to be a standard Gaussian distribution $\mathcal{N}(0, \mathbf{I})$ for a series of computational advantages in the following statistical inversion, which we will discuss later.

The objective function of the regular DGAN in Eq. (4) adopts the sigmoid cross entropy loss function. As stated by Mao et al. (2017), when training the generator, this loss function will cause the well-known problem of vanishing gradients for those generated samples that are on the correct side of the decision boundary, but are far away from the real samples. To mitigate the problem of vanishing gradients, we select to use the least squares DGAN (LSDGAN) (Mao et al., 2017) to train the generator. Instead of using the sigmoid cross entropy loss function, LSDGAN uses the a–b coding scheme for the discriminator, where a and b denote the labels for generated samples and real samples, respectively. The optimization problem that LSDGAN aims to solve is as follows:

$$\min_{\Theta_G} \max_{\Theta_D} \mathbb{E}_{\mathbf{m}\sim\mathcal{M}}[D(\mathbf{m}; \Theta_D) - a]^2 + \mathbb{E}_{\mathbf{x}\sim\mathcal{X}}[D(G(\mathbf{x}; \Theta_G); \Theta_D) - b]^2. \quad (5)$$

Since the objective function in Eq. (5) does not involve the sigmoid cross entropy loss function, training generator with this objective function less suffers from the problem of vanishing gradients. This allows LSDGAN to perform more stable during the learning process (Mao et al., 2017).

The trained generator $G(\mathbf{x}; \Theta_G)$ defines an n_l-manifold in the original $\mathbb{R}^{n_g}$ space, where $n_l \ll n_g$. With the generator $G(\mathbf{x}; \Theta_G)$, we can reformulate the original Bayesian inversion problem on the n_l-manifold as follows,

$$\rho_{\text{post}}(\mathbf{x}|\mathbf{d}_{\text{obs}}) \propto \rho_{\text{like}}(\mathbf{d}_{\text{obs}}|\mathbf{x})\rho_{\text{prior}}(\mathbf{x})$$
$$\propto \exp\left(-\frac{1}{2} \| F(G(\mathbf{x}; \Theta_G)) - \mathbf{d}_{\text{obs}} \|^2_{\Sigma^{-1}} - \frac{1}{2\sigma^2} \| \mathbf{x} \|^2\right). \tag{6}$$

The parameter σ is an ad-hoc scaling that might be set to 1, but its presence is an acknowledgment that the user might not have full confidence in the generality of the prior. If the prior is too restrictive, meaning it would only generate models close to the truth only at the expense of a very large $\|x\|$, then setting σ^2 to a correspondingly large value is a way to de-emphasize the importance of the prior. With a careful choice of σ, the prior is always trustworthy (in the worst case, it is just removed from consideration), and the error bars will not be under-estimated. Note that the presence of σ is completely dual to the presence of (a scalar version of) the data covariance matrix Σ in the data likelihood term, so the determination of σ is intimately tied to the user's ability to determine Σ in the first place.

Compared to the original posterior PDF in Eq. (1), studying the new posterior PDF $\rho_{\text{post}}(\mathbf{x}|\mathbf{d}_{\text{obs}})$ has four advantages. First, we significantly reduce the dimensionality from n_g to n_l, which makes the new Bayesian inversion problem less suffer from the curse of dimensionality. For example, according to Roberts, Rosenthal, et al. (2001), the convergence time of random walk Metropolis algorithm for a problem with a n_g-dimensional variable is $\mathcal{O}(n_g)$. Therefore, reducing the dimensionality of the unknown parameter can significantly reduce the computational cost. Second, the Gaussian prior distribution is directly from the definition of the DGAN generator without any hand designs. Third, images generated from the prior distribution share the same statistics as the training data set. Finally, the Gaussian prior distribution enables us to apply the computationally efficient McMC method— preconditioned Crank–Nicolson algorithm (pCN) to sample the posterior PDF, which we will introduce in the next subsection.

2.3 Sample the posterior PDF by pCN

The pCN algorithm is an efficient dimension-free McMC type method, whose efficiency has been proven for problems with Gaussian prior PDFs. The idea of this algorithm is that using the prior distribution to generate the random walking direction, followed by an accept–reject

Statistical seismic inversion 187

procedure. The pseudocode of the algorithm is shown in Algorithm 1. Indeed, the pCN algorithm is a modification of the conventional random walk algorithm, which generates a new proposed sample by $\mathbf{y}^{(k)} = \mathbf{x}^{(k-1)} + \beta \mathbf{r}^{(k)}$ with $\mathbf{r}^{(k)} \sim \mathcal{N}(0, \mathbf{I}_{n_l \times n_l})$. The pCN algorithm differs only slightly from the random walk method: the proposed sample is generated by $\mathbf{y}^{(k)} = \sqrt{1 - \beta^2}\mathbf{x}^{k-1} + \beta \mathbf{r}^{(k)}$. This slight modification results in that the pCN algorithm is robust to the increase of the dimensionality caused by the discretization of the physical domain. For this important property, we use the pCN algorithm to sample the posterior PDF in Eq. (6) in this work.

ALGORITHM 1 pCN

1. Set $k = 0$ and pick $\mathbf{x}^{(0)}$.
2. Propose $\mathbf{y}^{(k)} = \sqrt{1 - \beta^2}\mathbf{x}^{(k-1)} + \beta \mathbf{r}^{(k)}$, $\mathbf{r}^{(k)} \sim \mathcal{N}(0, \mathbf{I}_{n_l \times n_l})$.
3. Set $\mathbf{x}^{(k)} = \mathbf{y}^{(k)}$ with probability $a(\mathbf{x}^{(k-1)}, \mathbf{y}^{(k)}) = \min\left(1, \frac{\rho_{\text{like}}(\mathbf{d}_{\text{obs}}|\mathbf{y}^{(k)})}{\rho_{\text{like}}(\mathbf{d}_{\text{obs}}|\mathbf{x}^{(k-1)})}\right)$, or set $\mathbf{x}^{(k)} = \mathbf{x}^{(k-1)}$.
4. $k \rightarrow k + 1$.

---

3. Seismic inversion applications

We apply the proposed Bayesian approach with the deep GAN prior model generator to two seismic inversion applications—traveltime tomography and full waveform inversion.

3.1 Traveltime tomography

Seismic traveltime tomography has been serving as an important tool to extract elastic properties of the Earth interior in local, regional, and global scales from seismic data (Aki et al., 1977; Dziewonski, 1984; Nolet, 2012; Romanowicz, 1979). For the sake of efficiency, only the travel time information on the seismic recordings is used. The goal of the traveltime tomography is to seek a velocity model that can fit the observed travel time data by linearized or nonlinear inversion methods. Under the high frequency approximation, we can simplify the wave equation and compute the travel time field for a certain model by solving the following Eikonal equation:

$$||\nabla \tau(x, z)||^2 = \mathbf{m}(x, z), \tag{7}$$

where the operator ∇ denotes the gradient operator, $\tau(x, z)$ and $\mathbf{m}(x, z)$ are the travel time field and the squared slowness at the spatial coordinate (x, z), respectively. For a detailed review, readers should refer to Nolet (2012).

3.2 Full waveform inversion

Full waveform inversion (FWI) is another important seismic imaging approach in the fields of both global seismology and exploration geophysics. FWI utilizes all kinds of waveforms including the reflection and refraction waves to reconstruct the subsurface velocity model. Through iteratively comparing the observed and predicted data and updating the model, FWI tries to find the best data-fit model. The forward map of FWI requires to solve the following time-domain acoustic wave equation:

$$-\mathbf{m}(x, z)\frac{\partial^2 \mathbf{u}(x, z; t)}{\partial t^2} + \Delta \mathbf{u}(x, z; t) = \mathbf{q}(x, z; t), \qquad (8)$$

where the symbol Δ denotes the Laplacian operator, $\mathbf{u}(x, z; t)$ and $\mathbf{q}(x, z; t)$ denote the wavefield and source function at the time of t and space location of (x, z), respectively. After discretization, we can solve the acoustic wave equation and obtain the predicted data $\mathbf{d}_{\mathrm{pred}}$ corresponding to the current model parameter $\mathbf{m}$. See Virieux and Operto (2009) for an extensive overview of state-of-the-art approaches to FWI.

4. Numerical examples

We assess the effectiveness of the proposed Bayesian seismic inversion scheme with deep generator priors via applications of the traveltime tomography and the full waveform inversion. To conduct the experiments, we first train an eight-block generative network and a three-block discriminative network using velocity slices extracted from the 3D Overthrust model. With the trained deep generator prior, we conduct statistical inversions on both the Overthrust model and the Sigsbee model.

4.1 Training the network architecture

4.1.1 Network architectures

In this work, we first design three different generators with different numbers of layers to study the influence of the number of layers to the performance of the generator. The three different generative networks (GN) consist of eight-block layers, six block layers, and four block layers, respectively. To simplify the notation, we name them GN-8, GN-6, and

GN-4, respectively. We also design a discriminative network with three-block layers. The architectures of GN-8 and the discriminator are shown in Fig. 1. The generative network is a deep deconvolutional neural network that maps a 50-dimensional Gaussian random vector $\mathbf{x} \in \mathbb{R}^{50}$ with $\mathbf{x} \sim \mathcal{N}(0, \mathbf{I})$ to a 64×64 velocity image, where $\mathbf{I}$ is a 50×50 identity matrix. We first use a fully connected network to map the input data to 128 16×16 features, followed by the batch normalization and a leaky ReLU activation. The fully connected network, batch normalization, and leaky ReLU activation form the first block of the generative network. After the first block, we add six additional deconvolutional based blocks. Each block contains a deconvolutional neural network, a batch normalization operation, and a leaky ReLU activation. The output of the 6 blocks is 32 64×64 features. The last block of the generative network is a deconvolutional neural network with 5×5 kernels and the tanh activation function that generates the output 64×64 image. GN-6 shares a similar structure with GN-8 while does not include the forth and seventh blocks of GN-8. Similar to GN-6, GN-4 does not include the second, forth, sixth, and seventh blocks of GN-8.

The discriminative network starts with the 64×64 velocity image and outputs a real number that aims to fit the labels a and b for the real and generated samples. In this work, we follow Mao et al. (2017) to use the 0-1 binary coding scheme and select $a = 1$ and $b = 0$. The discriminative network is a three-block convolutional neural network. The first block contains a convolutional neural network with 5×5 kernels and 64 features, a leaky ReLU activation, and a dropout operator with a dropout ratio of 0.3. The second block consists of a convolutional neural network with 5×5 kernels and 128 features, a leaky ReLU activation, and a dropout operator with a dropout ratio of 0.3. The last block consists of a fully connected network that outputs the label value.

4.1.2 Training

We train the generative and discriminative networks with velocity models extracted from the 3D Overthrust model. The size of the 3D Overthrust model is 20 km $\times$ 20 km $\times$ 4.65 km with a spatial spacing of 0.025 km, yielding a 3D volume with the size of 801 $\times$ 801 $\times$ 187. We extract 19,762 velocity models with the size of 64×64 from the 3D volume. To increase the diversity of the training set, we also rotate the extracted velocity models with random angles, yielding 79,048 velocity models. We use the 79,048 models as the training set to train the generative and discriminative networks.

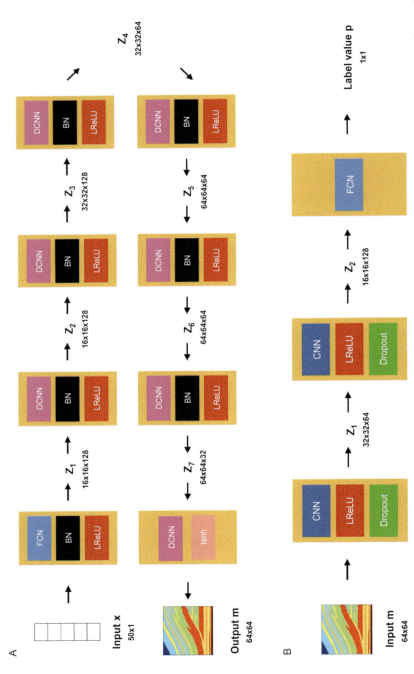

Fig. 1 The framework of DGAN. Architectures of (A) GN-8 and (B) the discriminative networks. *BN*, batch normalization; *CNN*, convolutional neural network; *DCNN*, deconvolutional neural network; *FCN*, fully connected network; *LReLU*, leaky ReLU; *tanh*, tanh activation function.

We use the built-in adaptive moment estimation (ADAM) optimizer (Kingma & Ba, 2014) of TensorFlow (Abadi et al., 2016) to train the networks. ADAM is a first-order gradient-based optimizer for stochastic optimizations, based on adaptive estimates of lower-order moments. ADAM has been widely used in many deep learning applications due to the fact that it is easy to implement, is computationally efficient, requires little memory cost, and is invariant to diagonal rescaling of the gradients. We select the momentum parameter $\gamma = 0.9$, mini-batch size $L_b = 256$, and initial rate $\mu = 1e\text{-}4$. We conduct the training with the TensorFlow interface (Abadi et al., 2016) and Keras toolbox (Chollet et al., 2015) on an NVIDIA GeForce GTX 1080 GPU with 12 GB RAM. We allow 1000 epochs that take around 23.9, 20.8, and 15.0 h for training GN-8, GN-6, and GN-4, respectively.

4.2 Quantitative performance evaluation of different generators

4.2.1 Comparison of generated samples

First, we compare the performance of the three generators in producing samples. Fig. 2 shows a comparison of the images from the training set and the three generative networks. Visually, it is very difficult to evaluate these images, since they all present some layered structures as shown in the training set. To further evaluate the three generators, we use each generator to produce 50,000 samples and compute the point-wise STD and probability density of the 50,000 samples. Fig. 3 shows the comparison of the STDs of the training samples and generated samples of the three generators. The STDs corresponding to GN-8 and GN-6 match with that of the training samples much better than the one of GN-4. Fig. 4 shows the comparison of the point-wise probability densities corresponding to the training samples and generated samples at three different locations. Clearly, the probability densities of GN-8 and GN-6 coincide with that of the training samples much better than the one of GN-4. Both comparisons imply that increasing the number of layers can yield a generator that is able to capture the distribution of the training set more accurately.

4.2.2 Quantitative evaluation criterion

Second, we quantitatively study the capability of the three generators in representing testing models. In general, the study of DGAN suffers from the lack of a quantitative criterion to evaluate the performance of trained generators. To design the evaluation criterion, we notice two facts about a good generator: (1) a good generator should be able to generate artificial

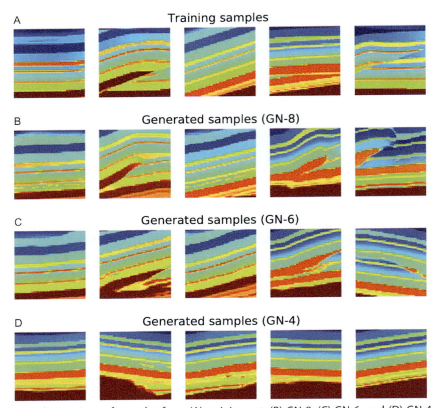

Fig. 2 Comparison of samples from (A) training set, (B) GN-8, (C) GN-6, and (D) GN-4.

models that are very close to the target testing models; (2) latent vectors for a good generator should have moderate magnitude to ensure that the generator is not overfitting as a result of ill-conditioning. Since the prior distribution is the standard Gaussian distribution, we can use the magnitude of standard Gaussian vectors as a reference for the magnitude of the latent vectors. We use both facts to design the evaluation criterion for the three generators.

We first select three testing models $\mathbf{m}_{t,i}$ ($1 \leq i \leq 3$) from the Overthrust model, which are not in the training set. Then for the jth generator ($1 \leq j \leq 3$) and each $\mathbf{m}_{t,i}$, we try to find the best latent vector $\mathbf{x}_{j,i}$ that can minimize the ℓ_2-norm difference between $\mathbf{m}_{t,i}$ and the generated model $\mathbf{m}_{g,j,i} = G_j(\mathbf{x}_{j,i}; \Theta_{G_j})$ by solving the following optimization problem:

$$\mathbf{x}_{j,i} = \arg\min_{\mathbf{x}} \| \mathbf{m}_{t,i} - G_j(\mathbf{x}; \Theta_{G_j}) \|_2^2. \tag{9}$$

Statistical seismic inversion

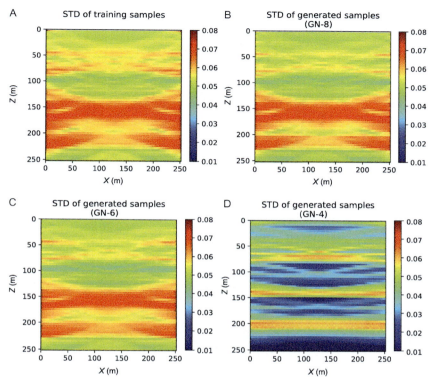

Fig. 3 Comparison of the point-wise STDs of (A) training samples, (B) samples generated by GN-8, (C) samples generated by GN-6, and (D) samples generated by GN-4.

To solve this optimization problem, we first generate 1000 samples and select the one $x_{ini,j,i}$ that produces the least ℓ_2-norm misfit. We start with $x_{ini,j,i}$ and use 100,000 gradient descent steps to find the optimal solution. Fig. 5 shows the three selected models and their corresponding best generated models $\{m_{g,j,i}\}_{1 \le i,j \le 3}$ produced by the three generators. Clearly, the generated models given by GN-8 are much closer to the testing models, compared to GN-6 and GN-4.

The first aforementioned criterion is easy to evaluate. We compute the relative model error $\frac{\|m_{t,i} - m_{g,j,i}\|_2}{\|m_{t,i}\|_2}$ between the testing models $\{m_{t,i}\}$ and the generated models $\{m_{g,j,i}\}$. We use this relative model error as the first evaluation quantity. To evaluate the second criterion, we first decide a Gaussian zone Z with respect to the ℓ_2-norm of the latent parameter x. We draw 10,000 random vectors p's from $\mathcal{N}(0, I)$ and compute their ℓ_2-norm. We compute $\|p\|_2$ for each p and compute the distribution of

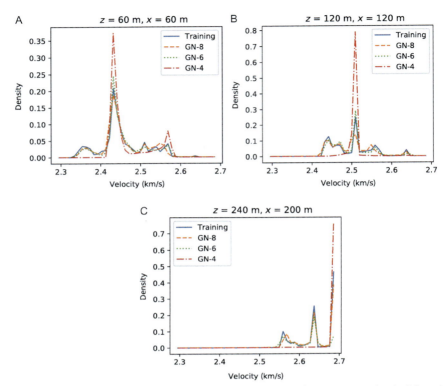

Fig. 4 Comparison of the point-wise probability densities of training samples (*solid*) and samples generated by GN-8 (*dashed*), GN-6 (*dotted*), and GN-4 (*dashdot*) at three locations: (A) $z = 60$m, $x = 60$m; (B) $z = 120$m, $x = 120$m; and (C) $z = 240$m, $x = 200$m.

the 10,000 $\|\mathbf{p}\|_2$'s. We define an interval along the ℓ_2-norm of the latent parameter $\|\mathbf{x}\|_2$ as a 95% Gaussian that contains 95% samples of $\|\mathbf{p}\|_2$'s. In this case, we have $Z = [5.29, 8.89]$. Given a latent parameter $\hat{\mathbf{x}}$, we can compute the distance between $\|\hat{\mathbf{x}}\|_2$ and Z as $\mathrm{dist}(\|\hat{\mathbf{x}}\|_2, Z) = \max\{0, \|\hat{\mathbf{x}}\|_2 - 8.89, 5.29 - \|\hat{\mathbf{x}}\|_2\}$. We compute the distance for each $\|\mathbf{x}\|_2$ and plot them as the gray background in Fig. 6, whose horizontal axis is $\log \|\mathbf{x}\|_2$. Fig. 6 plots the relative model error and logarithm ℓ_2-norm of the latent vector for each generated model. For a good prior generator, we expect that all the points should be at the bottom of the Gaussian zone (dark gray area). Clearly, GN-8 produces results with minimal relative model errors. Meanwhile, the $\|\mathbf{x}_{j,i}\|_2$ of the GN-8 results are closer to the Gaussian zone. This test illustrates that compared to GN-6 and GN-4, GN-8 has a stronger capability in representing testing models, which implies a stronger generality of GN-8.

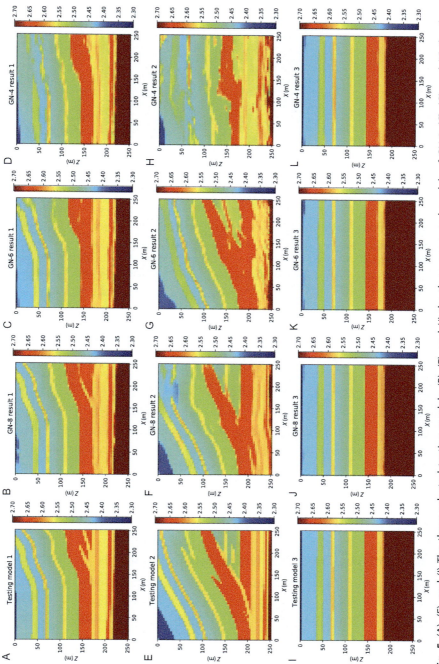

Fig. 5 (A), (E), and (I) The three selected testing models; (B), (F), and (J) the three generated models of GN-8; (C), (G), and (K) the three generated models of GN-6; (D), (H), and (L) the three generated models of GN-4.

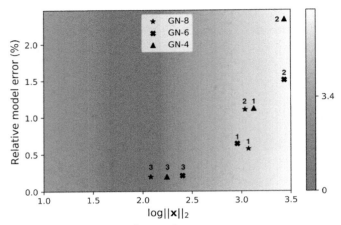

Fig. 6 ℓ_2-norm relative model error $\frac{\|\mathbf{m}_{t,i} - \mathbf{m}_{g,j,i}\|_2}{\|\mathbf{m}_{t,i}\|_2}$ between the testing models $\{\mathbf{m}_{t,i}\}$ and the generated models $\{\mathbf{m}_{g,j,i}\}$ of GN-8 (★), GN-6 (×), and GN-4 (△) vs. logarithm ℓ_2-norm of their latent vectors $\mathbf{x}_{j,i}$. The *gray scaled background* represents the distance between a latent vector and the standard Gaussian zone. The indices 1, 2, and 3 indicate the corresponding testing model indices.

With numerical experiments akin to those in Fig. 6, the user can make an informed choice of architecture and hyperparameters of their DGAN prior. The observed size of $\|\mathbf{x}\|$ needed for fitting representative test models informs the order of magnitude of their prior likelihood, and can help choose σ in Eq. (6), so that the prior's value $\exp\left(-\frac{1}{2\sigma^2}\|\mathbf{x}\|^2\right)$ is roughly comparable to the data likelihood value $\exp\left(-\frac{1}{2}\|F(G(\mathbf{x};\Theta_G)) - \mathbf{d}_{\text{obs}}\|^2_{\Sigma^{-1}}\right)$ (which in turn assumes that the user has some knowledge of the noise level, and of the uncertainties in the forward model F). If the size of $\|\mathbf{x}\|$ is too large for reasonable test models, then it is important to set σ to a correspondingly large value, so that the prior's effect is curtailed. Without tuning the σ, the risk is that the user will put too much faith in a prior that produces anomalously small error bars.

4.3 Statistical inversion with Overthrust models

We conduct the first Bayesian inversion with a testing velocity model $\mathbf{m}_t$ extracted from the 3D Overthrust model (shown in Fig. 7A) that does not belong to the training set. Since both the testing model and training models are extracted from the 3D Overthrust model, their spatial distributions should share a strong similarity and the prior generator should provide quite good prior information for the inversion.

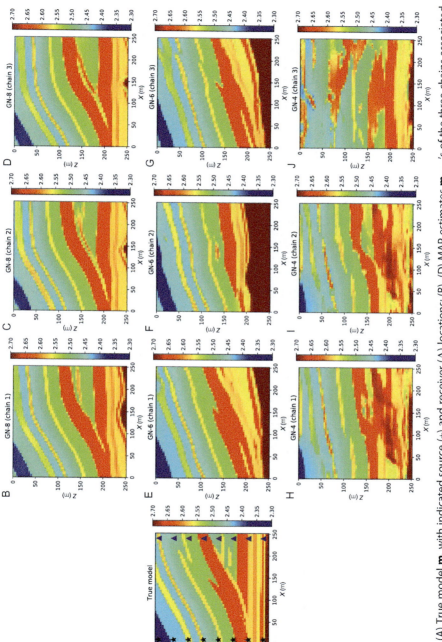

Fig. 7 (A) True model $\mathbf{m}_t$ with indicated source (⋆) and receiver (△) locations; (B)–(D) MAP estimates $\mathbf{m}_{MAP}$'s of the three chains associated with GN-8; (E)–(G) MAP estimates of the three chains associated with GN-6; (H)–(J) MAP estimates of the three chains associated with GN-4. The true model is the testing model 2 in Fig. 5.

4.3.1 Traveltime Tomography example

We first conduct the statistical inversion with the traveltime tomography. We mimic a cross-hole case, where 30 sources ($\star$) are located at the left side of the model and 60 receivers (Δ) are located at the right side of the model as shown in Fig. 7A, which is the testing model 2 in Fig. 5. The model size is 258 m × 258 m. We rescale the velocity model so that the range of the velocity is [2.3 km/s, 2.7km/s]. The observed data contain 0.5% additive Gaussian noise.

We conduct the Bayesian inversion with the three aforementioned different generators and compare their performances. For each generator, we use three different starting points to draw three different Markov chains with 100,000 samples. For McMC type methods, the selection of starting points will strongly affect the so-called "burn-in" period, after which the Markov chain reaches a stable status. To reduce the "burn-in" period, we need a starting point that is not too far from the true model. To meet this requirement, for each generator, we first generate 1000 samples and compute the ℓ_2-norm difference between each sample and the true model. Then we select the best models with the minimal ℓ_2-norm model difference from the first 100, 500, and 1000 samples as the starting points. With the starting points, we use the pCN method to produce the 9 Markov chains with a burn-in period of 30,000 steps. To ensure an appropriate acceptance ratio that can yield efficient sampling, we select $\beta = 0.01$, resulting in an average acceptance ratio of 0.4. Considering the similarity of the training models and the testing model, we select $\sigma = 1$.

Fig. 7B–J shows the MAP estimates of the 9 Markov chains, and Fig. 8 shows the relative ℓ_2-norm model error between each MAP estimate and the

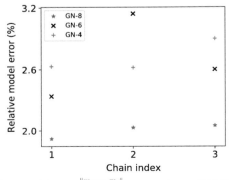

Fig. 8 ℓ_2-norm relative model error $\frac{\|\mathbf{m}_{\text{MAP}} - \mathbf{m}_t\|_2}{\|\mathbf{m}_t\|_2}$ comparison of MAP estimates associated with GN-8 ($\star$), GN-6 ($\times$), and GN-4 (+). Clearly, GN-8 produces MAP estimates with the minimal model errors.

true model. From Fig. 7, we can observe that the GN-8 outperforms GN-6 and GN-4 from the perspective of structure matching. Clearly, the MAP estimates of GN-8 matches most layered structures in the true model, despite some mismatches around the boundary of the model. Fig. 8 illustrates that GN-8 produces MAP estimates with minimal model errors. Additionally, the variation of the three MAP estimates of GN-8 is visually much smaller than those of GN-6 and GN-4.

We compute the point-wise STDs for all the 9 chains. For each generator, we also compute the point-wise STD from the ensemble of the three individual chains. Fig. 9 shows all the computed STDs. Overall, we can observe large STDs around the boundary of layers and small STDs inside layers. Comparing the STDs of individual chains with STDs of their corresponding ensembles, we can observe that the difference between STDs of individual chains and the ensemble chain associated with GN-8 is significantly smaller than those associated with GN-6 and GN-4. This result coincides with the observation that MAP estimates of GN-8 have the smallest variation.

Fig. 10 shows the comparison of the sample probability densities for the 9 chains at three different locations. Similar to the results of MAP estimates and STDs, the three probability density curves of GN-4 and GN-6 vary from each other, which implies that the three chains reach different stable statuses. On the other side, the three probability density curves of GN-8 share a strong similarity, which implies that the three chains reach a similar stable status.

The comparisons of MAP estimates, STDs, and probability densities imply that GN-8 outperforms GN-6 and GN-8 from both perspectives of model errors and the stability of the Markov chain. This comparison result coincides with the fact that GN-8 is the best one of the three networks in terms of the proposed evaluation criterion. Such out-performances may be due to the fact that GN-8 has more layers and network parameters. This strengthens the capability of GN-8 in producing more generic samples. Due to this out-performance of GN-8 over GN-6 and GN-4, we will only use GN-8 for all the following examples.

4.3.2 FWI example

We conduct the second statistical inversion with the application of FWI. We use the same velocity model as the one used in the first example. We also mimic a cross-hole case, in which we place eight sources at the left side of the model and place 64 receivers at the right side of the model. Fig. 11A shows the true model and locations of sources and receivers.

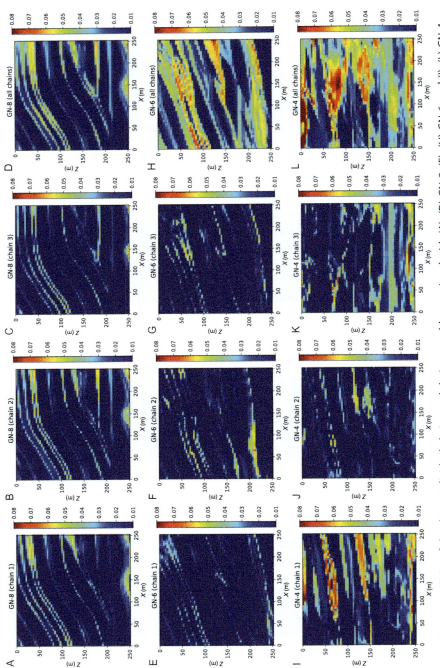

Fig. 9 Point-wise STDs of the three individual chains and their ensemble associated with (A)–(D) GN-8, (E)–(H) GN-6, and (I)–(L) GN-4.

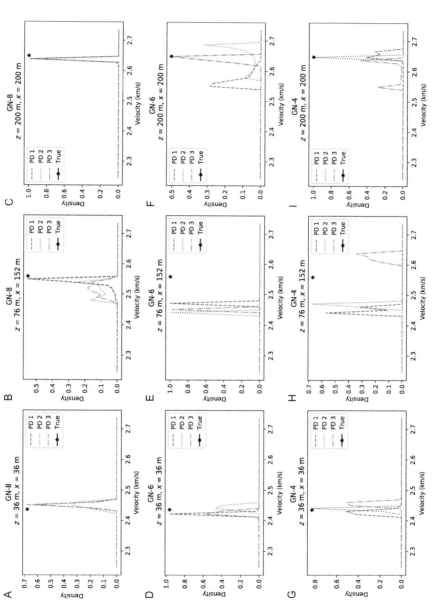

Fig. 10 Comparison of the probability densities associated with chains 1 (*dashed*), 2 (*dotted*), and 3 (*dashdot*) at three points: [z, x] = [36 m, 36 m], [76 m, 152 m], and [200 m, 200 m]. The symbol ⋆ indicates the true velocity at the target point. (A)–(C) Probability densities associated with the GN-8; (D)–(F) probability densities associated with GN-6; (G)–(I) probability densities associated with the GN-4.

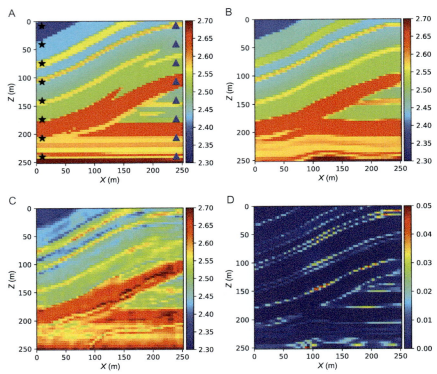

Fig. 11 (A) True model for FWI. (B) MAP of the statistical inversion. (C) Result of the conventional FWI. (D) STD of the 70,000 samples.

We use a Ricker wavelet centered at 50 Hz to simulate the data with 1% additive Gaussian noise. We use the same strategy as mentioned in the first example to select the starting point and use the pCN method to generate 100,000 samples with a burn-in period of 30,000 steps. We also select $\sigma = 1$. We compute statistics of interests with the remaining 70,000 samples. Fig. 11B shows the MAP estimate of the 70,000 samples. Clearly, the MAP estimate matches the true model quite well. As a comparison, Fig. 11C shows the result of the conventional FWI using the same starting point. We can observe that the MAP estimate obtained by the proposed method presents a better resolution compared to the conventional FWI that directly works on the physical domain. Despite the high-resolution image, the MAP estimate with the prior GN-8 does not reconstruct two faults at depths of $z = 70$ m and $z = 130$ m. This result is due to the fact that the corresponding latent parameter of the testing model is not in the predefined Gaussian zone, which has been observed from the previous result in Fig. 6. To improve the

Statistical seismic inversion 203

result, a larger σ would be necessary. Fig. 11D illustrates the point-wise STD of the 70,000 samples. Similar to the result of traveltime tomography, we can observe high STDs at boundaries of layers and low STDs inside layers. Fig. 12 shows probability density of samples at three positions. Again, velocities around the boundary of layers are more uncertain than those inside layers.

Fig. 13 depicts the history of $-\log \rho_{\text{post}}(\mathbf{x}|\mathbf{d}_{\text{obs}})$ vs. the iteration number of the Markov chain. We can observe after 30,000 iteration, $-\log \rho_{\text{post}}(\mathbf{x}|\mathbf{d}_{\text{obs}})$ starts to reach a relatively low and stable level (around 1000). Fig. 14 shows the generated samples drawn by McMC at iterations of 10,000, 30,000, 70,000, and 100,000. Clearly, after 30,000 iterations, the drawn samples converges to a stable status with small variations. This observation coincides with the observation of the history of $-\log \rho_{\text{post}}(\mathbf{x}|\mathbf{d}_{\text{obs}})$.

4.4 Statistical inversion with Sigsbee models

In this subsection, we conduct tests with a velocity model extracted from the Sigsbee model. This example is challenging, since the spatial distribution of the Sigsbee model is quite different from the Overthrust model.

4.4.1 Traveltime tomography example

We first conduct the traveltime tomography example. The settings for the experiment are the same with the ones in the previous example. We also select $\sigma = 1$. Fig. 15A–C shows the true model, MAP estimate and STD, respectively. We can observe clear differences between the true model and the MAP estimate. Fig. 17 shows the comparison of the true velocity (solid line) and the MAP estimate (dashed line) at horizontal positions of (A) $x = 40$ m, (B) $x = 120$ m, and (C) $x = 200$ m. The MAP estimates are shaded with the error bars corresponding to 99% confidence intervals. The velocities of the drawn samples show small variations at the horizontal locations of $x = 120$ m and $x = 200$ m. The error bars partially cover the true velocities and are partially under-estimated. However, if we compare the observed data and predicted data (Fig. 16), we can observe that the difference between them reaches the noise level (see Figs. 16C and 18). This implies that the priors are over-trusted and the proposed method finds a best data-fit solution restricted by the prior distribution. Considering the visible difference between the training models and the testing model, during the inversion we may select a larger σ to weaken our confidence on the priors.

We conduct an additional experiment with the selection of $\sigma = 10$. Fig. 19 shows the final MAP estimate and standard deviation, and Fig. 20

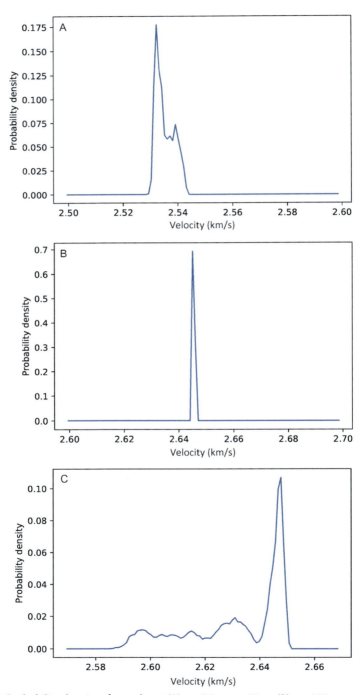

Fig. 12 Probability density of samples at (A) $z = 36$ m, $x = 76$ m, (B) $z = 188$ m, $x = 156$ m, and (C) $z = 244$ m, $x = 76$ m.

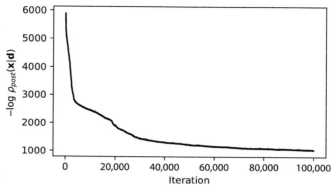

Fig. 13 $-\log \rho_{post}(\mathbf{x}|\mathbf{d}_{obs})$ vs. iteration number.

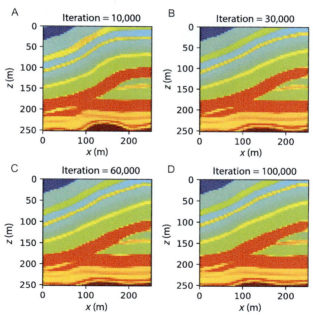

Fig. 14 Samples drawn by FWI McMC at iterations of (A) 10,000, (B), 30,000, (C) 60,000, and (D) 100,000.

shows the comparison of the true velocity (solid line) and the MAP estimate (dashed line) at horizontal positions of (A) $x = 40$ m, (B) $x = 120$ m, and (C) $x = 200$ m. The MAP estimates are shaded with the error bars corresponding to 99% confidence intervals. The change of σ yields a different MAP estimate. In addition, when we increase the parameter σ,

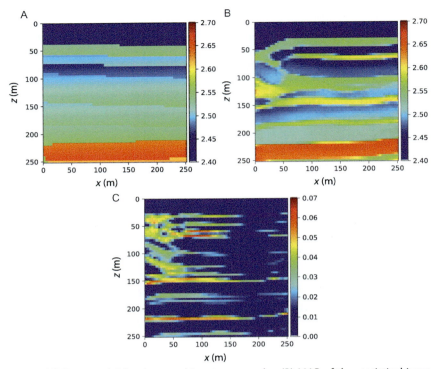

Fig. 15 (A) True model for the traveltime tomography. (B) MAP of the statistical inversion. (C) STD of the 70,000 samples.

the corresponding posterior standard deviation increases and the corresponding error bars expand. As shown in Fig. 20, error bars cover the true velocities, which indicates that the error bars might not be under-estimated.

4.4.2 FWI example

Then we conduct the statistical inversion with FWI. In this example, Fig. 21A shows the true model, whose size is 1.64 km × 1.64 km. We place 8 sources at the surface of the model and place 64 receivers at the same depth. We use a Ricker wavelet centered at 10 Hz to simulate the data with 1% additive Gaussian noise. The settings for McMC are the same as the previous example with $\sigma = 1$. Fig. 21B–D show the conventional FWI result, MAP estimate and STD, respectively. Conventional FWI presents a result that is a smoothed, blurred, and noisy version of the true model, while the proposed approach presents a sharp result with clear layers. However, we can observe

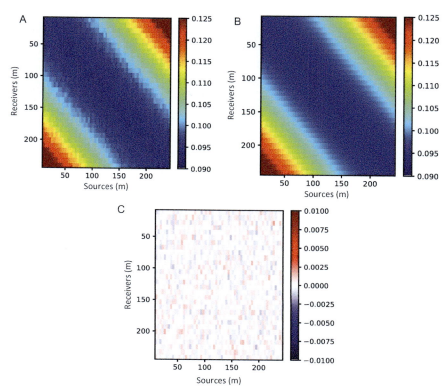

Fig. 16 (A) Observed data. (B) Predicted data obtained with the MAP estimate. (C) Difference between the observed and predicted data.

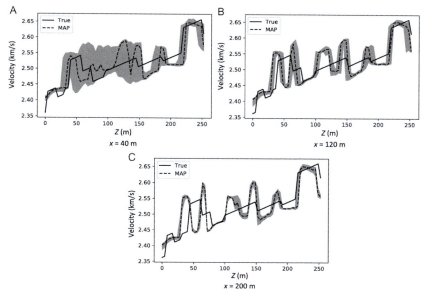

Fig. 17 Comparison of the true velocity (*solid line*) and the MAP estimate (*dashed line*) at horizontal positions of (A) $x = 40$ m, (B) $x = 120$ m, and (C) $x = 200$ m. The MAP estimates are *shaded* with the error bars corresponding to 99% confidence intervals.

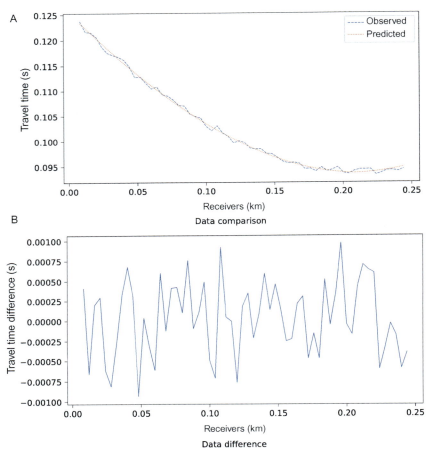

Fig. 18 (A) Observed data (*dashed line*) and predicted data (*dotted line*); (B) Difference between the observed and predicted data.

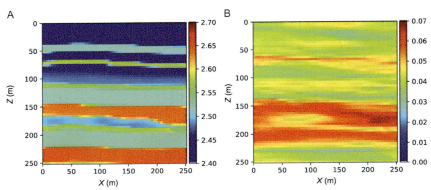

Fig. 19 (A) MAP of the statistical inversion. (B) STD of the 70,000 samples.

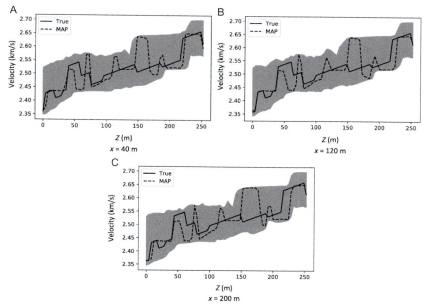

Fig. 20 Comparison of the true velocity (solid line) and the MAP estimate (*dashed line*) at horizontal positions of (A) $x = 40$ m, (B) $x = 120$ m, and (C) $x = 200$ m. The MAP estimates are shaded with the error bars corresponding to 99% confidence intervals.

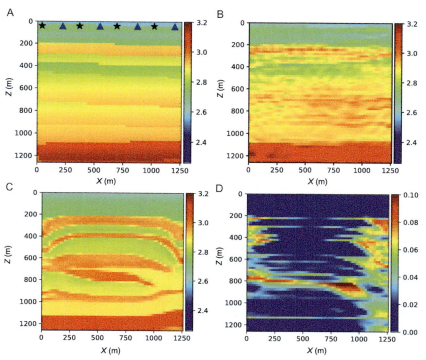

Fig. 21 (A) True model for FWI. (B) Result of the conventional FWI. (C) MAP of the statistical inversion. (D) STD of the 70,000 samples.

clear differences between the true model and the MAP estimate. Fig. 23 shows the comparison of the true velocity (solid line) and the MAP estimate (dashed line) at horizontal positions of (A) $x = 200$ m, (B) $x = 600$ m, and (C) $x = 1000$ m. The MAP estimates are shaded with the error bars corresponding to 99% confidence intervals. We can observe that the obtained error bars cover most of the true velocities while are under-estimated at some depths. To further analyze the inversion result, we compare the observed data and the predicted data obtained with the MAP estimate in Figs. 22 and 24. Clearly, the data differences have reached the noise level, which means the MAP estimate is also a best data-fit solution. All the observations implies that the priors are over trusted. To further improve the result, we may enlarge the parameter σ.

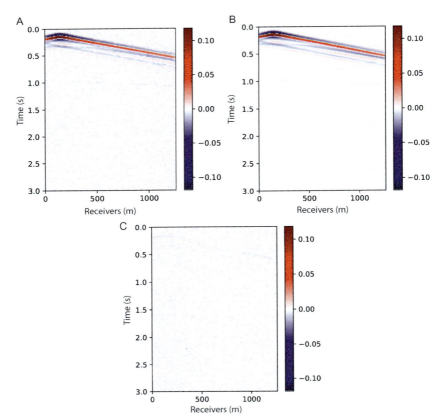

Fig. 22 (A) Observed data. (B) Predicted data obtained with the MAP estimate. (C) Difference between the observed and predicted data.

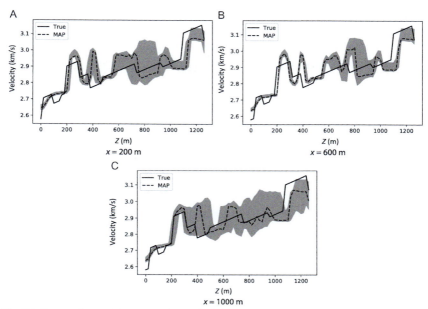

Fig. 23 Comparison of the true velocity (*solid line*) and the MAP estimate (*dashed line*) at horizontal positions of (A) $x = 200$ m, (B) $x = 600$ m, and (C) $x = 1000$ m. The MAP estimates are shaded with the error bars corresponding to 99% confidence intervals.

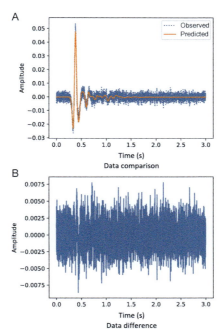

Fig. 24 (A) Observed data (*dotted line*) and predicted data (*solid line*); (B) Difference between the observed and predicted data.

5. Conclusions and discussion

We study the Bayesian statistical seismic inversion framework with a deep generator prior. Compared to conventional Bayesian inversion with hand-crafted prior information, using the deep generator prior has three major advantages. First, DGAN enables us to conduct inversions in the low-dimensional latent space instead of the original high-dimensional image space, which significantly reduces the computational complexity. Second, the trained DGAN is able to generate artificial models that share the similar spatial distributions as the training images, which provides us with the access to a complete description of the available prior information. Third, the prior distribution in the latent space is a normal distribution without any hand-crafted assumption. This enables us to apply the efficient pCN method to sample the posterior PDF. These benefits coincide with observations in recent works by Mosser et al. (2020) and Herrmann et al. (2019), which also study the application of GAN to statistical seismic inversion problems.

Given a training set, we find that the number of layers in the generative network plays an essential role in capturing the distribution and features of the training samples. As shown in the numerical examples, increasing the number of layers can strengthen the capability of the generative network in describing complete statistical information of the training set with a complicated distribution. As a direct consequence, Bayesian inversion using generative network with 8 block layers produces better inversion results than those with 6 and 4 block layers.

The quantitative evaluation of the quality of the trained generator is difficult for DGAN. We propose a quantitative evaluation criterion in this work. As we show in the numerical example, a good generator should be able generate artificial models that can match the target models quite well. Meanwhile the latent vectors of these artificial models should follow the predefined prior distribution used in the training procedure and have moderate magnitudes. We can use both the relative model error and the logarithm ℓ_2-norm of the latent vector as the two evaluation quantities. This provides us with a quantitative standard to evaluate the generator during the training procedure, and a way to calibrate the magnitude of its effect in the posterior. Based on the proposed evaluation criterion, we observe that the increase of the number of the neural network layers can strengthen the generality of the generator. Numerical examples show a strong consistence between the quality of the prior generator in terms of the proposed evaluation criterion and the quality of the statistical inversion result. The relative ℓ_2-norm model error used in this paper is the first attempt to realize the proposed evaluation criterion, which can be improved by more specific metrics for image quality quantification such as the structural similarity index metric (SSIM) (Wang, Bovik, Sheikh, & Simoncelli, 2004).

Since seismic inverse problems are typically very ill-posed, the noisy observed data can be explained by different models. Therefore, the selection of prior information has a strong influence to the final inversion result. Numerical examples illustrate the importance of the training dataset to the proposed Bayesian inversion with DGAN prior. Compared to the conventional hand-crafted priors, the generality of the DGAN prior straightforwardly depends on the diversity of the training set. When the target velocity model shares a similar spatial distribution with models in the training set, the proposed approach is able to produce an accurate best data-fit solution with a high resolution. While for the opposite situation, the proposed approach can only find a best data-fit solution in the range of the DGAN, which may differ from the true model. The parameter σ provides users the capability of controlling their confidence on the generality of the priors. A carefully selected σ can prevent the under-estimated error bars yielded by an over-trusted prior.

The selection of the starting point is essential to the efficiency and effectiveness of McMC type methods. Due to the complexity of the posterior distribution, we may not be able to sample the whole space given limited computational resources. As a consequence, the statistical analysis typically only capture local information associated with the starting point. Moreover, in practice, we do not have the access to the ground truth model and have to use more realistic strategies to select starting points. Indeed, there is one possible and tractable strategy based on the fact that the trained DGAN is able to efficiently generate high-quality artificial subsurface models. First, we can use the trained DGAN to generate a set of potential starting models with a relatively large size (e.g., 100 or 1000). Then we can select the model that maximizes the likelihood probability density function as the starting point for the McMC method. This strategy ensures us to start with a point that produces relatively high probability density value.

The deep generator provides us with a solution to automatically encode prior knowledge about the subsurface structures from existing subsurface models. The robustness of the deep generator prior depends on the generality and diversity of the training set. Through employing new subsurface models, the deep generator prior can keep on evolving to enhance its robustness. This property can help researchers automatically accumulate their knowledge about the subsurface structures in their research career. In practice, when applying a pretrained DGAN to a new study area, one possible approach to enriching the generality of DGAN is to apply DGAN and the conventional priors alternately, so that DGAN can account for new structures on the fly.

Acknowledgments

The authors acknowledge the funding and support provided by ExxonMobil Research and Engineering Company. Dr. Laurent Demanet is also supported by AFOSR grant FA9550-17-1-0316.

References

Abadi, M., Barham, P., Chen, J., Chen, Z., Davis, A., Dean, J., et al. (2016). Tensorflow: A system for large-scale machine learning. In *12th USENIX Symposium on Operating Systems Design and Implementation (OSDI 16)*, pp. 265–283.

Aki, K., Christoffersson, A., & Husebye, E. S. (1977). Determination of the three-dimensional seismic structure of the lithosphere. *Journal of Geophysical Research, 82*(2), 277–296.

Araya-Polo, M., Jennings, J., Adler, A., & Dahlke, T. (2018). Deep-learning tomography. *The Leading Edge, 37*(1), 58–66.

Arridge, S., Maass, P., Öktem, O., & Schönlieb, C.-B. (2019). Solving inverse problems using data-driven models. *Acta Numerica, 28*, 1–174.

Bui-Thanh, T., Ghattas, O., Martin, J., & Stadler, G. (2013). A computational framework for infinite-dimensional Bayesian inverse problems. Part I: The linearized case, with application to global seismic inversion. *SIAM Journal on Scientific Computing, 35*(6), A2494–A2523.

Chollet, F., et al. (2015). In *Keras*. https://keras.io.

Clapp, R. G. (2000). *Geologically constrained migration velocity analysis*. Citeseer.

Cotter, S. L., Roberts, G. O., Stuart, A. M., White, D., et al. (2013). MCMC methods for functions: Modifying old algorithms to make them faster. *Statistical Science, 28*(3), 424–446.

Duijndam, A. J. W. (1988). Bayesian estimation in seismic inversion. Part II: Uncertainty Analysis. *Geophysical Prospecting, 36*(8), 899–918.

Dziewonski, A. M. (1984). Mapping the lower mantle: Determination of lateral heterogeneity in P velocity up to degree and order 6. *Journal of Geophysical Research: Solid Earth, 89*(B7), 5929–5952.

Ely, G., Malcolm, A., & Poliannikov, O. V. (2017). Assessing uncertainties in velocity models and images with a fast nonlinear uncertainty quantification method. *Geophysics, 83*(2), R63–R75.

Fang, Z., Da Silva, C., Kuske, R., & Herrmann, F. J. (2018). Uncertainty quantification for inverse problems with weak partial-differential-equation constraints. *Geophysics, 83*(6), R629–R647.

Goodfellow, I., Pouget-Abadie, J., Mirza, M., Xu, B., Warde-Farley, D., Ozair, S., & Bengio, Y. (2014). Generative adversarial nets. In *Advances in neural information processing systems*, pp. 2672–2680.

Haber, E., Ascher, U. M., & Oldenburg, D. (2000). On optimization techniques for solving nonlinear inverse problems. *Inverse Problems, 16*(5), 1263.

Herrmann, F. J., Siahkoohi, A., & Rizzuti, G. (2019). Learned imaging with constraints and uncertainty quantification. *In 33rd Neural Information Processing Systems (NeurIPS)*.

Kaipio, J., & Somersalo, E. (2006). *Statistical and computational inverse problems*. Springer Science & Business Media.

Kingma, D. P., & Ba, J. (2014). Adam: A method for stochastic optimization. *In 6th International Conference on learning representations*.

Krizhevsky, A., Sutskever, I., & Hinton, G. E. (2012). Imagenet classification with deep convolutional neural networks. In *Advances in neural information processing systems*. pp. 1097–1105.

Lailly, P., et al. (1983). The seismic inverse problem as a sequence of before stack migrations. In Conference on inverse scattering.

LeCun, Y., Bengio, Y., & Hinton, G. (2015). Deep learning. *Nature*, *521*(7553), 436.

Ledig, C., Theis, L., Huszár, F., Caballero, J., Cunningham, A., Acosta, A., et al. (2017). Photo-realistic single image super-resolution using a generative adversarial network. In *Proceedings of the ieee conference on computer vision and pattern recognition*. pp. 4681–4690.

Li, X., Aravkin, A. Y., van Leeuwen, T., & Herrmann, F. J. (2012). Fast randomized full-waveform inversion with compressive sensing. *Geophysics*, *77*(3), A13–A17.

Li, Y., Biondi, B., Clapp, R., & Nichols, D. (2016). Integrated VTI model building with seismic data, geologic information, and rock-physics modeling—Part 1: Theory and synthetic test. *Geophysics*, *81*(5), C177–C191.

Lunz, S., Öktem, O., & Schönlieb, C.-B. (2018). Adversarial regularizers in inverse problems. In *Advances in neural information processing systems*, pp. 8507–8516.

Mao, X., Li, Q., Xie, H., Lau, R. Y. K., Wang, Z., & Paul Smolley, S. (2017). Least squares generative adversarial networks. In *Proceedings of the IEEE international conference on computer vision*, pp. 2794–2802.

Matheron, G. (2012). *Estimating and choosing: An essay on probability in practice*. Springer Science & Business Media.

Mosser, L., Dubrule, O., & Blunt, M. J. (2020). Stochastic seismic waveform inversion using generative adversarial networks as a geological prior. *Mathematical Geosciences*, *52*(1), 53–79.

Nolet, G. (2012). *Seismic tomography: With applications in global seismology and exploration geophysics*. Springer Science & Business Media.

Osypov, K., Nichols, D., Woodward, M., Zdraveva, O., Yarman, C. E., et al. (2008). Uncertainty and resolution analysis for anisotropic tomography using iterative eigendecomposition. In *2008 SEG annual meeting*. Society of Exploration Geophysicists.

Osypov, K., Yang, Y., Fournier, A., Ivanova, N., Bachrach, R., Yarman, C. E., et al. (2013). Model-uncertainty quantification in seismic tomography: Method and applications. *Geophysical Prospecting*, *61*(6), 1114–1134.

Pratt, R. G. (1999). Seismic waveform inversion in the frequency domain, Part 1: Theory and verification in a physical scale model. *Geophysics*, *64*(3), 888–901.

Rick Chang, J. H., Li, C.-L., Poczos, B., Vijaya Kumar, B. V. K., & Sankaranarayanan, A. C. (2017). One network to solve them all–solving linear inverse problems using deep projection models. In *Proceedings of the IEEE international conference on computer vision*, pp. 5888–5897.

Roberts, G. O., Rosenthal, J. S., et al. (2001). Optimal scaling for various Metropolis-Hastings algorithms. *Statistical Science*, *16*(4), 351–367.

Romanowicz, B. A. (1979). Seismic structure of the upper mantle beneath the United States by three-dimensional inversion of body wave arrival times. *Geophysical Journal International*, *57*(2), 479–506.

Sambridge, M., Gallagher, K., Jackson, A., & Rickwood, P. (2006). Trans-dimensional inverse problems, model comparison and the evidence. *Geophysical Journal International*, *167*(2), 528–542.

Scales, J. A., & Tenorio, L. (2001). Prior information and uncertainty in inverse problems. *Geophysics, 66*(2), 389–397.

Siahkoohi, A., Louboutin, M., & Herrmann, F. J. (2019). The importance of transfer learning in seismic modeling and imaging. *Geophysics, 84*(6), A47–A52.

Sun, H., & Demanet, L. (2018). Low frequency extrapolation with deep learning. In *SEG technical program expanded abstracts 2018*. Society of Exploration Geophysicists, pp. 2011–2015.

Sun, J., Slang, S., Elboth, T., Greiner, T. L., McDonald, S., & Gelius, L.-J. (2020). A convolutional neural network approach to deblending seismic data. *Geophysics, 85*(4), WA13–WA26.

Tarantola, A., & Valette, B. (1982a). Generalized nonlinear inverse problems solved using the least squares criterion. *Reviews of Geophysics, 20*(2), 219–232. https://doi.org/10.1029/RG020i002p00219.

Tarantola, A., & Valette, B. (1982b). Inverse problems = quest for information. *Journal of Geophysics, 50*, 159–170.

Tromp, J. (2019). Seismic wavefield imaging of Earth's interior across scales. *Nature Reviews Earth & Environment*, 1–14.

Tu, N., & Herrmann, F. J. (2015). Fast imaging with surface-related multiples by sparse inversion. *Geophysical Journal International, 201*(1), 304–317.

Virieux, J., & Operto, S. (2009). An overview of full-waveform inversion in exploration geophysics. *Geophysics, 74*(6), WCC1–WCC26. https://doi.org/10.1190/1.3238367.

Wang, Z., Bovik, A. C., Sheikh, H. R., & Simoncelli, E. P. (2004). Image quality assessment: From error visibility to structural similarity. *IEEE Transactions on Image Processing, 13*(4), 600–612.

Yeh, R. A., Chen, C., Lim, T. Y., Hasegawa-Johnson, M., Do, M. N., & Schwing, A. G. (2017). Semantic image inpainting with deep generative models. *Proceedings of the IEEE conference on computer vision and pattern recognition*, 5485–5493.

Ying, L., Demanet, L., & Candes, E. (2005). 3D discrete curvelet transform. In *Wavelets XI. 5914* (p. 591413). International Society for Optics and Photonics.

Yu, S., Ma, J., & Wang, W. (2019). Deep learning for denoising. *Geophysics, 84*(6), V333–V350.

Zhu, H., Li, S., Fomel, S., Stadler, G., & Ghattas, O. (2016). A Bayesian approach to estimate uncertainty for full-waveform inversion using a priori information from depth migration. *Geophysics, 81*(5), R307–R323.

CHAPTER SIX

An introduction to the two-scale homogenization method for seismology

Yann Capdeville[a,*], Paul Cupillard[b], and Sneha Singh[a]
[a]LPG, CNRS, Université de Nantes, Nantes, France
[b]GeoRessources, Université de Lorraine, Nancy, France
*Corresponding author: e-mail address: yann.capdeville@univ-nantes.fr

Contents

1. Introduction — 218
2. Mathematical notions and notations — 222
   2.1 Periodic functions — 223
   2.2 Two-variable functions — 223
   2.3 Linear filtering and Fourier domain — 223
3. A numerical introduction to the subject — 226
   3.1 A simple example in 1-D — 227
   3.2 Small scales heterogeneities and solution smoothness — 230
   3.3 The Backus solution — 231
   3.4 Beyond the Backus solution — 235
4. Two-scale homogenization: the 1-D periodic case — 237
   4.1 Resolution of the homogenization problem — 240
   4.2 Convergence theorem — 246
   4.3 External point sources — 248
   4.4 An example — 249
5. Two-scale homogenization: The 1-D nonperiodic case — 252
   5.1 The Backus solution — 252
   5.2 Naive solutions to the two-scale homogenization periodicity limitation — 253
   5.3 New functional spaces — 254
   5.4 Cell properties in the nonperiodic case — 257
   5.5 Nonperiodic homogenized equations and resolution — 257
   5.6 Construction of the cell properties E^{ε_0} and ρ^{ε_0} — 260
   5.7 Two examples — 263
6. Two-scale homogenization: Higher dimensions — 266
   6.1 The 2-D and 3-D elastic case — 266
   6.2 The 2-D and 3-D acoustic case — 270
7. What we skipped — 273
   7.1 Spatial filters — 273
   7.2 Boundary conditions — 274
   7.3 Other aspects to address — 275

8. Examples of applications	276
8.1 2-D refracted waves	276
8.2 A solid with a fluid inclusion	280
8.3 A failing case: Helmholtz resonators	282
8.4 A 3-D geological model	285
8.5 Rotational receivers corrector	290
8.6 Full waveform inversion (FWI)	292
9. Discussion and conclusions	299
Acknowledgments	302
References	302

1. Introduction

The Earth is a multiscale medium: whatever the scale we place ourselves at, there are always heterogeneities of smaller scales. Looking at a cross section of the global Earth, we would probably see the main layers (crust, mantle, core) along with slabs and plumes. Zooming in on a subduction zone, we would see an accretion prism, magma chambers below volcanoes and more details within the crust. Zooming in again on the shallow crust, we would see thinner layers along with faults and fractures. Zooming in again, we would see grains of many different materials, pores, and so on and so forth. So, the Earth is definitely multiscale, but it is also a deterministic medium (as opposed to a stochastic medium): there is only one Earth.

The fact that the Earth is a multiscale body could be a serious problem for many Earth sciences, including seismology. If one had to know all details about the Earth to model body wave arrival times or frequency band limited seismograms, seismology would not be very useful. Fortunately, it is well known that, somehow, simple Earth models such as PREM (Dziewonski & Anderson, 1981) can model body wave time arrivals and low-frequency seismic data with a surprising accuracy. Therefore, it is clear that an underlying homogenization process (also called upscaling or effective process) exists for seismic wave propagation, which makes it possible to model the real Earth with a simple effective media.

For decades, implicitly knowing that a homogenization process exists but having no mathematical theory for this process has been enough. Nowadays, with the emergence of more and more precise modeling tools and inversion techniques, the lack of this theory is becoming a problem. For the forward modeling case, the numerical cost of a simulation in a smooth 3-D medium is directly linked to the maximum frequency emitted at the source (or, equivalently, the inverse of the minimum wavelength) to the power 4. But in a rough medium, the numerical cost is proportional to the inverse of the smallest scale size to the power 4, regardless of the maximum frequency of the source. Therefore, scales smaller than the minimum wavelength can be extremely expensive numerically. They can even make the modeling impossible when the mesh cannot be designed because of the medium complexity. To overcome these issues, the ability to remove small scales to work with a smooth effective medium, whatever the complexity of the original medium, would be a great advantage. For the inverse problem, we know that the media we are trying to image always contain scales below the resolution limit. Understanding the relationship between the small scales and the scales we are can image is a necessity.

A practical example of the reason why this missing tool can be a problem is the following: when working on the very early stage of global scale full waveform inversion (FWI) (Capdeville, Gung, & Romanowicz, 2005), the first author of this work faced a difficulty because of the thin layers of the crust. An important ingredient for a successful inversion is a good quality starting model. Because of computing limitations in the early 2000s, the inversion was carried out with very long period data (150 s and above) that led to very long wavelengths (about 600 km and longer). For such long period data, PREM is an appropriate starting model. The problem is that PREM has a 9 km thick lower-crust, which is very small compared to the minimum wavelength of 600 km. This difference in scales poses a serious problem for both the forward modeling (the small layer drastically increases the numerical cost necessary to solve the wave equation with modern solvers) and the inverse problem (the small layer imposes a very fine parameterization, leading to an ill-posed inverse problem).

When facing such a difficulty, it is tempting to remove the small layer or to average it ("it is so small, it should not matter..."). However, these trivial solutions have a surprisingly strong effect on the waveforms, especially on surface wave phase velocities, making them an invalid option (Capdeville & Marigo, 2008). When looking for a solution (which was the starting point of the present work actually), it appeared that an entire scientific field is dedicated to finding effective or homogenized solutions for different types of equations. The science of dealing with small scales to find effective behaviors, also called homogenization, has been in existence since the 1960s. The literature on this subject is very vast and gathers thousands of publications.

We do not attempt to give an exhaustive review of this field here but we provide some broad outlines instead.

There are many different approaches to solve the problem of finding effective behaviors. The first one is the two-scale homogenization, or formal asymptotic homogenization analysis (Babuška, 1976; Bensoussan, Lions, & Papanicolaou, 1978; Murat & Tartar, 1985; Sanchez-Palencia, 1980). It is the method on which the present work is based upon. It involves an explicit scale separation between large and small scales and the introduction of two space variables, one for the large scales and one for the small scales. It is often considered as a mathematical approach because theorems of convergence exist (e.g., Allaire, 1992). However, this method is not exact: it is asymptotic and, as such, it implies some approximations to be applied to practical cases. An advantage of this method is that it can deal with periodic, quasi-periodic, and stochastic (Papanicolaou & Varadhan, 1979) heterogeneous media. A second approach is the variational approach, or the Willis approach (Hashin, 1972; Willis, 1981, 1983). It does not involve an explicit scale separation and it is often dedicated to finding bounds of effective elastic properties of complex material. It does not rely on any specific approximation but it often leads to nonlocal effective behavior, both in space and time, which makes it difficult to use in practice (Willis, 1985, 2009). Many other approaches, such as the volume average approach (Pride, Gangi, & Morgan, 1992) which is more intuitive and physical, the theory of mixtures (Bowen, 1976) and the numerical homogenization (Engquist & Souganidis, 2008), also exist. Finally, several works have been dedicated to comparing and connecting some of these approaches (see, for instance, Davit et al., 2013; Meng & Guzina, 2018; Nassar, He, & Auffray, 2016).

Many classes of problems are tackled with those various methods, like obtaining bounds of composite materials (e.g., Francfort & Murat, 1986; Hashin & Shtrikman, 1962; Hill, 1965; Willis, 1981) or porous media (Auriault, Borne, & Chambon, 1985; Auriault & Sanchez-Palencia, 1977; Boutin & Auriault, 1990; Burridge & Keller, 1981; Pride et al., 1992), elasto-dynamics (e.g., Boutin & Auriault, 1993; Sanchez-Palencia, 1980; Willis, 1997), elastic rupture (e.g., Abdelmoula & Marigo, 2000), and so on.

The Earth science communities, including seismology, have long remained mostly aloof from these methods. Nevertheless, some important contributions have been made. One of the main ones is Backus (1962) (often named the "Backus averaging method"), which is dedicated to the upscaling of finely layered media for elastic waves in the low-frequency (long-wavelength) regime. Among other results, Backus has shown that a finely layered isotropic medium gives, in general, an anisotropic effective medium. This is the apparent anisotropy of layered media, which is often seen in nature. Interestingly, the Backus method does not require any hypothesis apart from handling a layered medium (no periodicity requirement, no scale separation hypothesis, etc.). Since then, some works have been dedicated to improving and attempting to generalize Backus' results (Gold, Shapiro, Bojinski, & Müller, 2000; Grechka, 2003), mostly for exploration geophysics applications (see Tiwary, Bayuk, Vikhorev, and Chesnokov (2009) for a review of layered media upscaling in that domain), without trying to reach the rigor or the quality of the two-scale periodic homogenization method. Similarly, effective elastic properties of mineral media for geophysical applications have been derived from the classical Voigt-Reuss averages (e.g., Thomsen, 1972) and from the Hashin–Shtrikman bounds (e.g., Watt, 1988).

To sum up, keeping in mind that geological media are neither periodic nor stochastic and heterogeneities have no natural scale separation, the situation in the early 2000s offered no option to upscale general geological media for the seismic (elastic or acoustic) wave equation. Going back to the global scale FWI example mentioned above, the first nonnaive upscaling attempt was to apply the Backus averaging. While the Backus averaging gives good results for the volume and body waves, it gives deceivingly poor results for shallow layers and the surface waves. Unfortunately, the Backus approach does not provide any direction to go beyond its results. Moreover, solutions provided by the two-scale homogenization community are limited to periodic heterogeneities (Dumontet, 1990; Marigo & Pideri, 2011). The first successful attempt to mix the periodic homogenization approach and the Backus method (at least the fact that it does not require a periodicity hypothesis) was made by Capdeville and Marigo (2007, 2008) for layered media. It shows that the order 0 two-scale homogenization falls back to Backus' results (which was known for long, see for instance Sanchez-Palencia (1980), or more recently Guillot, Capdeville, and Marigo (2010) and C. Lin, Saleh, Milkereit, and Liu (2017)) and it makes it possible to go beyond the leading order. This was an improvement, but not yet the desired general solution, as it was limited to the layered case.

To go further, a more general solution has been proposed in 1-D (Capdeville, Guillot, & Marigo, 2010 a), then in 2-D (Capdeville, Guillot, & Marigo, 2010 b), and finally in 3-D (Capdeville, Zhao, & Cupillard, 2015; Cupillard & Capdeville, 2018). Note that alternative approaches have been developed by Fichtner and Hanasoge (2017) in 1-D and Jordan (2015). Homogenization of a rapid topography has been treated by Capdeville and Marigo (2013) and the effect of small scale heterogeneities on seismic sources, more specifically on explosions, by Burgos, Capdeville, and Guillot (2016). Finally, an important emerging topic is the link between the seismic inversion problem and homogenization for full waveform imaging techniques (Afanasiev, Boehm, May, & Fichtner, 2016; Capdeville & Métivier, 2018; Capdeville, Stutzmann, Wang, & Montagner, 2013) and for downscaling and interpreting of seismic results (Alder et al., 2017; Bodin, Capdeville, Romanowicz, & Montagner, 2015; Faccenda et al., 2019; Fichtner, Kennett, & Trampert, 2013; Wang, Montagner, Fichtner, & Capdeville, 2013).

The objective of this work is to propose an introduction to the two-scale homogenization method, in the periodic case first, then in the nonperiodic case, and demonstrate its applications to seismology. Homogenization is a nonintuitive approach that can be difficult to step into for someone with a background in geophysics. Not everyone likes it; for example, Pride leaves the following comment about it in Pride et al. (1992): "However, the entire (possibly confusing) notion of having all quantities functionally depend on two independent length scales and of performing asymptotic expansions is unnecessary when direct volume averaging proceeds so directly." Nevertheless, we hope to convince the readers that the gain of such an approach is worth the effort. The first part is dedicated to some mathematical notions that will be useful for the rest of the paper. The second part is dedicated to a numerical experiment in 1-D. This 1-D introduction has no real application but it is meant to show where the idea of the two-scale variables comes from and why it is useful to introduce it. It is also the occasion to introduce Backus' solution. The third part is dedicated to the formal two-scale homogenization in the 1-D periodic case. It is the simplest case and it makes the first contact to the classical two-scale homogenization as simple as it can be. We then move to the 1-D nonperiodic case. Once again, there is no real application of this 1-D case, but it is the simplest way to introduce the necessary concepts. The sixth part is dedicated to the 2-D and 3-D cases. Finally, the last part is dedicated to examples and applications of nonperiodic homogenization.

2. Mathematical notions and notations

Before getting started, let us introduce some simple mathematical notions and notations that will be useful throughout the paper.

2.1 Periodic functions

Homogenization strongly relies on periodic functions, at least in its classic form. In 1-D, a T-periodic function h is such that, for any x, $h(x + T) = h(x)$. In d dimensions, where d is 2 or 3 for 2-D or 3-D respectively, a function h is $\mathbf{T}$-periodic if, for any $\mathbf{x}$ and for any $i \in \{1, ..., d\}$, $h(\mathbf{x} + T_i\hat{\mathbf{i}}) = h(\mathbf{x})$ with no implicit summation on i and where the $\hat{\mathbf{i}}$ are the space basis unit vectors.

2.2 Two-variable functions

Homogenization makes the scale separation explicit by using two spatial variables: one for the large scales and one for the small scales. In practice, it implies the use of two-scale functions. A two-scale function $h(x, y)$ is a function that depends on two space variables x and y. It is usually periodic for the second variable.

Assuming h is T-periodic in y, we define the "cell average" as

$$\langle h \rangle (x) = \frac{1}{T} \int_0^T h(x, y) dy. \tag{1}$$

At this stage, the notion of "cell" is not yet defined. Here, it is one periodicity T. The following properties of h will be useful:

$$\partial_y h = 0 \Leftrightarrow h(x, y) = \langle h \rangle (x) \tag{2}$$

and

$$\langle \partial_y h \rangle = 0. \tag{3}$$

Both properties can be demonstrated using the T-periodicity and an integration by parts.

Finally we define $\mathcal{T}$, the functional space of two-variable functions $\lambda_{\min}$-periodic for the second variable, where $\lambda_{\min}$ is the minimum wavelength of the wavefield (defined in the next section).

2.3 Linear filtering and Fourier domain

It is useful, for any function $h(x)$, to define its wavenumber domain version:

$$\bar{h}(k) = \int_{-\infty}^{+\infty} h(x) e^{i2\pi kx} dx, \tag{4}$$

where k is the spatial frequency and $\lambda = 1/k$ its corresponding wavelength. $\bar{h}$ is the Fourier transform of h; it is a complex function.

One of the important physical scales is the wavefield minimum wavelength λ_{min} to which we can associate the maximum spatial frequency $k_{max} = 1/\lambda_{min}$. In the following, we will often use an arbitrarily user-defined wavelength λ_0 and we will measure its position relative to λ_{min} with

$$\varepsilon_0 = \frac{\lambda_0}{\lambda_{min}}. \tag{5}$$

λ_0 is used to define what can be considered fine scales and what can be considered large scales. It is usually small with respect to 1.

We will also make extensive use of linear filtering, mainly low-pass filtering, mostly in the space domain. We introduce a low-pass filter $\mathcal{F}^{\varepsilon_0}$ to be applied in the space domain such that, for any function $h(x)$,

$$\mathcal{F}^{\varepsilon_0}(h)(x) = \int_{-\infty}^{+\infty} h(x')w_{\varepsilon_0}(x - x')dx', \tag{6}$$

where w_{ε_0} is the filter wavelet. There are many types of wavelets with different properties. For low-pass filters, the wavelet is such that its convolution with h removes all spatial variations smaller than λ_0 in $\mathcal{F}^{\varepsilon_0}(h)$. In our notation, i.e., $\mathcal{F}^{\varepsilon_0}$, we emphasize the dependency of the filter to ε_0 instead of λ_0 because we will be working with a fixed λ_{min}.

In the wavenumber domain, the linear low-pass filtering is simply a product:

$$\bar{\mathcal{F}}^{\varepsilon_0}(h)(k) = \bar{h}(k)\bar{w}_{\varepsilon_0}(k). \tag{7}$$

The low-pass filter wavelet is defined such that its amplitude spectrum $|\bar{w}_{\varepsilon_0}|(k)$ is equal to 1 in $[0, k_0]$ (where $k_0 = 1/\lambda_0$) and then goes down to 0 more or less sharply (apodization) depending on the desired support size of the wavelet in the space domain (Fig. 1).

Based on this low-pass filter, we can introduce the notion of "smooth functions" and "rough functions":
— h is said to be smooth if $\mathcal{F}^{\varepsilon_0}(h) = h$. In other words, a smooth function has no spatial variations smaller than λ_0;
— h is said to be rough if $\mathcal{F}^{\varepsilon_0}(h) \neq h$.

We will also make use of the following properties:
(i) partial derivatives and $\mathcal{F}^{\varepsilon_0}$ commute, meaning that

$$\mathcal{F}^{\varepsilon_0}\left(\frac{\partial h}{\partial x}\right) = \frac{\partial \mathcal{F}^{\varepsilon_0}(h)}{\partial x}, \tag{8}$$

(ii) if h is smooth, then for any function g we have

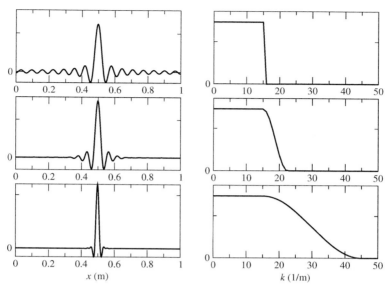

Fig. 1 Low-pass filter wavelet w_{ε_0} in the space domain (*left*) and its spectrum norm $|\bar{w}_{\varepsilon_0}|$ in the spatial frequency domain (*right*) for different apodizations. The smoother is the spectral apodization, the more compact is the spatial support of the wavelet.

$$\mathcal{F}^{\varepsilon_0}(hg) \simeq h\mathcal{F}^{\varepsilon_0}(g). \qquad (9)$$

Note that, for two rough functions h and g, $\mathcal{F}^{\varepsilon_0}(hg) \neq \mathcal{F}^{\varepsilon_0}(h)\mathcal{F}^{\varepsilon_0}(g)$.

In the case of a two-scale function $h(x, y)$, low-pass filters are always applied to the second variable. Even though x and y are independent variables, as we will see later on, in the homogenization theory, y is built from x with the relation $y = x/\varepsilon_0$. This relation also applies in the spectral domain and therefore, to obtain a similar low-pass filtering effect in the y domain like in the x domain, the λ_0 cutoff (for a function of x) becomes $\lambda_{\min}$ (for a function of y). Indeed, in the y domain, the filter wavelet w_1 is related to w_{ε_0} by $w_1(y) = \varepsilon_0 w_{\varepsilon_0}(\varepsilon_0 y)$. The cutoff of w_1 is therefore $\lambda_1 = \lambda_0/\varepsilon_0 = \lambda_{\min}$ (the corresponding ε_0 is indeed 1). For a two-scale variable, the filtering is therefore independent of ε_0 and is simply noted.

$$\mathcal{F}(h)(x, y) = \int_{-\infty}^{+\infty} h(x, y')w_1(y - y')\, dy'. \qquad (10)$$

To exemplify the equivalence of the two domain filtering, let us take a simple function $g(y) = h(\varepsilon_0 y)$. Then

$$\mathcal{F}(g)(y) = \int_{-\infty}^{+\infty} g(y')w_1(y - y') \, dy' \tag{11}$$

$$= \int_{-\infty}^{+\infty} g\left(\frac{x'}{\varepsilon_0}\right) w_1\left(\frac{x - x'}{\varepsilon_0}\right) \frac{dx'}{\varepsilon_0} \tag{12}$$

$$= \int_{-\infty}^{+\infty} h(x')w_{\varepsilon_0}(x - x') \, dx' \tag{13}$$

$$= \mathcal{F}^{\varepsilon_0}(h)(x) \tag{14}$$

$$= \mathcal{F}^{\varepsilon_0}(h)(\varepsilon_0 y). \tag{15}$$

Applying $\mathcal{F}$ to g indeed gives the same result as applying $\mathcal{F}^{\varepsilon_0}$ to h.

2.3.1 Arithmetic and harmonic averages

For any T-periodic function $h(x)$, the arithmetic average is

$$\langle h \rangle = \frac{1}{T} \int_0^T h(x) \, dx \tag{16}$$

and the harmonic average is

$$\langle h \rangle_H = \left(\frac{1}{T} \int_0^T \frac{1}{h}(x) \, dx\right)^{-1}. \tag{17}$$

These two averages can give significantly different results. For instance, let us assume a 2-periodic function $h(x)$ such that $h(x) = a$ for $x \in [0, 1)$ and $h(x) = b$ for $x \in [1, 2)$. The arithmetic average of this function is $\langle h \rangle = (a + b)/2$ whereas the harmonic average is $\langle h \rangle_H = \frac{2ab}{a + b}$. Taking some values for a and b, we see how different $\langle h \rangle$ and $\langle h \rangle_H$ can be:

- $a = 1.01;\ b = 0.99 \Rightarrow \langle h \rangle = 1;\ \langle h \rangle_H = 0.99990$
- $a = 1.10;\ b = 0.90 \Rightarrow \langle h \rangle = 1;\ \langle h \rangle_H = 0.990$
- $a = 1.20;\ b = 0.80 \Rightarrow \langle h \rangle = 1;\ \langle h \rangle_H = 0.960$
- $a = 1.50;\ b = 0.50 \Rightarrow \langle h \rangle = 1;\ \langle h \rangle_H = 0.750$

Even though the low-pass filtering $\mathcal{F}^{\varepsilon_0}$ we are going to use in this article is different from the cell average $\langle\ \rangle$, the conclusion we are drawing holds for $\mathcal{F}^{\varepsilon_0}$.

3. A numerical introduction to the subject

This section is dedicated to a numerical experiment in 1-D. It has little equivalence in the real world but it aims at gradually bringing the reader, through observations of numerical results in a simple setting, to ideas behind the two-scale homogenization method.

3.1 A simple example in 1-D

Now that the mathematical notions have been exposed, we present some trivial numerical experiments to introduce the subject. Let us take a very simple example: a 1-D elastic wave propagating in a 1-D bar with periodic heterogeneities. Let Ω be a fine and long cylinder of length L and section A (Fig. 2). We assume that the radius of the section is very small compared to the wavelength of the wave we are about to consider so that any displacement of any particle in Ω which is not parallel to the bar can be neglected. Let $u(x, t)$ be the displacement of the particle (relative to a position at rest) along the $\hat{x}$ direction as function of x and time t. If an internal force $F(x, t)$ is applied in the bar along the $\hat{x}$ axis, we define the stress σ as

$$\sigma(x, t) = \frac{F(x, t)}{A} \tag{18}$$

We assume that we are in the linear elasticity regime so that a linear relationship between the stress and the strain $\epsilon(x, t) = \partial_x u(x, t)$ exits:

$$\sigma(x, t) = E \partial_x u(x, t), \tag{19}$$

where E is the Young modulus (or the elastic coefficient). This last equation is the material constitutive equation, also called the Hooke's law. Based on Newton's second law, the dynamics equation in the bar is

$$\rho(x) \partial_{tt} u(x, t) - \partial_x \sigma(x, t) = f(x, t), \tag{20}$$

where ρ is the mass per unit length of the bar, $\partial_{tt} u$ the second time derivative of u and f an external force applied to the bar. We assume free stress conditions at both ends

$$\sigma(0, t) = \sigma(L, t) = 0, \tag{21}$$

and we assume the bar is at rest at $t = 0$. Eqs. (20) and (19) together with the boundary conditions (21) make the elastic wave equation in the 1-D bar Ω.

We use a point source located in x_0:

$$f(x, t) = \delta(x - x_0) g(t), \tag{22}$$

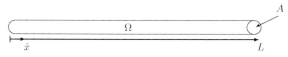

Fig. 2 A 1-D bar of length L and section A along the $\hat{x}$ axis.

where $g(t)$ is the source time wavelet. Here, we use a Ricker function (second derivative of a Gaussian) which can be characterized by its central frequency f_0. The maximum frequency of such a wavelet can be estimated as $f_{max} \simeq 3f_0$.

Before moving forward, we need to introduce the notion of minimum wavelength λ_{min} associated to f_{max}. Assuming homogeneous mechanical properties and a monochromatic wave of type $u(x, t) = Ue^{i(2\pi kx - \omega t)}$, where ω is the angular frequency and U is the amplitude, Eqs. (19) and (20) leads to the dispersion relation

$$k = \frac{\omega}{2\pi\alpha}, \tag{23}$$

where $\alpha = \sqrt{E/\rho}$ is the wave velocity. Note that, in this work, k is the spatial frequency and not the more standard wavenumber (if k was the wavenumber, then the dispersion relation would be $k = \omega/\alpha$). Using the last equation along with $k = 1/\lambda$ and $\omega = 2\pi f$, we find

$$\lambda_{min} = \frac{\alpha}{f_{max}}. \tag{24}$$

This last equation shows that, for a source with a limited frequency band (i.e., with a maximum frequency f_{max}), the wavefield does not have oscillations smaller than λ_{min}. Note that this is only true in the far-field, i.e., a few minimum wavelengths away from the point source. For heterogeneous media, the dispersion relation might be difficult to establish analytically, but it always exists and it is bounded (with a few exceptions in some very particular media, such as metamaterials, see Section 8.3 for instance). In general, we can obtain an estimate of the lower bound of λ_{min} using

$$\lambda_{min} = \frac{\alpha_{min}}{f_{max}}, \tag{25}$$

where α_{min} is the minimum wave velocity. The minimum wavelength is very useful for solving the wave equation numerically as it provides an estimate of the regularity of the solution, which is always necessary to calibrate the grid spacing of the mesh that comes with the numerical method. It is also very important for the homogenization process because what is small and what is large is determined with respect to λ_{min}.

To solve the wave Eqs. (19)–(21) in the heterogeneous bar, we rely on the spectral-element method (SEM) (Chaljub et al., 2007; Komatitsch & Vilotte, 1998). A good introduction to the method can be found in Igel (2017).

SEM is a type of high-degree finite-element method. It has many advantages but the one that interests us the most is its ability to accurately take into account material discontinuities if they are matched by an element boundary. By meshing all the bar mechanical discontinuities, we can obtain a solution accurate enough to be used as a reference solution.

We now assume that the mechanical properties $\rho(x)$ and $E(x)$ vary periodically with a ℓ-periodicity along x and we assume ℓ is smaller than λ_{min}. We take $\ell = 0.01$m and we set up the properties such that the wave velocity $\alpha(x)$ jumps from 1.25km/s to 1.875km/s with a ℓ-periodicity (Fig. 3). The corresponding average velocity is $\langle \alpha \rangle = 1.562$ km/s. We use a source at $x_0 = 0.3$m and with $f_0 = 20$kHz corresponding to $\lambda_{min} = 0.02$m. Finally, we choose $L = 1$m. In Fig. 3, two snapshots of the displacement along $\hat{x}$ for a time $t = 0.16$ms are shown: one computed in the periodic heterogeneous bar and one in the average velocity bar. We observe that the wave pulses have a very similar shape in the two cases, even though they seem to propagate at a different speed. It appears that, for the heterogeneous case, the waves propagate the same way as in a homogeneous bar: no scattering or dispersion are observed. Therefore, it seems that an effective propagation occurs. The second observation is that this effective wave propagation has a different wave speed than the average wave speed, which is counterintuitive. It is interesting to note that the true effective wave propagation

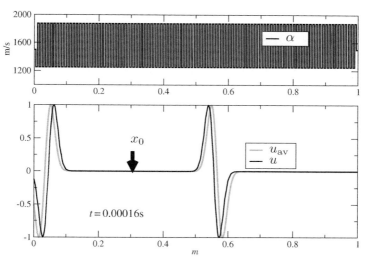

Fig. 3 *Top*: periodic wave speed α along the bar. *Bottom*: snapshots of the displacement u computed in the bar with the periodic heterogeneity and of the displacement u_{av} computed in a 1-D bar with the average velocity $\langle \alpha \rangle$ for $t = 0.16$ms.

is slower than the average wave speed, which is consistent with the "velocity shift" experimentally observed when comparing time arrivals of waves propagating in random media to time arrivals computed with the corresponding average velocity (Shapiro, Schwarz, & Gold, 1996).

From this simple example, we conclude that an effective wave propagation occurs, but it does not correspond to a propagation in a simple average velocity model. We need to understand this mismatch; this is the starting point of this work.

3.2 Small scales heterogeneities and solution smoothness

Based on the observations from the previous section and before moving forward, we make a slight change in notations. In this work, we have two types of scales: the small scales, that are much smaller than λ_{min}, and the large scales, that are comparable or larger than λ_{min}. In reality, it is not obvious to tell what small and large scales are. Only the periodic case makes this distinction simple. We introduce ε, a parameter that somehow measures how small the small scales are compared to λ_{min}. For periodic heterogeneities, it is simply defined as

$$\varepsilon = \frac{\ell}{\lambda_{min}}, \tag{26}$$

where ℓ is still the medium heterogeneity periodicity. For more general heterogeneities (nonperiodic), there is no simple definition of ε; let us say for now that ε is just a symbol indicating that a given quantity depends on the small scales. A priori, if the mechanical properties contain small scales, all quantities of the problem also depend on the small scales. We, therefore, rewrite the wave equation as

$$\rho^\varepsilon \partial_{tt} u^\varepsilon - \partial_x \sigma^\varepsilon = f^\varepsilon, \tag{27}$$

$$\sigma^\varepsilon = E^\varepsilon \partial_x u^\varepsilon. \tag{28}$$

Based on the example in the previous section, it appears that effective quantities exist: we will mark them with an $*$. For example, u^* is the effective displacement. Using the "averaging operator" $\mathcal{F}^{\varepsilon_0}$ we can define the effective solution to the wave equation:

$$u^* = \mathcal{F}^{\varepsilon_0}(u^\varepsilon), \tag{29}$$

$$\sigma^* = \mathcal{F}^{\varepsilon_0}(\sigma^\varepsilon), \tag{30}$$

$$\epsilon^* = \mathcal{F}^{\varepsilon_0}(\epsilon^\varepsilon). \tag{31}$$

In the following, we choose $\varepsilon_0 > \varepsilon$ such that, for any purely periodic function h^ε, $\mathcal{F}^{\varepsilon_0}(h^\varepsilon)$ is a constant function in x.

3.3 The Backus solution

To analyze the observation made in the previous section, we use the Backus approach to obtain effective media. Backus (1962) indeed developed an elegant solution to upscale finely layered media and obtained widely used results. While his work has been developed for layered media for wave propagation in 3-D, we will apply it to our 1-D case. His method is based on applying a low-pass filter to the wave equation. To do so, Backus determines whether the parameters u^ε, $\partial_x u^\varepsilon$ and σ^ε are smooth or rough based on mathematical arguments. Here, we rely on a simple numerical observation instead of mathematical arguments: in Fig. 4, snapshots of displacement, strain, and stress for an elastic wave propagating in the finely layered bar used in Section 3.1 are plotted. One can notice that the displacement and the stress are smooth, meaning that their variations are on scales larger than those of the heterogeneities. The strain, however, displays a large scale variation on which changes at the scale of the heterogeneity layering is superimposed.

In the next step, Backus applies $\mathcal{F}^{\varepsilon_0}$ to the dynamic Eq. (20) without its second member f:

$$\mathcal{F}^{\varepsilon_0}\left(\rho^\varepsilon \partial_{tt} u^\varepsilon - \partial_x \sigma^\varepsilon\right) = 0. \tag{32}$$

Using the linearity of the low-pass filter operator, we have

$$\mathcal{F}^{\varepsilon_0}\left(\rho^\varepsilon \partial_{tt} u^\varepsilon\right) - \mathcal{F}^{\varepsilon_0}\left(\partial_x \sigma^\varepsilon\right) = 0. \tag{33}$$

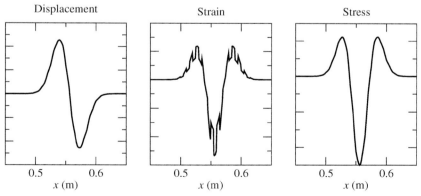

Fig. 4 Snapshots of displacement u^ε (*left*), strain $\partial_x u^\varepsilon$ (*middle*) and stress σ^ε (*right*) for an elastic wave propagating in the periodic heterogeneities used in Section 3.1.

We know that u^ε is a smooth function (see Section 2.3 for a definition of smooth and rough functions), therefore

$$u^* = \mathcal{F}^{\varepsilon_0}(u^\varepsilon) = u^\varepsilon. \tag{34}$$

Using this last equation along with (9), it comes

$$\mathcal{F}^{\varepsilon_0}(\rho^\varepsilon \partial_{tt} u^\varepsilon) = \mathcal{F}^{\varepsilon_0}(\rho^\varepsilon)\partial_{tt} u^*. \tag{35}$$

Note that, to obtain the last equation, we ignored the time derivatives. This is expected because the filter only acts on space.

Using (8) and the fact that σ^ε is smooth, we have

$$\mathcal{F}^{\varepsilon_0}(\partial_x \sigma^\varepsilon) = \partial_x \mathcal{F}^{\varepsilon_0}(\sigma^\varepsilon), \tag{36}$$
$$= \partial_x \sigma^*. \tag{37}$$

Finally, putting together (33), (35) and (37), we obtain

$$\mathcal{F}^{\varepsilon_0}(\rho^\varepsilon)\partial_{tt} u^* - \partial_x \sigma^* = 0. \tag{38}$$

The above equation is the effective dynamic equation, and we can see that the effective density is simply

$$\rho^* = \mathcal{F}^{\varepsilon_0}(\rho^\varepsilon). \tag{39}$$

We now move to the constitutive relation (28). Applying the same recipe like the one we just applied, it comes

$$\mathcal{F}^{\varepsilon_0}(\sigma^\varepsilon) = \mathcal{F}^{\varepsilon_0}(E^\varepsilon \partial_x u^\varepsilon). \tag{40}$$

We get stuck here because the right-hand term is the product of E^ε and $\partial_x u^\varepsilon$ which are two rough functions. In that case, property (9) does not apply and we cannot obtain an effective equation. To get around this difficulty, Backus noticed that, in the 1-D or layered case, the constitutive relation can be rewritten as

$$\partial_x u^\varepsilon = \frac{1}{E^\varepsilon}\sigma^\varepsilon. \tag{41}$$

We now have, on the right-hand term of the last equation, the product of a rough function ($1/E^\varepsilon$) with a smooth function (σ^ε). Therefore, property (9) can be used, and applying the filter to the last equation leads to

$$\partial_x u^* = \mathcal{F}^{\varepsilon_0}\left(\frac{1}{E^\varepsilon}\right)\sigma^*. \tag{42}$$

We finally obtain the effective constitutive relation:

$$\sigma^* = E^* \partial_x u^* \tag{43}$$

with

$$E^* = \left(\mathcal{F}^{\varepsilon_0} \left(\frac{1}{E^\varepsilon} \right) \right)^{-1}. \tag{44}$$

Contrary to the density, the effective Young modulus is not the trivial linear filtering of the fine-scale Young modulus. This nontrivial relation is responsible for the difference of wave speeds observed in Fig. 3 and it comes from the difference between the arithmetic and the harmonic average (see Section 2.3.1). In the example presented in Section 2.3.1, we see that the larger the contrast between a and b, the larger the difference between the two averages. Moreover, we see that the harmonic average is systematically smaller than the arithmetic average. This fact is at the root of the observation showing that true wave propagation is always slower than the propagation in the average velocity media. Similarly, the difference between these two averages is at the origin of the effective anisotropy often observed in nature for finely layered media.

To wrap-up this section, with simple linear filtering rules, Backus found that the effective displacement u^* and stress σ^* are driven by

$$\rho^* \partial_{tt} u^* - \partial_x \sigma^* = 0, \tag{45}$$

$$\sigma^* = E^* \partial_x u^*, \tag{46}$$

with

$$\rho^* = \mathcal{F}^{\varepsilon_0}(\rho)$$

$$E^* = \left(\mathcal{F}^{\varepsilon_0} \left(\frac{1}{E^\varepsilon} \right) \right)^{-1}. \tag{47}$$

Note that Backus does not say anything about the source.

If we apply the Backus formula (47) to the example of the periodic bar in Section 3, using $\varepsilon_0 = 0.5$, we obtain a constant effective model with an effective velocity $\alpha^* = \sqrt{E^*/\rho^*}$ which is plotted in Fig. 5A . As expected, it can be seen that α^* is not the average velocity (it is not the average slowness either). A snapshot of the reference displacement u^ε computed in the periodic bar and of the Backus effective displacement u^* computed in the

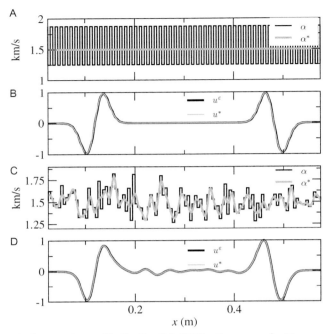

Fig. 5 Numerical example used in Section 3.3. (A) Elastic wave velocities α and α^* in the same periodic model as for Fig. 3 and in the corresponding Backus effective model, respectively. (B) Displacement snapshot for $t = 2$ms computed in the original periodic model (u^ε) and in the corresponding Backus effective model (u^*). (C) Elastic wave velocities in the random model (α) and in the corresponding Backus effective model (α^*) computed for $\varepsilon_0 = 0.5$. (D) Displacement snapshot for $t = 2$ms computed in the original random model (u^ε) and in the corresponding Backus effective model (u^*).

Backus effective bar (Fig. 5B) display a good agreement. The time shift observed in Fig. 3 has gone. Because the Backus solution does not require a periodicity hypothesis of the mechanical properties, we can go further and generate a bar with random mechanical properties (it is, nonetheless, a deterministic case: the bar is generated only once). The resulting velocity and its Backus effective version are shown in Fig. 5C. A snapshot of the reference displacement u^ε computed in the random bar and of the Backus effective displacement u^* computed in the Backus effective bar (Fig. 5D) display, once again, a good agreement.

The Backus (1962) method described here can be applied to a more general case, namely layered media (i.e., a 1-D variation of the properties and 3-D wave propagation) with transversely isotropic elastic properties. Such media carry P and S waves. Beyond the technical aspect of obtaining effective media, Backus (1962) shows that the long-wavelength effective medium of fine isotropic layers is anisotropic in general: this is the so-called apparent anisotropy. For the S waves, two effective parameters come out:

$$L = \left\langle \frac{1}{\mu} \right\rangle^{-1}, \tag{48}$$

$$N = \langle \mu \rangle, \tag{49}$$

where μ is the second Lamé coefficient and the elastic parameters L and N can be related to the vertically (SV) and horizontally (SH) polarized S waves, respectively, with wave speeds $\alpha_{SV} = \sqrt{L/\rho}$ and $\alpha_{SH} = \sqrt{N/\rho}$. Due to the difference between harmonic and arithmetic averages mentioned earlier, α_{SH} and α_{SV} are different in general: this is the shear wave anisotropy. The P wave case is more complex, as shown in Backus (1962).

3.4 Beyond the Backus solution

If we take a closer look at the Backus effective solution and compare it to the reference solution, some differences can still be observed. In Fig. 6, on the left panel, a zoom on the displacement propagating in the 1-D periodic bar, for two different ℓ-periodicities (and, consequently, for two different ε), is plotted. One can notice that the observed differences depend upon ε. In Fig. 6, on the right panel, the residuals $u^\varepsilon - u^*$ for $\varepsilon = 0.5$ and $\varepsilon = 0.25$ are plotted. We see that each residual is the product of a smooth function and a function with fine-scale variations. The amplitude of the smooth

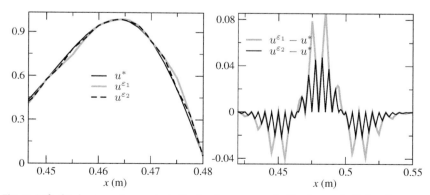

Fig. 6 *Left*: displacement snapshots for $t = 2$ ms computed in the periodic bar for $\varepsilon = \varepsilon_1 = 0.5$ (u^{ε_1}) and $\varepsilon = \varepsilon_2 = 0.25$ (u^{ε_2}) and in the Backus effective model (u^*). *Right*: residual snapshot $u^\varepsilon - u^*$ for the two ε values.

function looks proportional to ε, and its variation in x seems independent of ε. On the contrary, the small-scale variations in x look directly tied to ε. Moreover, they look periodic. Translating these observations to an equation, we can write

$$u^\varepsilon(x, t) - u^*(x, t) \simeq \varepsilon r\left(x, \frac{x}{\varepsilon}, t\right), \tag{50}$$

where r is a function smooth in x and almost periodic in $\frac{x}{\varepsilon}$.

Let us now observe snapshots of the residual at different times (Fig. 7). It appears that the residual r is a product of two functions, as mentioned previously, and that only the smooth function is propagating and time dependent. Mathematically, this means that

$$r\left(x, \frac{x}{\varepsilon}, t\right) = \chi\left(\frac{x}{\varepsilon}\right) v(x, t), \tag{51}$$

where $\chi\left(\frac{x}{\varepsilon}\right)$ is probably locally periodic and v is a function smooth in space and time dependent. We will see from the theoretical homogenization sections that this observation is confirmed by a mathematical development.

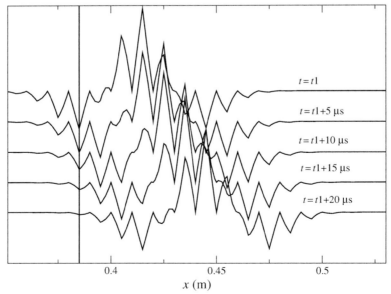

Fig. 7 Residual snapshots $u^\varepsilon - u^*$ for $\varepsilon = 0.5$ at five different times. The vertical line at $x = 0.38$ m marks one of the sharp elastic property changes in the bar model.

An introduction to the two-scale homogenization method 237

To conclude, even if Backus' results are accurate, visible differences remain between the true and the Backus solutions. These differences argue in favor of the following relation:

$$u^\varepsilon(x,t) \simeq u^*(x,t) + \varepsilon r(x, \frac{x}{\varepsilon}, t), \tag{52}$$

where $r(x, ., t)$ is ℓ-periodic. This could be the first terms of a series expansion such as

$$u^\varepsilon(x,t) = u^0(x, \frac{x}{\varepsilon}, t) + \varepsilon u^1(x, \frac{x}{\varepsilon}, t) + \varepsilon^2 u^2(x, \frac{x}{\varepsilon}, t) + \cdots \tag{53}$$

This is the starting point of the **two-scale periodic homogenization** that we develop in the next section.

4. Two-scale homogenization: the 1-D periodic case

In this section, we follow the classical two-scale periodic homogenization technique as, for example, developed by Sanchez-Palencia (1980). Two-scale periodic homogenization technique is based on a mathematical construction that explicitly makes use of two space variables, one for the small scales and one for the large scales. This construction can be puzzling at first: there is of course only one space-variable in nature. The mathematical construction we are about to present could just be an interesting exercise with no connection to the original one space-variable problem. It would not be very useful in the end. Fortunately, a theorem shows that the original one scale problem converges to the two-scale mathematical construction as ε goes to 0. Thanks to this theorem, the results of the two-scale periodic homogenization apply to our original one scale wave propagation problem and make this development useful. The distinction between small and large scales is made simple by the periodicity of the mechanical properties: everything that is periodic, or quasi-periodic, is a small scale and all the rest are large scales.

As one can guess from the examples in the previous sections, the solution to the wave equation in a bar with small periodic heterogeneities depends on two scales: x and $\frac{x}{\varepsilon}$. To explicitly take small-scale heterogeneities into account when solving the wave equation, the small space-variable is introduced (see Fig. 8):

$$y = \frac{x}{\varepsilon}, \tag{54}$$

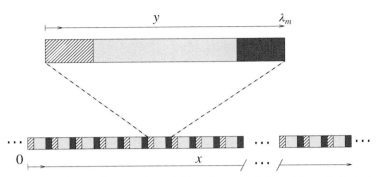

Fig. 8 Sketch displaying the x and y variables along a 1-D bar with periodic heterogeneities. y corresponds to a zoom on a single periodic cell.

where $\varepsilon = \frac{\ell}{\lambda_{\min}}$, as defined in (26). In many publications, y is often named the microscopic variable and x the macroscopic variable. In the following, we will also use "fine-scale," "microscale" or "small scale" for y and "large scale" or "macroscale" for x. When $\varepsilon \to 0$, that is the periodicity is infinitely small compared to $\lambda_{\min}$, any change in y induces a very small change in x. This leads to the idea of separation of scales: *y and x are treated as independent variables*. As already mentioned, there is, of course only one space-variable in the original problem. The introduction of the small scale variable y is a pure mathematical trick that makes sense in the end thanks to the convergence theorem.

The observations made in the previous section suggests that the solution to the wave equation could be sought as an asymptotic power series in ε:

$$u^\varepsilon(x,t) = u^0(x,y,t) + \varepsilon u^1(x,y,t) + \varepsilon^2 u^2(x,y,t) + \cdots, \qquad (55)$$
$$\sigma^\varepsilon(x,t) = \frac{1}{\varepsilon}\sigma^{-1}(x,y,t) + \sigma^0(x,y,t) + \varepsilon\sigma^1(x,y,t) + \cdots,$$

in which coefficients u^i and σ^i depend on both space variables x and y and are $\lambda_{\min}$-periodic in y (they belong to $\mathcal{T}$, see Section 2.2). This ansatz- the x and y dependence of the solution-explicitly incorporates our observations, that the sought solution depends on the wavefield at the large scale, and also on the small variations of elastic properties locally. As we will see it soon, the constitutive relation between the stress and the displacement involves a $1/\varepsilon$. Therefore, if the displacement power series starts at i_0, the stress power series should start at i_0-1. Starting the displacement series at $i=0$, as expected, implies to start the stress expansion at $i=-1$.

We now need to build the equations that drive the u^i and σ^i coefficients. The x and y relation (54) implies that partial derivatives with respect to x should become:

$$\frac{\partial}{\partial x} \rightarrow \frac{\partial}{\partial x} + \frac{1}{\varepsilon}\frac{\partial}{\partial y}. \tag{56}$$

We now introduce ρ and E, the cell mechanical properties:

$$\begin{aligned} \rho(y) &= \rho^\varepsilon(\varepsilon y), \\ E(y) &= E^\varepsilon(\varepsilon y), \end{aligned} \tag{57}$$

E and ρ are the unit cell elastic modulus and density. It is important to note that, with such a definition, ρ and E are $\lambda_{\min}$-periodic and independent of ε. Indeed, for any function ℓ-periodic h^ε, the function $h(y) = h^\varepsilon(\varepsilon y)$ is $\lambda_{\min}$-periodic and independent of ε:

$$\begin{aligned} h(y + \lambda_{\min}) &= h^\varepsilon(\varepsilon y + \varepsilon \lambda_{\min}), \\ &= h^\varepsilon(\varepsilon y + l), \\ &= h^\varepsilon(\varepsilon y), \\ &= h(y), \end{aligned} \tag{58}$$

where we have use the definition (26) of ε. It shows that h is $\lambda_{\min}$-periodic whatever the value of ε.

The definitions (57) of the cell properties are important: it implies that all spatial variation of the mechanical properties belong to the small scale domain. This is trivial in the periodic case, nevertheless, in the nonperiodic case, this definition will be one of the most difficult but critical points. One of the reasons explaining the simplicity of this definition lies in the outstanding properties of periodic functions: periodic functions remain periodic after most mathematical nonlinear manipulations such as taking the product, the square, the square root, or the inverse. For example, it can happen that the cell properties need to be defined (57) on (ρ, α) and not (ρ, E). Nevertheless, due to the periodic nature of the mechanical properties, it does not matter and any other choice than (ρ, E) works. We will see that this simplicity does not hold anymore in the nonperiodic case.

At this stage, we do not say much about the source f. In the following development, we treat f as a constant term, even if it is obviously not the case, and we postpone the discussion about the source.

Introducing expansions (55) in Eqs. (27)–(28), using (56) we obtain:

$$\rho \partial_{tt} \sum_{i\geq 0} \varepsilon^i u^i - \left(\partial_x + \frac{1}{\varepsilon}\partial_y\right) \sum_{i\geq -1} \varepsilon^i \sigma^i = f\delta_{i,0} \tag{59}$$

$$\sum_{i\geq -1} \varepsilon^i \sigma^i = E\left(\partial_x + \frac{1}{\varepsilon}\partial_y\right) \sum_{i\geq 0} \varepsilon^i u^i \tag{60}$$

where δ is the Kronecker symbol. It leads to

$$\rho \partial_{tt} \sum_{i\geq 0} \varepsilon^i u^i - \partial_x \sum_{i\geq -1} \varepsilon^i \sigma^i - \partial_y \sum_{i\geq -1} \varepsilon^{i-1} \sigma^i = f\delta_{i,0} \tag{61}$$

$$\sum_{i\geq -1} \varepsilon^i \sigma^i = E\sum_{i\geq 0} \varepsilon^i \partial_x u^i + E\sum_{i\geq 0} \varepsilon^{i-1} \partial_y u^i \tag{62}$$

Renaming $i - 1$ as i on the last sums, identifying term by term in ε^i we obtain, $\forall i \geq 0$:

$$\rho \partial_{tt} u^i - \partial_x \sigma^i - \partial_y \sigma^{i+1} = f\delta_{i,0}, \tag{63}$$

$$\sigma^i = E(\partial_x u^i + \partial_y u^{i+1}), \tag{64}$$

The last equations are the ones that drive the homogenization expansion coefficients. They have to be solved for each i. In this context, for any function $h(x, y)$ $\lambda_{\min}$-periodic in y, the cell average defined in Section 2 is

$$\langle h\rangle(x) = \frac{1}{\lambda_{\min}} \int_0^{\lambda_{\min}} h(x, y)dy. \tag{65}$$

4.1 Resolution of the homogenization problem

We now solve Eqs. (63) and (64). In the following, the time dependence t is dropped to ease the notations.

4.1.1 Resolution of the homogenized equations, step 1

Eqs. (63) for $i = -2$ and (64) for $i = -1$ give

$$\partial_y \sigma^{-1} = 0,$$
$$\sigma^{-1} = E\partial_y u^0, \tag{66}$$

which implies

$$\partial_y(E\partial_y u^0) = 0. \tag{67}$$

Multiplying the last equation by u^0, integrating over the unit cell, using integration by parts and taking into account the periodicity of u^0 and $E\partial_y u^0$, we find

$$\int_0^{\lambda_{min}} (\partial_y u^0)^2 E \, dy = 0. \tag{68}$$

$E(y)$ being a strictly positive function, the unique solution to the above equation is $\partial_y u^0 = 0$. This implies that u^0 does not depend on y and we therefore have

$$u^0 = \langle u^0 \rangle, \tag{69}$$

$$\sigma^{-1} = 0. \tag{70}$$

The fact that the order 0 solution in displacement u^0 is independent of the microscale variable y is an important result, which confirms the observation made in Fig. 4, that the displacement is a smooth function to the leading order.

4.1.2 Resolution of the homogenized equations, step 2

Eqs. (63) for $i = -1$ and (64) for $i = 0$ give

$$\partial_y \sigma^0 = 0, \tag{71}$$

$$\sigma^0 = E(\partial_y u^1 + \partial_x u^0). \tag{72}$$

Eq. (71) implies that $\sigma^0(x, y) = \langle \sigma^0 \rangle(x)$ and, with (72), that

$$\partial_y (E\partial_y u^1) = -\partial_y E \, \partial_x u^0. \tag{73}$$

The last equation is a partial differential equation in y. For a fixed x, knowing that u^0 only depends on x, $\partial_x u^0$ is a constant in y. (73) is therefore an equation of type

$$\partial_y (E(y)\partial_y q(y)) = \gamma f(y), \tag{74}$$

where γ is a constant. The last equation is linear with respect to the second member and a general solution takes the form

$$q(y) = \gamma q_0(y) + \beta \tag{75}$$

where q_0 is the solution of (74) for $\gamma = 1$ and a constant β. Applying this observation to (73), we can separate the variables and look for a solution of the form

$$u^1(x, y) = \chi^1(y)\partial_x u^0(x) + \langle u^1 \rangle(x) \tag{76}$$

where $\chi^1(y)$ is called the first-order periodic corrector (γ is $\partial_x u^0$ and β is $\langle u^1 \rangle(x)$ in (74) and (75)). To enforce the uniqueness of the solution, we impose $\langle \chi^1 \rangle = 0$. Introducing (76) into (73), we obtain the equation of the so-called cell problem:

$$\partial_y \left[E(1 + \partial_y \chi^1) \right] = 0, \tag{77}$$

χ^1 being λ_{min}-periodic and verifying $\langle \chi^1 \rangle = 0$. Introducing (76) into (72), taking the cell average and using the fact that u^0 and σ^0 do not depend upon y, we find the order 0 constitutive relation,

$$\sigma^0 = E^* \partial_x u^0, \tag{78}$$

where E^* is the order 0 homogenized elastic coefficient,

$$E^* = \langle E(1 + \partial_y \chi^1) \rangle. \tag{79}$$

4.1.3 Resolution of the homogenized equations, step 3

Eqs. (63) for $i = 0$ and (64) for $i = 1$ give

$$\rho \partial_{tt} u^0 - \partial_x \sigma^0 - \partial_y \sigma^1 = f, \tag{80}$$

$$\sigma^1 = E(\partial_y u^2 + \partial_x u^1). \tag{81}$$

Applying the cell average on (80), using the property (3), the fact that u^0 and σ^0 do not depend on y and gathering the result with (78), we find the order 0 wave equation:

$$\begin{aligned} \rho^* \partial_{tt} u^0 - \partial_x \sigma^0 &= f \\ \sigma^0 &= E^* \partial_x u^0, \end{aligned} \tag{82}$$

where

$$\rho^* = \langle \rho \rangle \tag{83}$$

is the effective density and E^* is defined by Eq. (99). This a remarkable result: the effective wave equation is still a classical wave equation meaning that it can be solved using classical wave equation solvers without any change.

An introduction to the two-scale homogenization method 243

Knowing that ρ^* and E^* are constant, solving the wave equation for the order 0 homogenized medium is a much simpler task than solving for the original medium and no numerical difficulty related to the rapid variation of the properties of the bar arises. One of the important results of the homogenization theory is to show that u^ε "converges" toward u^0 when ε tends toward 0 (the so-called convergence theorem, see Sanchez-Palencia (1980) and Section 4.2).

Once u^0 is found, the first-order correction, $\chi^1(x/\varepsilon)\partial_x u^0(x)$, can be computed. To obtain the complete order 1 solution u^1 using (76), $\langle u^1 \rangle$ remains to be found. Subtracting (82) from (80) we have,

$$\partial_y \sigma^1 = (\rho - \langle \rho \rangle)\partial_{tt} u^0, \tag{84}$$

which, together with (81) and (76) gives

$$\partial_y \left(E \partial_y u^2 \right) = -\partial_y (E \partial_x u^1) + (\rho - \langle \rho \rangle)\partial_{tt} u^0, \tag{85}$$

$$= -\partial_y E \, \partial_x \langle u^1 \rangle - \partial_y \left(E \chi^1 \right)\partial_{xx} u^0 + (\rho - \langle \rho \rangle)\partial_{tt} u^0. \tag{86}$$

Using the linearity of the last equation we can separate the variables and look for a solution of the form

$$u^2(x, y) = \chi^2(y)\partial_{xx} u^0(x) + \chi^1(y)\partial_x \langle u^1 \rangle(x) + \chi^\rho(y)\partial_{tt} u^0 + \langle u^2 \rangle(x), \tag{87}$$

where χ^2 and χ^ρ are solutions of

$$\partial_y \left[E(\chi^1 + \partial_y \chi^2) \right] = 0, \tag{88}$$

$$\partial_y \left[E \partial_y \chi^\rho \right] = \rho - \langle \rho \rangle, \tag{89}$$

with χ^2 and χ^ρ λ_{min}-periodic and where we impose $\langle \chi^2 \rangle = \langle \chi^\rho \rangle = 0$ to ensure the uniqueness of the solutions. Introducing (87) into (81) and taking the cell average, we find the order 1 constitutive relation:

$$\langle \sigma^1 \rangle = E^* \partial_x \langle u^1 \rangle + E^{1*} \partial_{xx} u^0 + E^{\rho *} \partial_{tt} u^0 \tag{90}$$

with

$$E^{1*} = \left\langle E(\chi^1 + \partial_y \chi^2) \right\rangle \tag{91}$$

$$E^{\rho *} = \left\langle E \partial_y \chi^\rho \right\rangle \tag{92}$$

The periodicity condition on χ^2 imposes $\partial_y \chi^2 = -\chi^1$ and therefore $E^{1*} = 0$.

Finally using (76) and taking the average of Eq. (63) for $i = 1$ gives the order 1 wave equation

$$\langle \rho \rangle \partial_{tt} \langle u^1 \rangle + \langle \rho \chi^1 \rangle \partial_x \partial_{tt} u^0 - \partial_x \langle \sigma^1 \rangle = 0$$
$$\langle \sigma^1 \rangle = E^* \partial_x \langle u^1 \rangle + E^{\rho*} \partial_{tt} u^0. \tag{93}$$

It can shown (see Capdeville et al., 2010a, appendix A) that $E^{\rho*} = \langle \rho \chi^1 \rangle$ and therefore, renaming $\langle \widetilde{\sigma}^1 \rangle = \langle \sigma^1 \rangle - E^{\rho*} \partial_{tt} u^0$, the last equations can be simplified to

$$\langle \rho \rangle \partial_{tt} \langle u^1 \rangle - \partial_x \langle \widetilde{\sigma}^1 \rangle = 0$$
$$\langle \widetilde{\sigma}^1 \rangle = E^* \partial_x \langle u^1 \rangle. \tag{94}$$

This is once again a classical wave equation, but with no second member. Knowing that such equations have a unique solution and that knowing that 0 is solution, we conclude that $\langle u^1 \rangle = 0$. Summing u^0 and u^1, and using the fact that $\langle u^1 \rangle = 0$, we obtain the complete order 1 solution:

$$\widehat{u}^1(x, y) = u^0(x) + \varepsilon \chi(y) \partial_x u^0(x). \tag{95}$$

We stop the resolution of (63) and (64) here but it is possible to go up to a higher order (see Fish, Chen, & Nagai, 2002 for a 1-D periodic case up to the order 2).

4.1.4 Analytical solution to the cell problem

It is useful to note that a general analytical solution to the cell problem (77) exists and is

$$\chi^1(y) = -y + a \int_0^y \frac{1}{E(y')} dy' + b. \tag{96}$$

The periodicity condition imposes

$$a = \left\langle \frac{1}{E} \right\rangle^{-1}, \tag{97}$$

and b can be found using $\langle \chi^1 \rangle = 0$. We therefore have

$$\partial_y \chi^1(y) = -1 + \left\langle \frac{1}{E} \right\rangle^{-1} \frac{1}{E(y)} \tag{98}$$

An interesting follow up to the last equation is that, once introduced in (79), we have

$$E^* = \left\langle \frac{1}{E} \right\rangle^{-1}. \tag{99}$$

An introduction to the two-scale homogenization method 245

This solution for the effective Young modulus is very similar to the Backus solution (44) and both solutions give the same results if applied to the periodic media. There is nevertheless an important difference: the two-scale homogenization solution only applies to periodic media while the Backus solution applies to any media.

4.1.5 Strain leading order

What about strain? The strain can also be expanded as a power series in ε:

$$\epsilon^\varepsilon(u^\varepsilon)(x) = \sum_i \varepsilon^i \epsilon^i\left(x, \frac{x}{\varepsilon}\right). \tag{100}$$

Using (56) and (55) we have

$$\epsilon^\varepsilon(u^\varepsilon) = \left(\partial_x + \frac{1}{\varepsilon}\partial_y\right) \sum_i \varepsilon^i u^i \tag{101}$$

and therefore

$$\epsilon^\varepsilon(u^\varepsilon) = \sum_i \varepsilon^i\left(\partial_x u^i + \partial_y u^{i+1}\right) \tag{102}$$

For the leading order (using $\partial_y u^0(x) = 0$), we have

$$\epsilon^0 = \partial_x u^0 + \partial_y u^1. \tag{103}$$

Using $\partial_y u^1 = \partial_y \chi^1 \partial_x u^0$, we finally find

$$\epsilon^0(x, y) = \left(1 + \partial_y \chi^1(y)\right)\partial_x u^0(x). \tag{104}$$

The last equation is interesting: on the contrary to the displacement, the strain depends on y to the leading order. This implies that the strain is not smooth and depends strongly on the small scale, which has been already observed in Fig. 4. Another example is given in Fig. 9, which shows that the amplitude on the small-scale effect does not change with ε: no matter how small the heterogeneity, if it has a large contrast, it will have a strong effect on strain. This has important consequences for any wavefield gradient measurements such as rotation and strain (see Section 6.1.1).

4.1.6 Summary

To summarize, at this stage, we have shown that:

- u^0 and σ^0 do not depend on the small-scale variable y;
- u^0 and σ^0 are solution of effective equations (82) which are the same as the original wave equation, but for the effective coefficients;

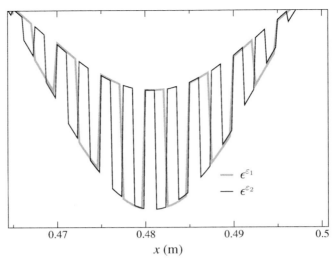

Fig. 9 Strain snapshots computed in the same periodic bar as for Section 3.1, but for two ε, $\varepsilon_1 = 0.5$, and $\varepsilon_2 = 0.25$. One can notice that if the local periodicity of the strain changes with ε as for the displacement residual, the oscillations' amplitude are independent of ε and does not scale as ε as for the displacement residual.

- to the order 1, we have

$$u^\varepsilon(x,t) = u^0(x,t) + \varepsilon \chi^1\left(\frac{x}{\varepsilon}\right) \partial_x u^0(x,t) + O(\varepsilon^2), \quad (105)$$

where χ^1 is the first-order periodic corrector. ;
- χ^1 is solution of the cell Eq. (77), a static time independent equation;
- the effective Young modulus E^* is obtained though (79).

It is interesting to note that Eq. (105) confirms the numerical observation made in Section 3.4, in particular in the Eq. (51).

4.2 Convergence theorem

So far, we have extensively used the ansatz

$$u^\varepsilon(x,t) = u^0\left(x, \frac{x}{\varepsilon}, t\right) + \varepsilon u^1\left(x, \frac{x}{\varepsilon}, t\right) + \varepsilon^2 u^2\left(x, \frac{x}{\varepsilon}, t\right) + \cdots \quad (106)$$

with $u^i(x,y)$ $\lambda_{\min}$-periodic for the second variable and the separation of scale (x and y are treated as independent variables) but with no formal proof or justification.

The proof of the homogenization theorem

$$\begin{cases} u^\varepsilon \xrightarrow[\varepsilon \to 0]{} u^0 & \text{weakly in the appropriate space} \\ E^\varepsilon \partial_x u^\varepsilon \xrightarrow[\varepsilon \to 0]{} E^* \partial_x u^0 & \text{weakly in the appropriate space} \end{cases}$$

is an important concern in the homogenization community and has yielded many works. Among them, we can find

- the oscillating test function method (sometimes called the *energy method*) derived by Tartar (Tartar & Murat, 1997; Tartar, 1978). This method does not require the periodicity hypothesis to work. It is also the base of some numerical homogenization methods (Owhadi & Zhang, 2008);
- the two-scale convergence (Allaire, 1992; Nguetseng, 1989) designed for the periodic case. It justifies the scale separation and shows strong convergence when higher order terms of the expansion are used;
- the variational homogenization and the Γ-convergence methods (De Giorgi, 1984);
- the convergence issues are also addressed in the general G-convergence and H-convergence theories (De Giorgi & Spagnolo, 1973; Tartar & Murat, 1997).

A weak convergence means that the convergence is true on average (integral form), but not for every individual point. For example, the true strain never converges to the order 0 strain, but its average does. Strong convergence (for every individual point) can be obtained if the first-order corrector is accounted for (Allaire, 1992).

The convergence theorems justify the ansatz (106) and the explicit scale separation of the two-scale method: as ε becomes smaller and smaller, the true problem converges to the two-scale problem. It implies that the strange two-scale construction makes sense and is useful. Note that, in most standard convergence problems, the approximate solution converges toward the true problem. Here, it is the other way around: we have the solutions to a series of true problems that converge toward the asymptotic solution. In the series of true problems, only one corresponds to the real periodicity of the bar. This has an important consequence: if the periodicity of the real bar corresponds to an ε that is not small, or not small enough, then the effective solution may not be accurate. If it is the case, unfortunately, nothing can be done to improve the solution in the two-scale homogenization framework, besides using a purely numerical solution in the real bar. We will see that it is different for the non-periodic homogenization.

4.3 External point sources

In practice, the external source process often occurs over an area much smaller than the smallest wavelength λ_{min} allowing us to ideally consider it as a point source: $f(x, t) = g(t)\delta(x - x_0)$ or $f(x, t) = g(t)\partial_x\delta(x - x_0)$. There are two ways to deal with a point source or with multiple point sources.

In the first approach, we assume that we are not interested in the near-field of the source and we would like to keep the source as a point source. Indeed, forgetting about the near-field, point sources are simple and effective to implement into solvers such as SEM that use the weak form of the wave equation (it is more complex for solvers such as finite differences that use the strong form). Then, two potential issues arise:

1. in the vicinity of x_0, there is no such a thing as a minimum wavelength. The asymptotic development presented here is therefore only valid far away enough from x_0;
2. a point source has a local interaction with the microscopic structure that needs to be accounted for.

The first point is not an issue as we assumed that we are not interested in the near-field. One should nevertheless keep in mind that, very close to x_0, the solution is not accurate, but not less than any standard numerical methods used to solve the wave equation. For example, this is the case for the spectral-element method for the element containing the source and some times for the elements next to it (Nissen-Meyer, Fournier, & Dahlen, 2007). The second point is more important and can be addressed in the following way: the hypothetical point source is just a macroscopic representation of a more complex physical process, and what is relevant is to ensure the conservation of the energy released at the source. Therefore we need to find a corrected source f^ε that preserves the energy associated with the original force f up to the desired order (here 1 as an example). We therefore need

$$(u^\varepsilon, f) = (u^0, f^\varepsilon) + O(\varepsilon^2), \tag{107}$$

where $(\cdot, \cdot)$ is the L^2 inner product, which for any function g and h is:

$$(g, h) = \int_{\mathbb{R}} g(x)h(x)dx. \tag{108}$$

Using (105) we have, to the order 1,

$$(u^0, f) + \varepsilon(\chi^1(y)\partial_x u^0(x, t), f) = (u^0, f^\varepsilon) + O(\varepsilon^2), \tag{109}$$

and, using integration by parts, we find

$$f^\varepsilon(x, t) = \left[1 - \varepsilon\chi^1\left(\frac{x}{\varepsilon}\right)\partial_x\right]\delta(x - x_0)g(t). \tag{110}$$

f^ε is the corrected source that needs to be used in the homogenized wave Eq. (82) instead of f.

If the source term has the form $f(x, t) = g(t)\partial_x\delta(x - x_0)$, which is in the case of earthquake moment tensors, to the leading order,

$$f^\varepsilon(x, t) = \left[1 + \partial_y\chi^1\left(\frac{x}{\varepsilon}\right)\right]\partial_x\delta(x - x_0)g(t). \tag{111}$$

It can be seen from the last equation that no ε appears in front of the corrector, which means source term is strongly modified by the local structure. This leads to significant distortion on the apparent moment tensor in realistic situations and can lead to wrong interpretation of the inverted source tensors (Burgos et al., 2016; Capdeville et al., 2010b).

If we are interested in the near-field, we need a different approach. The second approach implies the two-scale construction of the external source term $f(x, y)$ from the original source term $f(x)$. This can only be done in the nonperiodic framework presented in Section 5. We will not develop this approach here, it is the subject of publication in preparation. Nevertheless, the near-field can be modeled as another corrector θ, and this corrector appears at the order 0 in the displacement when sources appear as a derivative of a Dirac function, such as for earthquake moment tensors. Therefore, displacement depends on y at the zero order in the vicinity of the source, which is very unusual in homogenization, but consistent with both numerical and real observations. Similarly, an effective source appears, which is not a point source anymore, but a volumetric force map instead.

4.4 An example

A numerical experiment in a bar of periodic properties shown in Fig. 10 is performed. The periodicity of the structure is $l = 6$ mm. First, the cell problem (77) is solved with periodic boundary conditions using a finite-element method based on the same mesh and quadrature like the one that will be used to solve the wave equation. Although this is not necessary for this simple 1-D case (we could use the analytical solution), it is a convenient solution and makes it simple to access E^*, the correctors χ and well as the external source term f^ε. Then, the homogenized wave equation,

$$\rho^*\partial_{tt}u - \partial_x(E^*\partial_x u) = f, \tag{112}$$

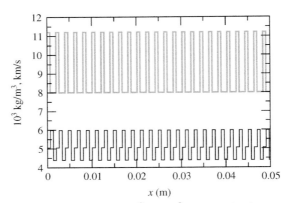

Fig. 10 A 5 cm sample of the density in 10^3 kg m^{-3} (*gray*) and velocity in km s^{-1} (*black*) for the bar.

where $u = u^0$ and $f = f^\varepsilon$ is solved using the SEM (see Capdeville (2000) or Igel (2017) for a complete description of the 1-D SEM).

The point source is located at $x = 2$ m. The time wavelet $g(t)$ is a Ricker with a central frequency of 50 kHz (which gives a corner frequency of about 125 kHz) and a central time shift $t_0 = 6.4 \times 10^{-5}$ s. In the far-field, this wavelet gives a minimum wavelength of about 4 cm which corresponds to a wave propagation with $\varepsilon = 0.15$. In practice, the bar is of course not infinite, but its length (5 m) and the time at which the displacement is recorded (4.9×10^{-4} s) is such that the wave pulse does not reach the extremity of the bar. To be accurate, the reference solution is computed with a SEM mesh matching all interfaces with an element boundary (7440 elements for the 5 m bar). To make sure that only the effect of homogenization is seen in the simulations, the mesh and time step used to compute the reference solution are also used to compute the homogenized solution. Once the simulation is done for the time step corresponding to $t = 4.9 \times 10^{-4}$ s, the complete order 1 solution can be computed with (95). We can also compute the incomplete homogenized solution at the order 2 as shown in Capdeville et al. (2010b).

The results of the simulation are shown in Fig. 11. On the upper left plot (Fig. 11A) the reference solution (bold red line), the order 0 solution (black line) and a solution obtained in the bar with a $E^* = \langle E^\varepsilon \rangle$ ("E average," dashed line) for $t = 4.9 \times 10^{-4}$ s are shown as a function of x. As expected, the "E average" solution is not in phase with the reference solution and shows that this "natural" filtering is not accurate. On the other hand, the order 0 homogenized solution is already in excellent agreement with the reference solution. On Fig. 11B the residual between the order 0 homogenized solution

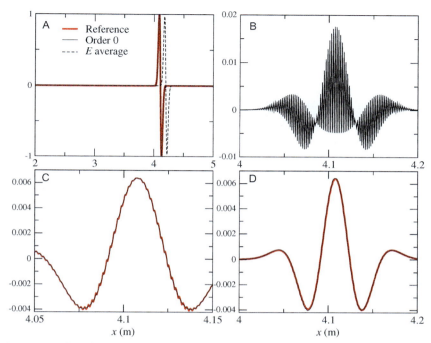

Fig. 11 Results of the numerical example in Section 4.4. (A) displacement $u^\varepsilon(x, t)$ at $t = 4.9 \times 10^{-3}$ s computed in the reference model shown in Fig. 10 (*red*), the order 0 homogenized solution $\hat{u}^0(x, t)$ (*black*) and the solution computed in a model obtained by averaging the elastic properties ($\langle \rho^\varepsilon \rangle$ and $\langle E^\varepsilon \rangle$) (*dashed*). (B) Order 0 residual ($u^0 - u^\varepsilon$). (C) Order 1 residual ($u^0 + \varepsilon u^1 - u^\varepsilon$) (*red*) and partial order 2 residual ($u^0 + \varepsilon u^1 + \varepsilon^2 u^2 - u^\varepsilon(x, t)$) (*black*). (D) Partial order 2 residual for $\varepsilon = 0.15$ (*red*) and partial order 2 residual for $\varepsilon = 0.075$ with amplitude multiplied by 4 (*black*).

and the reference solution $u^0 - u^\varepsilon$ is shown. The error amplitude reaches 2% and contains fine-scale variations. In Fig. 11C the order 1 residual $\hat{u}^1 - u^\varepsilon$ (bold red line) and the partial order 2 residual $\hat{u}^2(x, t) - u^\varepsilon(x, t)$ (see Capdeville et al., 2010a for the order 2 calculus), where $\hat{u}^i$ are the sum of homogenized terms from 0 to i is shown. Comparing Fig. 11B and C, it can be seen that the order 1 periodic corrector removes most of the fine-scale variations present in the order 0 residual. The remaining variations disappear with the partial order 2 residual. The remnant smooth residual is due to the $\langle u^2 \rangle$ that has not computed. In order to check that it is indeed an ε^2 residual, the residual computed for $\varepsilon = 0.15$ is overlapped with a residual computed for $\varepsilon = 0.075$ (which corresponds to $l_0 = 3$ mm) and the amplitude is multiplied by a factor 4. The fact that these two signals overlap is consistent with a ε^2 residual.

5. Two-scale homogenization: The 1-D nonperiodic case

We now treat the nonperiodic case. In the periodic case, we were able to define a series of ε indexed problems by varying the periodicity. Here, this is not possible anymore as the mechanical properties ($\rho(x)$, $E(x)$) can contain all possible scales with no restriction. No ε can be defined simply and here we do not use the superscript ε for the mechanical properties and solution of the wave equation. The wave equations are

$$\rho \partial_{tt} u - \partial_x \sigma = f,$$
$$\sigma = E \partial_x u. \qquad (113)$$

We assume that the source time function has a maximum frequency f_{max} and the a minimum wavelength λ_{min} exists. The solution u, σ to the wave equations implicitly depend upon f_{max}.

5.1 The Backus solution

As shown in Fig. 5C and D, the Backus solution (47) can be applied to nonperiodic settings and provide an accurate order 0 solution, at least in 1-D and for layered media. If we take a closer look at the difference between the reference solution and the Backus solution shown in Fig. 12, it can be seen that, similar to the periodic case, the residual is not zero and contains small scale oscillations. The Backus solution does not explain this difference.

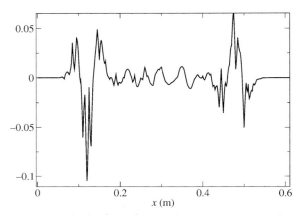

Fig. 12 Displacement residual ($u^\varepsilon - u^*$) snapshot at $t = 2$ ms in the same random model as for Fig. 5, where u^* is obtained in the Backus effective medium. The residual is normalized by the maximum amplitude of the reference solution.

5.2 Naive solutions to the two-scale homogenization periodicity limitation

At this stage, we have two different approaches to deal with small scales: the Backus approach and the two-scale homogenization method. On the one hand, the Backus method applies to both periodic and nonperiodic heterogeneities and it gives an accurate effective solution. However, as seen above, it does not go beyond the leading order of the effective solution and it does not explain the small oscillations around it (or large oscillations for strain) and it does not propose a solution for the source. Furthermore, it cannot be extended to higher dimensional 2-D or 3-D cases. On the other hand, the two-scale homogenization method is a complete theory that gives solutions for leading order effective solution, corrector, strain, and external sources. Moreover, as we will see in Section 6, it extends naturally to dimensions higher than 1-D. However, it only applies to media with periodic heterogeneities, which is an unacceptable limitation for geological media.

It is tempting to directly apply the two-scale periodic method to nonperiodic media. But, to do so, what should the periodic cell be? It could be the entire bar. In that case, the effective media is constant all along the bar, with $\rho^* = \frac{1}{L} \int_0^L \rho(x) dx$ and $\frac{1}{E^*} = \frac{1}{L} \int_0^L \frac{1}{E}(x) dx$. In a homogeneous bar, a coda wave after the ballistic wave cannot appear, but we know from Fig. 5D that coda wave is present. Therefore, this solution does not work. Another option is to use a moving window of a size comparable to the wavelength and perform many homogenizations along the bar. Unfortunately, in that case, the cell window acts as a convolution with a boxcar function. In such a case, zero order discontinuities (jump) are transformed into first-order discontinuities (kink). The latter are as difficult to mesh as the zero order discontinuities and thus, no gain is obtained.

Another tempting idea is to mix the two methods naively. It is interesting that the filters in the two methods are different. The two-scale homogenization uses a cell average (65) whereas the Backus solution use a low-pass filter average (6). Why not just replacing the cell average in the two-scale homogenization development (65) by the Backus low-pass filter average (6)? This would work for the effective density ρ^*, from (39) to (83). Nevertheless, for the effective Young modulus, starting from (96) and applying the periodic condition $\chi^1(0) = \chi^1(\lambda_{\min})$, we have

$$0 + b + 0 = -\lambda_{\min} + a \int_0^{\lambda_{\min}} \frac{1}{E(y')} dy' + b, \qquad (114)$$

which leads to

$$a = \lambda_{\min} \left(\int_0^{\lambda_{\min}} \frac{1}{E(y')} dy \right)^{-1}, \tag{115}$$

$$= \left\langle \frac{1}{E} \right\rangle^{-1}. \tag{116}$$

We therefore see that (99) remains unchanged. Changing $\langle . \rangle$ by $\mathcal{F}^{\varepsilon_0}$ in (79) leads to

$$E^* = \mathcal{F}^{\varepsilon_0} \left(E(1 + \partial_y \chi^1) \right), \tag{117}$$

which, using (98) implies

$$E^* = \mathcal{F}^{\varepsilon_0}(a). \tag{118}$$

a is constant and therefore $\mathcal{F}^{\varepsilon_0}(a) = a$ and we fall back on

$$E^* = \left\langle \frac{1}{E} \right\rangle^{-1}. \tag{119}$$

There is no way to avoid this result and we fail to replace $\langle . \rangle$ by $\mathcal{F}^{\varepsilon_0}$ in the last equation. This naive approach is clearly not a solution and we propose a more successful solution in the next section.

5.3 New functional spaces

We want to keep the theoretical results obtained from the two-scale periodic homogenization method while preserving the flexibility of the Backus solution. After a few attempts, we knew there is no trivial way to mix those two methods. Therefore, we need to set up a good theoretical framework to achieve our objective.

We first need to define the functional space to which our solutions should belong. In the periodic case, each two-scale expansion coefficients $(u^i(x, y), \sigma^i(x, y) \ldots)$ belongs to $\mathcal{T}$, the space of two-scale functions $\lambda_{\min}$-periodic for the second variable (see Section 2.2). For the classical periodic homogenization, the periodicity makes it possible to simply and efficiently define the fine-scales. Moreover, as already mentioned, $\mathcal{T}$ has some very interesting properties: one of them is that the product of two functions $(g, h) \in \mathcal{T}^2$ also belongs to $\mathcal{T}$. Actually, a function h in $\mathcal{T}$ still belongs to $\mathcal{T}$ after a nonlinear operation, such as taking its inverse. These unusual properties make the periodic homogenization very special: it warranties that fine scales remain in the fine-scale domain through mathematical manipulations and make sure scales are separated. Unfortunately, it is not possible to keep these properties in the nonperiodic case.

For the nonperiodic case, we now need to define a new functional space V that would be the equivalent of T for the periodic case. The idea is to define a two-scale functional space such that only fine-scale variations exist on the y variable in the Fourier sense. To do so, we use $\mathcal{F}$, the spatial filter on the y variable, defined in (10). It is the same as the $\mathcal{F}^{\varepsilon_0}$ filter defined earlier, but it applies to y instead of x. Let λ_0 be the user-defined spatial wavelength scale separation and $k_0 = \lambda_0^{-1}$ the spatial frequency separation. λ_0 is always defined in the large scale domain (the x domain) because it is the physical domain. In the y domain, because of the relation $y = x/\varepsilon_0$, the separation of scale is always $\lambda_{\min}$. So, in the x domain, any spatial oscillation smaller than λ_0 is considered as small scale. In the spectral domain, for a function h, any value of $\bar{h}(k)$ for $|k| > k_0$ is considered as a small scale. Practically, for a given function $h(x)$, in order to tell if it has small scale variations, we need to compute its Fourier transform (4) and check if $\bar{h}(k)$ has non zero values for $|k| > k_0$. Even so, we can guess it makes sense to choose λ_0 smaller than $\lambda_{\min}$, λ_0 is defined by the user, and therefore, the ratio ε_0 (5) is user-defined.

Based on this definition of small scales, we can define a function space of two-scale functions that only contain fine-scale on the second variable:

$$V = \{h(x, y), Y - \text{periodic in } y \text{ such that } \mathcal{F}(h)(x, y) = \langle h \rangle(x)\}. \quad (120)$$

In this definition, for the sake of simplicity, we have omitted to define precisely Y. Y is a segment in y, centered on x/ε_0 and it has to be wide enough to fit the filter wavelet w_m support. A precise definition can be found in Capdeville et al. (2010 a). Graphic examples of a function h not in V and one in V are given in Fig. 13.

To try to visualize the concept of functions in V more, let us assume we need to build a two-scale function $g(x, y)$ in V from a function $h(x)$ with the constrain $h(x) = g(x, \frac{x}{\varepsilon_0})$. For example, h could be the function plotted in Fig. 13A. The periodic solution $g(y) = h(\varepsilon_0 y)$ would not work because in such a case $\mathcal{F}(g)(y)$ would not be constant (smooth variations would still be present). Instead, for any $(x, y) \in \partial\Omega \times Y$, we define

$$g(x, y) = \mathcal{F}^{\varepsilon_0}(h)(x) + (1 - \mathcal{F}^{\varepsilon_0})(h)(\varepsilon_0 y). \quad (121)$$

Obviously, setting $y = x/\varepsilon_0$ in the last equation leads to $g(x, \frac{x}{\varepsilon_0}) = h(x)$. Moreover, one can check that $\mathcal{F}(g) = \langle g \rangle = \mathcal{F}^{\varepsilon_0}(h)(x)$. Indeed, assuming

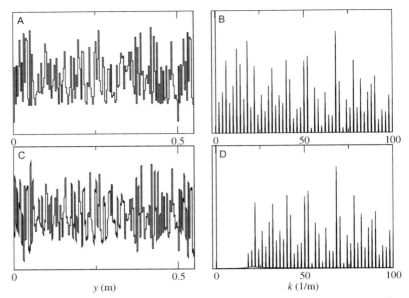

Fig. 13 Example of a function $h(x,y)$ not in $\mathcal{V}$ (graph A) and $h(x,y) \in \mathcal{V}$ (graph C) for a $k_0 = 16m^{-1}$ plotted for a given x as a function of y and their respective power spectra (graphs B and D) for positive wavenumber (k). It can be seen that, for $h(x,y) \in \mathcal{V}$ the power spectrum is 0 is the range $]0m^{-1}, 16m^{-1}]$. Both functions are periodic with a periodicity of 0.5m.

that $\mathcal{F}^{\varepsilon_0} \circ \mathcal{F}^{\varepsilon_0} = \mathcal{F}^{\varepsilon_0}$ (which is only an approximation because of the apodization of w_{ε_0}), we have

$$\mathcal{F}((1-\mathcal{F}^{\varepsilon_0})(h)(\varepsilon_0 y)) = \mathcal{F}^{\varepsilon_0}((1-\mathcal{F}^{\varepsilon_0})(h)(x)), \quad (122)$$
$$= (\mathcal{F}^{\varepsilon_0} - \mathcal{F}^{\varepsilon_0} \circ \mathcal{F}^{\varepsilon_0})(h)(x)), \quad (123)$$
$$= (\mathcal{F}^{\varepsilon_0} - \mathcal{F}^{\varepsilon_0})(h)(x)), \quad (124)$$
$$= 0, \quad (125)$$

where, for (122), we have proceeded in a similar way to (12)–(15). Therefore, g is indeed in $\mathcal{V}$. We will see more of these type of constructions later on.

To summarize,

- in the periodic case, $(u^i, \sigma^i) \in \mathcal{T}^2$, i.e., the expansion coefficients are $\lambda_{\min}$-periodic in y;
- in the nonperiodic case, $(u^{\varepsilon_0,i}, \sigma^{\varepsilon_0,i}) \in \mathcal{V}^2$, i.e., the expansion coefficients contain only fine-scale in the Fourier sense.

Note that, in the nonperiodic case, the expansion coefficients implicitly depend on ε_0, whereas in the periodic case, they do not depend upon ε.

5.4 Cell properties in the nonperiodic case

We need a similar version of the cell properties (57) to move on. In the periodic case, the construction of cell properties is trivial thanks to the amazing properties of periodic functions mentioned earlier (nonlinear manipulation of periodic functions are periodic functions). In the nonperiodic case, this construction is not trivial. It is actually the main difficulty of the nonperiodic case. We need to find two scales cell properties $\rho^{\varepsilon_0}(x, y)$, $E^{\varepsilon_0}(x, y)$ such that

$$\rho(x) = \rho^{\varepsilon_0}\left(x, \frac{x}{\varepsilon_0}\right),$$

$$E(x) = E^{\varepsilon_0}\left(x, \frac{x}{\varepsilon_0}\right),$$

(126)

and such that the expansion coefficients

$$\left(u^{\varepsilon_0, i}, \sigma^{\varepsilon_0, i}\right) \in \mathcal{V}^2.$$

(127)

The idea is to assume it is possible to find cell properties such that (126) and (127) are fulfilled, solve the asymptotic equations and find the necessary condition on $\left(\rho^{\varepsilon_0}, E^{\varepsilon_0}\right)$ so that a solution exists and then construct them.

5.5 Nonperiodic homogenized equations and resolution

The solution to the wave equations (113) is again sought as an asymptotic expansion in ε_0, but this time we look for $u^{\varepsilon_0, i}$ and $\sigma^{\varepsilon_0, i}$ in $\mathcal{V}$:

$$u(x, t) = \sum_{i=0}^{\infty} \varepsilon_0^i u^{\varepsilon_0, i}(x, x/\varepsilon_0, t),$$

$$\sigma(x, t) = \sum_{i=-1}^{\infty} \varepsilon_0^i \sigma^{\varepsilon_0, i}(x, x/\varepsilon_0, t).$$

(128)

Note that imposing $u^{\varepsilon_0, i}$ and $\sigma^{\varepsilon_0, i}$ in $\mathcal{V}$ is a strong condition that means that only large scale variations must appear in x and only fine-scale variations in y. One can notice that, in the above equations, the left terms do not depend on ε_0 but the right terms do. For the periodic case, in (55), this is the opposite. Actually, for a consistent mathematical development in the nonperiodic case, two epsilons are necessary, ε_0 and ε. If we do so, the ε dependency appears the same way in (128) like in (55). Nevertheless, the role of ε is in that case purely formal and we skip this difficulty here. The complete development can be found in Capdeville et al. (2010a, 2010b) and Guillot et al. (2010).

In the following, we work at ε_0 fixed and to simplify the notation further, we drop the explicit ε_0 dependency of u^i and σ^i.

In the nonperiodic case, the equations for the two-scale expansion remain unchanged :

$$\rho^{\varepsilon_0}(x, y)\partial_{tt}u^i(x, y) - \partial_x\sigma^i(x, y) - \partial_y\sigma^{i+1}(x, y) = f(x)\delta_{i,0}, \tag{129}$$

$$\sigma^i(x, y) = E^{\varepsilon_0}(x, y)(\partial_x u^i(x, y) + \partial_y u^{i+1}(x, y)). \tag{130}$$

The only difference from the periodic case is that ρ^{ε_0} and E^{ε_0} depend on ε_0 and both on (x, y) while they depend only on y in the periodic case. To solve these equations, we follow the same procedure as for the periodic case. Because the y periodicity is kept in $\mathcal{V}$, the resolution of the homogenized equations is almost the same as in the periodic case.

- As for the periodic case, Eq. (129) for $i = -2$ and (130) for $i = -1$ gives $\sigma^{-1} = 0$ and $u^0 = \langle u^0 \rangle$.
- Eqs. (129) for $i = -1$ and (130) for $i = 0$ implies $\sigma^0 = \langle \sigma^0 \rangle$ and

$$\partial_y\left(E^{\varepsilon_0}\partial_y u^1\right) = -\partial_y E^{\varepsilon_0} \partial_x u^0. \tag{131}$$

Using the linearity of the last equation we can separate the variables and look for a solution of the form

$$u^1(x, y) = \chi^{\varepsilon_0}(x, y)\partial_x u^0(x) + \langle u^1 \rangle(x). \tag{132}$$

As $u^1 \in \mathcal{V}$ and $u^0 = \langle u^0 \rangle$, χ^{ε_0} must lie in $\mathcal{V}$ and satisfies

$$\partial_y\left[E^{\varepsilon_0}(1 + \partial_y\chi^{\varepsilon_0})\right] = 0, \tag{133}$$

with periodic boundary conditions. We impose $\langle \chi^{\varepsilon_0} \rangle(x) = 0$. A solution in $\mathcal{V}$ to the last equation exists only if E^{ε_0} has been correctly built, i.e., using the general solution (96), $1/E^{\varepsilon_0}$ must lie in $\mathcal{V}$. If this condition is met, $\chi^{\varepsilon_0}(x, y)$ is in $\mathcal{V}$ and

$$\partial_y\chi^{\varepsilon_0}(x, y) = -1 + \left\langle \frac{1}{E^{\varepsilon_0}} \right\rangle^{-1}(x) \ \frac{1}{E^{\varepsilon_0}(x, y)}. \tag{134}$$

As for the periodic case, we find the order 0 constitutive relation

$$\sigma^0(x) = E^{*,\varepsilon_0}(x)\partial_x u^0(x), \tag{135}$$

with

$$E^{*,\varepsilon_0}(x) = \langle E^{\varepsilon_0}(1 + \partial_y \chi^{\varepsilon_0}) \rangle(x), \tag{136}$$

$$= \left\langle \frac{1}{E^{\varepsilon_0}} \right\rangle^{-1}(x). \tag{137}$$

- Eqs. (129) for $i = 0$ and (130) for $i = 1$ give

$$\rho^{\varepsilon_0} \partial_{tt} u^0 - \partial_x \sigma^0 - \partial_y \sigma^1 = f, \tag{138}$$

$$\sigma^1 = E^{\varepsilon_0}(\partial_y u^2 + \partial_x u^1). \tag{139}$$

From (138) we see that the equation that drives σ^1 has the form

$$\partial_y \sigma^1(x, y) = a(x)\rho^{\varepsilon_0}(x, y) + b(x), \tag{140}$$

where $a(x)$ and $b(x)$ are constants as a function of y. Therefore, to obtain σ^1 in $\mathcal{V}$, (138) implies that ρ^{ε_0} must lie in $\mathcal{V}$. Taking the average of (138) together with (135) leads to the order 0 wave equation:

$$\rho^{*,\varepsilon_0} \partial_{tt} u^0 - \partial_x \sigma^0 = f \tag{141}$$

$$\sigma^0 = E^{*,\varepsilon_0} \partial_x u^0, \tag{142}$$

where $\rho^{*,\varepsilon_0} = \langle \rho^{\varepsilon_0} \rangle$.

We stop the development here, but we could compute order 2 correctors similarly to the periodic case. This is done in Capdeville et al. (2010a). Nevertheless, it is found that, in general, some of the order 2 correctors do not belong to $\mathcal{V}$, which implies there is no general nonperiodic solution beyond the order 1, at least in this form. In practice, this is not a strong limitation as only the leading and first order are necessary for most applications. And we will see in the example that the order 2 corrector can still be computed, ignoring the theoretical limitation, and good results can be obtained, at least in that example.

To summarize, we have seen that the nonperiodic homogenization is very similar to the periodic case. The main difference is that the effective properties $(\rho^{*,\varepsilon_0}, E^{*,\varepsilon_0})$ depend on the position x, and the corrector χ^{ε_0} depends on both x and y. We have also found the necessary conditions so that a solution in $\mathcal{V}$ is possible. Constructing the cell properties such that these conditions are fulfilled is the purpose of the next section.

5.6 Construction of the cell properties E^{ε_0} and ρ^{ε_0}

Here, we present two ways of building $E^{\varepsilon_0}(x, y)$ and $\rho^{\varepsilon_0}(x, y)$ in $\mathcal{T}$ with the following constraints obtained in the previous sections:
1. ρ^{ε_0} and χ^{ε_0} must lie in $\mathcal{V}$ (see Eqs. 134 and 140);
2. ρ^{ε_0} and E^{ε_0} must be positive functions;
3. $\rho^{\varepsilon_0}(x, x/\varepsilon_0) = \rho(x)$ and $E^{\varepsilon_0}(x, x/\varepsilon_0) = E(x)$.

The first constraint is necessary to obtain solutions in $\mathcal{V}$, at least up to the order 1.

5.6.1 Direct construction

Let us start with the simplest: the cell density property. The condition is that ρ^{ε_0} must be in $\mathcal{V}$. We propose the following construction, for a given x and any $y \in Y_x$:

$$\rho^{\varepsilon_0}(x, y) = \mathcal{F}^{\varepsilon_0}(\rho)(x) + (\rho - \mathcal{F}^{\varepsilon_0}(\rho))(\varepsilon_0 y), \tag{143}$$

and then extended to $\mathbb{R}$ in y by periodicity. Thanks to the fact that, for any h

$$\mathcal{F}^{\varepsilon_0}(\mathcal{F}^{\varepsilon_0}(h)) = \mathcal{F}^{\varepsilon_0}(h), \tag{144}$$

we have

$$\mathcal{F}^{\varepsilon_0}((\rho - \mathcal{F}^{\varepsilon_0}(\rho))) = \mathcal{F}^{\varepsilon_0}(\rho) - \mathcal{F}^{\varepsilon_0}(\rho)$$
$$= 0. \tag{145}$$

$\mathcal{F}^{\varepsilon_0}(\rho^{\varepsilon_0})$ is therefore a constant in y: it contains only fine-scale variations on y, which is the necessary condition to belong to $\mathcal{V}$. It can be easily checked that the condition (3) is met by construction. Finally, it is always possible to find a filter wavelet w such that ρ^{ε_0} is positive. Note that the property (144) is not completely true in practice because the filter w does not have a sharp cutoff, which implies that (144) is not fully accurate. We consider this side effect as being negligible.

For the elastic parameter, it is in general more difficult. Nevertheless, the 1-D case is interesting because it gives an explicit formula to obtain χ^{ε_0} in $\mathcal{V}$. It implies that $1/E^{\varepsilon_0}$ should be in $\mathcal{V}$ (constraint (1)) so that a solution to the nonperiodic homogenized problem exists (see Eq. 134). Thanks to this explicit constraint, similar to the density construction, we can propose, for a given x and any $y \in Y_x$,

$$E^{\varepsilon_0}(x, y) = \left[\mathcal{F}^{\varepsilon_0}\left(\frac{1}{E}\right)(x) + \left(\frac{1}{E} - \mathcal{F}^{\varepsilon_0}\left(\frac{1}{E}\right)\right)(\varepsilon_0 y) \right]^{-1}, \qquad (146)$$

and then extended to $\mathbb{R}$ in y by periodicity. Similar to the density case, it can easily be checked than $1/E^{\varepsilon_0}$ is in $\mathcal{V}$. This is only true for $1/E^{\varepsilon_0}$ and, for example, E^{ε_0} is not in $\mathcal{V}$.

One can check that $\rho^{\varepsilon_0}(x, x/\varepsilon_0) = \rho(x)$ and $E^{\varepsilon_0} = E(x, x/\varepsilon_0)$ such that the constraint number 3 is fulfilled.

For most standard applications, ρ^{ε_0} and E^{ε_0} are positive functions for any filter wavelet w. Nevertheless, for some extreme cases (e.g., a single discontinuity with several orders of magnitude of elastic modulus contrast), some filter wavelet w designs could lead to a negative E^{ε_0}. In such an extreme case, one should make sure that the design of w allows ρ^{ε_0} and E^{ε_0} to be positive functions.

Finally, we can check that

$$\frac{1}{E^{*,\varepsilon_0}} = \left\langle \frac{1}{E^{\varepsilon_0}} \right\rangle = \mathcal{F}^{\varepsilon_0}\left(\frac{1}{E}\right)(x), \qquad (147)$$

which is the Backus solution obtained in Section 3.3. We also have

$$\rho^{*,\varepsilon_0} = \mathcal{F}^{\varepsilon_0}(\rho)(x) \qquad (148)$$

5.6.2 Implicit construction

For dimensions higher than 1-D, a cell problem, similar to (133), arises (see, for example, Sanchez-Palencia, 1980). Unfortunately, there is no explicit solution to this cell problem that leads to an analytical solution equivalent to (134) (there is one for layered media, but it can be considered as a 1-D case). The direct solution explained above to build E^{ε_0} is therefore not available for higher dimensions (it still is for ρ^{ε_0}). Here, we propose a procedure that gives a similar result to the explicit construction without the knowledge that the construction should be done on $1/E$. The main interest of the procedure is that it can be generalized to higher space dimensions. It is based on the work of Papanicolaou and Varadhan (1979) on the homogenization for random media. They suggest working with the gradients of corrector rather than the corrector directly. If we call

$$G^{\varepsilon_0} = \partial_y \chi^{\varepsilon_0} + 1, \qquad (149)$$

$$H^{\varepsilon_0}(x, y) = E^{\varepsilon_0}(x, y) G^{\varepsilon_0}(x, y), \qquad (150)$$

a solution to our problem in $\mathcal{V}$ up to the order 1 can be found if we can build $E^{\varepsilon_0}(x, y)$ such that $(H^{\varepsilon_0}, G^{\varepsilon_0}) \in \mathcal{V}^2$ and $\langle G^{\varepsilon_0} \rangle = 1$. To do so, we propose the following procedure:

1. Build a start $E_s^{\varepsilon_0}$ defined as, for a given x and for any $y \in Y_x$, $E_s^{\varepsilon_0}(x, y) = E(\varepsilon_0 y)$ and then extended to $\mathbb{R}$ in y by periodicity ($E_s^{\varepsilon_0}$ is therefore in $\mathcal{T}$). Y_x is a $\mathbb{R}$ segment centered on x/ε_0, at least large enough to contain the support of w_1. Then solve (133) with periodic boundary conditions on Y_x to find $\chi_s^{\varepsilon_0}(x, y)$.

2. Compute $G_s^{\varepsilon_0} = \partial_y \chi_s^{\varepsilon_0} + 1$, then $H_s^{\varepsilon_0}(x, y) = E_s^{\varepsilon_0}(x, y) G_s^{\varepsilon_0}(x, y)$ and finally

$$G^{\varepsilon_0}(x, y) = \frac{1}{\mathcal{F}(G_s^{\varepsilon_0})(x, x/\varepsilon_0)} \left(G_s^{\varepsilon_0} - \mathcal{F}(G_s^{\varepsilon_0})\right)(x, y) + 1,$$

$$H^{\varepsilon_0}(x, y) = \frac{1}{\mathcal{F}(G_s^{\varepsilon_0})(x, x/\varepsilon_0)} \left[\left(H_s^{\varepsilon_0} - \mathcal{F}(H_s^{\varepsilon_0})\right)(x, y) \right. \qquad (151)$$

$$\left. + \mathcal{F}(H_s^{\varepsilon_0})(x, x/\varepsilon_0)\right].$$

At this stage, we have, by construction, $(H^{\varepsilon_0}, G^{\varepsilon_0}) \in \mathcal{V}^2$ and $\langle G^{\varepsilon_0} \rangle = 1$.

3. From (150) and (136), we have

$$E^{\varepsilon_0}(x, y) = \frac{H_s^{\varepsilon_0}}{G_s^{\varepsilon_0}}(x, y),$$ (152)

$$E^{*, \varepsilon_0}(x) = \langle H^{\varepsilon_0} \rangle(x) = \frac{\mathcal{F}(H_s^{\varepsilon_0})}{\mathcal{F}(G_s^{\varepsilon_0})}(x, x/\varepsilon_0).$$ (153)

4. Once $E^{\varepsilon_0}(x, y)$ is known, following the homogenization procedure to find the different correctors can be pursued.

Once again, we insist on the fact that the main interest of this procedure is that obtaining an explicit solution to the cell problem is not required and it can be extended to 2-D or 3-D.

Remarks:

- in practical cases, the bar is finite and Y_x can be chosen to enclose the whole bar. In that case, the dependence to the macroscopic location x in $\chi_s^{\varepsilon_0}$, $G_s^{\varepsilon_0}$, $H_s^{\varepsilon_0}$, and $E_s^{\varepsilon_0}$ disappears.
- the step (1) of the implicit construction procedure involves solving (133) with periodic boundary conditions on Y_x. This step implies the use of a finite-element solver on a single large domain (if Y_0 is set as the whole bar) or on a set of smaller domains (Y_x) and this implies that a mesh,

An introduction to the two-scale homogenization method 263

or a set of meshes, of the elastic properties in the Y_x domain must be designed. Therefore, even if the meshing problem for the elastic wave propagation in the order 0 homogenized model is much simpler than for the original model, the problem is still not mesh free. Indeed, fine meshes must still be designed to solve the homogenization problem. Nevertheless, these meshes can be based on tetrahedra even if the wave equation solver is based on hexahedra. Moreover, as the homogenization problem is time independent, the consequences of very small or badly shaped elements on the computing time are limited.

We can check that this procedure gives a correct result on our 1-D case:

1. Taking Y_x as $\mathbb{R}$, the first step allows finding $\partial_y \chi_s^{\varepsilon_0}(y) = \frac{C}{E}(\varepsilon_0 y) - 1$, where $C = \left(\lim_{T \to \infty} \frac{1}{2T} \int_{-T}^{T} \frac{1}{E}(x) dx \right)^{-1}$

2. H^{ε_0} and G^{ε_0} are straight forward to compute from step (i). We have

$$G^{\varepsilon_0}(x, y) = \frac{1}{\mathcal{F}^{\varepsilon_0}\left(\frac{1}{E}\right)}(x)\left(\frac{1}{E} - \mathcal{F}^{\varepsilon_0}\left(\frac{1}{E}\right)\right)(\varepsilon_0 y) + 1 \tag{154}$$

where the fact that, for any h, $\mathcal{F}(h)(x/\varepsilon_0) = \mathcal{F}^{\varepsilon_0}(h)(x)$. We also find $H^{\varepsilon_0}(x, y) = \left(\mathcal{F}^{\varepsilon_0}\left(\frac{1}{E}\right)(x)\right)^{-1}$.

3. The third step allows to find

$$E^{\varepsilon_0}(x, y) = \left(\left(\frac{1}{E} - \mathcal{F}^{\varepsilon_0}\left(\frac{1}{E}\right)\right)(\varepsilon_0 y) + \mathcal{F}^{\varepsilon_0}\left(\frac{1}{E}\right)(x)\right)^{-1} \tag{155}$$

$$E^{*, \varepsilon_0}(x) = \left(\mathcal{F}^{\varepsilon_0}\left(\frac{1}{E}\right)(x)\right)^{-1} \tag{156}$$

which are the desired results.

5.7 Two examples

5.7.1 1-D bar with random properties

In Fig. 14, using the same random bar as for Fig. 5, we present some results obtained using the nonperiodic method developed above. In Fig. 14A, we compare the reference solution with the order 0 nonperiodic homogenized solution and a solution obtained using a naive upscaling (with $E^* = \mathcal{F}^{\varepsilon_0}(E)$, the "E average" solution). Note the strong coda wave trapped in the random model on the left of the ballistic pulse which was not at all present in the periodic case. As expected, the "E average" solution is not in phase with the reference solution and shows that this "natural" filtering is not accurate. On the other hand, the order 0 homogenized solution is already in excellent

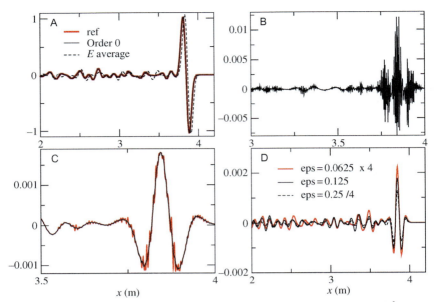

Fig. 14 Results of the numerical example in Section 5.7.1. (A) displacement $u^{ref}(x, t)$ at $t = 4.9 \times 10^{-3}$ s computed in a random bar model (*red*), the order 0 homogenized solution $u^0(x, t)$ (*black*) and the solution computed in a model obtained by averaging the elastic properties ($\mathcal{F}^{\varepsilon_0}(\rho)$ and $\mathcal{F}^{\varepsilon_0}(E)$) (*dashed*). (B) Order 0 residual, $u^0(x, t) - u^{ref}(x, t)$. (C) Order 1 residual (*red*), $u^0(x, t) + u^1(x, t) - u^{ref}(x, t)$. Partial order 2 residual (it is partial because $\langle u^2 \rangle$ is not computed) (*black*). (D) Partial order 2 residual for $\varepsilon_0 = 0.125$ mm (*black*), partial order 2 residual for $\varepsilon_0 = 0.0625$ mm with amplitude multiplied by 4 (*red*) and partial order 2 residual for $\varepsilon_0 = 0.25$ mm with amplitude divided by 4 (*dashed*).

agreement with the reference solution, as we have already seen with the Backus solution. In Fig. 14B the residual between the order 0 homogenized solution and the reference solution $u^0(x, t) - u^{ref}(x, t)$ is shown. The error amplitude reaches 1% and contains fast variations. In Fig. 14C the order 1 residual $(u^0 + u^1)(x, t) - u^{ref}(x, t)$ (bold red line) and the partial order 2 residual (it is partial because the $\langle u^2 \rangle$ is not computed) is shown. The order 2 solution development has not been presented, but it can be computed similarly as the order 1 (Capdeville et al., 2010a). By comparing Fig. 14B and Fig. 14C, it can be seen that the order 1 periodic corrector removes most of the fine-scale variations present in the order 0 residual. The remaining fine-scale variations residual disappear with the partial order 2 residual. The remnant smooth residual is due to the $\langle u^2 \rangle$ that is not computed. To check that this residual is indeed a ε_0^2 residual, the same residual, computed for $\varepsilon_0 = 0.125$ is compared to the partial order 2 residual computed for $\varepsilon_0 = 0.0625$ (multiplying its amplitude by 4) and for $\varepsilon_0 = 0.25$ (dividing its amplitude by 4). It can be seen that these three signals overlap but not completely. This is consistent with a ε_0^2 residual but it shows that the approximations made, mainly the fact that support of the filters w_{ε_0} has been truncated to make their support finite, has some effect on the convergence rate (it is not completely an order 2 convergence).

5.7.2 1-D bar with random properties but no smooth variations

From the 1-D bar with random properties used in the previous section, using the homogenization principle, we can create a heterogeneous bar that behaves as a homogeneous bar. For that, we build a medium (ρ^n, E^n) such that

$$\rho^n(x) = \rho^0 + (\rho - \mathcal{F}^{\varepsilon_0}(\rho))(x), \tag{157}$$

$$E^n(x) = \left[\frac{1}{E^0} + \left(\frac{1}{E} - \mathcal{F}^{\varepsilon_0}\left(\frac{1}{E}\right)\right)(x)\right]^{-1}, \tag{158}$$

where (ρ, E) are the random properties used in the previous example and (ρ^0, E^0) are two constant mechanical properties. In Fig. 15, a snapshot of the wavefield computed in (ρ, E) and in (ρ^n, E^n) is plotted. Even if (ρ^n, E^n) is strongly heterogeneous and nonperiodic, no scattering occurs during the wave propagation and the waves appears to propagate in a homogeneous medium. (ρ^n, E^n) contains only small scales variation and $(\rho^n)^* = \rho^0$ and $(E^n)^* = E^0$.

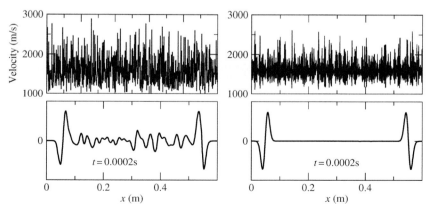

Fig. 15 *Top*: elastic coefficient $E(x)$ along x (*left*) and elastic coefficient $E^n(x)$ along x following (158) (*right*). *Bottom*: displacement $u(x, t)$ snapshot for $t = 2.10^{-4}$s in the heterogeneous bar (ρ, E) (*left*) and displacement $u(x, t)$ snapshot for the same time in the heterogeneous bar (ρ^n, E^n) (*right*).

6. Two-scale homogenization: Higher dimensions

On the one hand, the asymptotic expansion, the equations and how the equations are solved in higher dimensions are very similar to the 1-D case. On the other hand, many things that can be solved simply in 1-D, such as the cell problem, are more complex in 2-D and 3-D. Indeed, as we shall see later, the cell problem has no analytical solution in dimensions higher than 1.

In the following, we distinguish two cases: the elastic and the acoustic case.

6.1 The 2-D and 3-D elastic case

The elastic wave equations are

$$\rho \partial_{tt} \mathbf{u} - \nabla \cdot \boldsymbol{\sigma} = f, \tag{159}$$

$$\boldsymbol{\sigma} = \mathbf{c} : \boldsymbol{\epsilon}(\mathbf{u}), \tag{160}$$

where $\rho(\mathbf{x})$ is the material density and $\mathbf{c}(\mathbf{x})$ the elastic tensor for any $\mathbf{x}$ in a domain Ω with boundary $\partial\Omega$ and $\mathbf{c} : \boldsymbol{\epsilon} = \sum_{kl} c_{ijkl} \epsilon_{kl}$. We use the free boundary condition: $\boldsymbol{\sigma}(\mathbf{x}) \cdot \mathbf{n}(\mathbf{x}) = 0$ for $\mathbf{x} \in \partial\Omega$ and where $\mathbf{n}(\mathbf{x})$ is the outward unit normal vector to $\partial\Omega$ in $\mathbf{x}$.

$\mathbf{c}$, which is a 4th order elastic tensor such that in d–dimensions we have

$$\mathbf{c}(\mathbf{x}) = \{c_{ijkl}(\mathbf{x})\}, \ (i, j, k, l) \in \{1, ..., d\}^4,$$

is positive-definite and satisfies the following symmetries:

$$c_{ijkl} = c_{jikl} = c_{ijlk} = c_{klij}, \tag{161}$$

These symmetries reduce the maximum number of independent parameters necessary to characterize $\mathbf{c}$ to 6 in 2-D and to 21 in 3-D. The fact that $\mathbf{c}$ is positive-definite is a necessary condition to ensure the uniqueness of the solution of the wave equation. Breaking this hypothesis leads to unforeseeable, generally wrong, results. This happens when one tries to model acoustic wave setting $V_S = 0$ in $\mathbf{c}$ within the solid wave equations. It is also an important condition for the homogenization.

6.1.1 Total anisotropy and isotropic projection

In the following, we will need represent some components of general, potentially anisotropic, elastic tensors $\mathbf{c}$. To do so, we choose to first project the general anisotropic $\mathbf{c}$ to the nearest isotropic tensor $\mathbf{c}^{\text{iso}}$ following Browaeys and Chevrot (2004). Then, P and S wave velocities are defined as

$$V_P(\mathbf{x}) = \sqrt{(c_{1111}^{\text{iso}}(\mathbf{x})/\rho(\mathbf{x}))}, \tag{162}$$

and, for the 2-D case,

$$V_S(\mathbf{x}) = \sqrt{(c_{1212}^{\text{iso}}(\mathbf{x})/\rho(\mathbf{x}))}, \tag{163}$$

and the total anisotropy, or the anisotropy index, as

$$\text{aniso}(\mathbf{x}) = \sqrt{\frac{\sum_{ijkl}\left(c_{ijkl}^{\text{iso}}(\mathbf{x}) - c_{ijkl}(\mathbf{x})\right)^2}{\sum_{ijkl}\left(c_{ijkl}^{\text{iso}}(\mathbf{x})\right)^2}}. \tag{164}$$

6.1.2 Asymptotic expansion

We jump directly to the nonperiodic case, skipping the periodic case. The procedure is very similar to the 1-D case and, therefore, we show less details. They can nevertheless be found in Guillot et al. (2010), Capdeville et al. (2010b), and Cupillard and Capdeville (2018).

We first assume that we have been able to define $(\rho^{\varepsilon_0}(\mathbf{x}, \mathbf{y}), \mathbf{c}^{\varepsilon_0}(\mathbf{x}, \mathbf{y}))$ in $\mathcal{T}^2$ with the conditions

$$\begin{aligned} \rho^{\varepsilon_0}(\mathbf{x}, \mathbf{x}/\varepsilon_0) &= \rho(\mathbf{x}), \\ \mathbf{c}^{\varepsilon_0}(\mathbf{x}, \mathbf{x}/\varepsilon_0) &= \mathbf{c}(\mathbf{x}), \end{aligned} \tag{165}$$

and such that solutions to the following development has a solution.

The solution to the wave equations (159) and (160) is then sought as an asymptotic expansion in ε_0 with $\mathbf{u}^i$ and $\boldsymbol{\sigma}^i$ in $\mathcal{V}$:

$$\begin{aligned} \mathbf{u}(\mathbf{x}, t) &= \Sigma_{i=0}^{\infty}\varepsilon_0^i\mathbf{u}^i(\mathbf{x}, \mathbf{x}/\varepsilon_0, t) = \Sigma_{i=0}^{\infty}\varepsilon_0^i\mathbf{u}^i(\mathbf{x}, \mathbf{y}, t), \\ \boldsymbol{\sigma}(\mathbf{x}, t) &= \Sigma_{i=-1}^{\infty}\varepsilon_0^i\boldsymbol{\sigma}^i(\mathbf{x}, \mathbf{x}/\varepsilon_0, t) = \Sigma_{i=-1}^{\infty}\varepsilon_0^i\boldsymbol{\sigma}^i(\mathbf{x}, \mathbf{y}, t). \end{aligned} \tag{166}$$

where the superscript i is a power on ε_0 but not on $\mathbf{u}^i$ and $\boldsymbol{\sigma}^i$. Once again, the condition for $\mathbf{u}^i$ and $\boldsymbol{\sigma}^i$ to be in $\mathcal{V}$ is a strong condition which means that only large variations in $\mathbf{x}$ and fine variations in $\mathbf{y}$ are allowed. It is equivalent to the $\mathbf{y}$ periodic condition in the periodic case. Like in the 1-D case, both $\mathbf{u}^i$ and $\boldsymbol{\sigma}^i$ depend on ε_0. Note that, as already said for the 1-D case, to be fully consistent, we should define two ε (ε and ε_0) as it is done in Capdeville et al. (2010b). We skip that difficulty for the sake of simplicity.

6.1.3 Series of equations

Introducing expansions (166) in the wave equations (159) and (160) and using

$$\nabla \to \nabla_{\mathbf{x}} + \frac{1}{\varepsilon_0}\nabla_{\mathbf{y}}, \tag{167}$$

we obtain:

$$\rho^{\varepsilon_0}\partial_{tt}\mathbf{u}^i - \nabla_{\mathbf{x}} \cdot \boldsymbol{\sigma}^i - \nabla_{\mathbf{y}} \cdot \boldsymbol{\sigma}^{i+1} = f\delta_{i,0}, \tag{168}$$

$$\boldsymbol{\sigma}^i = \mathbf{c}^{\varepsilon_0} : \left(\boldsymbol{\epsilon}_x\left(\mathbf{u}^i\right) + \boldsymbol{\epsilon}_y\left(\mathbf{u}^{i+1}\right)\right). \tag{169}$$

The external source is built as independent of $\mathbf{y}$.

6.1.4 Resolution

Solving (168) and (169) for $i = -2$ and $i = -1$, respectively, makes it possible to show that $\boldsymbol{\sigma}^{-1} = 0$ and that $\mathbf{u}^0 = \langle\mathbf{u}^0\rangle$. Then, Eq. (168) for $i = -1$ and (169) for $i = 0$ give

$$u_i^1(\mathbf{x}, \mathbf{y}) = \chi_i^{\varepsilon_0, kl}(\mathbf{x}, \mathbf{y})\epsilon_{x,kl}^0(\mathbf{x}) + \langle u_i^1\rangle(\mathbf{x}). \tag{170}$$

where $\boldsymbol{\epsilon}_x^0 = \boldsymbol{\epsilon}_x(\mathbf{u}^0)$ and $\boldsymbol{\chi}^{\varepsilon_0}$ is the first-order corrector, a third-order tensor. It is the solution in $\mathcal{V}$ of cell equation

$$\partial_{y_i}H_{ijkl}^{\varepsilon_0} = 0, \tag{171}$$

with

$$H_{ijkl}^{\varepsilon_0} = c_{ijmn}^{\varepsilon_0}G_{mnkl}^{\varepsilon_0}, \tag{172}$$

$$G_{ijkl}^{\varepsilon_0} = \frac{1}{2}\left(\delta_{ik}\delta_{jl} + \delta_{jk}\delta_{il} + \partial_{y_j}\chi_i^{\varepsilon_0, kl} + \partial_{y_i}\chi_j^{\varepsilon_0, kl}\right), \tag{173}$$

where, to enforce the uniqueness of the solution, $\langle\boldsymbol{\chi}^{\varepsilon_0}\rangle = 0$ is imposed. Like in the 1-D case, we find the order 0 constitutive relation:

$$\langle\boldsymbol{\sigma}^0\rangle = \mathbf{c}^{*,\varepsilon_0} : \boldsymbol{\epsilon}_x(\mathbf{u}^0), \tag{174}$$

where the effective elastic tensor is

$$\mathbf{c}^{*,\varepsilon_0}(\mathbf{x}) = \langle\mathbf{H}^{\varepsilon_0}\rangle(\mathbf{x}). \tag{175}$$

To the contrary to the 1-D case, $\boldsymbol{\sigma}^0$, the order 0 stress, depends on $\mathbf{y}$ and therefore, $\langle\boldsymbol{\sigma}^0\rangle \neq \boldsymbol{\sigma}^0$.

Eq. (168) for $i = 0$ gives

$$\rho^{\varepsilon_0}\partial_{tt}\mathbf{u}^0 - \boldsymbol{\nabla}_x \cdot \boldsymbol{\sigma}^0 - \boldsymbol{\nabla}_y \cdot \boldsymbol{\sigma}^1 = f. \tag{176}$$

Taking the cell average of the last equation, using property (3) and taking the f independence on $\mathbf{y}$ into account, we find

$$\rho^{*,\varepsilon_0}\partial_{tt}\mathbf{u}^0 - \boldsymbol{\nabla} \cdot \langle \boldsymbol{\sigma}^0 \rangle = f, \tag{177}$$

where $\rho^{*,\varepsilon_0} = \langle \rho^{\varepsilon_0} \rangle$. The last equation together with the order 0 constitutive relation (174) are the order 0 effective wave equations. They form a classical elastic wave equation for the effective elastic model $(\rho^{*,\varepsilon_0}, \mathbf{c}^{*,\varepsilon_0})$.

6.1.5 Construction of the effective elastic tensor and density

To build the two-scale elastic tensor $\mathbf{c}(\mathbf{x}, \mathbf{y})$ and the effective elastic tensor $\mathbf{c}^{*,\varepsilon_0}$, we follow the procedure described in Section 5.6.2.

Step 1. For a given elastic model defined by its elastic tensor $\mathbf{c}(\mathbf{x})$, we solve the following set of problems (6 in 3-D), called the cell problem, to find the initial guess for the corrector χ_s^{kl}:

$$\boldsymbol{\nabla} \cdot \boldsymbol{\sigma}^{kl} = -\boldsymbol{\nabla} \cdot (\mathbf{c} : (\hat{\mathbf{k}} \otimes \hat{\mathbf{l}})) \tag{178}$$

$$\boldsymbol{\sigma}^{kl} = \frac{1}{2}\mathbf{c} : \left(\boldsymbol{\nabla}\chi_s^{kl} + {}^t\boldsymbol{\nabla}\chi_s^{kl} \right) \tag{179}$$

with periodic boundary conditions and $\hat{\mathbf{k}}$ and $\hat{\mathbf{l}}$ being the space basis unit vectors.

This is a simple static elastic equation with a set of loading. Currently, we have two solvers, both in 2-D and 3-D, to solve them:

- Based on a finite-element method. This method requires a finite-element mesh based on tetrahedron but is accurate. It is the approach used in Capdeville et al. (2010b) and Cupillard and Capdeville (2018).
- Based on fast Fourier transform (FFT) method (Michel, Moulinec, & Suquet, 1999; Moulinec & Suquet, 1998). This method is easy to implement and does not need a mesh. It may present some limitations in the case of discontinuous media with very strong contrast, but experience shows that it gives good results, even in that case (Capdeville et al., 2015).

Step 2. Once the initial corrector χ_s^{kl} has been obtained,

- We compute the following tensors:

$$\begin{aligned} \mathbf{G}_s(\mathbf{x}) &= \mathbf{I} + \frac{1}{2}\left(\boldsymbol{\nabla}\chi_s + {}^T\boldsymbol{\nabla}\chi_s \right) \\ \mathbf{H}_s(\mathbf{x}) &= \mathbf{c} : \mathbf{G}_s \end{aligned} \tag{180}$$

- The ε_0 effective tensor can be computed as

$$\mathbf{c}^{*,\varepsilon_0}(\mathbf{x}) = \mathcal{F}^{\varepsilon_0}(\mathbf{H}_s) : \mathcal{F}^{\varepsilon_0}(\mathbf{G}_s)^{-1}(\mathbf{x}) \tag{181}$$

- The effective density is then simply:

$$\rho^{*,\varepsilon_0} = \mathcal{F}^{\varepsilon_0}(\rho)$$

- Finally, the two-scale tensor $\mathbf{G}^{\varepsilon_0}(\mathbf{x}, \mathbf{x}/\varepsilon_0)$, $\mathbf{H}^{\varepsilon_0}(\mathbf{x}, \mathbf{x}/\varepsilon_0)$ and $\boldsymbol{\chi}^{\varepsilon_0}(\mathbf{x}, \mathbf{x}/\varepsilon_0)$ can also be built.

Step 3.

- The solution $\mathbf{u}^0$ can be obtained by solving the effective wave equation:

$$\rho^{\varepsilon_0} \cdot \partial_{tt} \mathbf{u}^0 - \boldsymbol{\nabla} \cdot \langle \boldsymbol{\sigma}^0 \rangle = f \tag{182}$$

$$\langle \boldsymbol{\sigma}^0 \rangle = \mathbf{c}^{*,\varepsilon_0} : \boldsymbol{\epsilon}(\mathbf{u}^0) \tag{183}$$

- Once the order 0 solution is obtained, we can access to the order 0 stress and deformation with:

$$\boldsymbol{\sigma}^0(\mathbf{x}, \mathbf{x}/\varepsilon_0) = \mathbf{H}^{\varepsilon_0}(\mathbf{x}, \mathbf{x}/\varepsilon_0) : \boldsymbol{\epsilon}(\mathbf{u}^0)(\mathbf{x}) \tag{184}$$

$$\boldsymbol{\epsilon}^0(\mathbf{x}, \mathbf{x}/\varepsilon_0) = \mathbf{G}^{\varepsilon_0}(\mathbf{x}, \mathbf{x}/\varepsilon_0) : \boldsymbol{\epsilon}(\mathbf{u}^0)(\mathbf{x}) \tag{185}$$

- The partial order 1 displacement is then obtained as:

$$\mathbf{u}(\mathbf{x}) = \mathbf{u}^0(\mathbf{x}) + \varepsilon_0 \boldsymbol{\chi}^{\varepsilon_0}(\mathbf{x}, \mathbf{x}/\varepsilon_0) : \boldsymbol{\epsilon}(\mathbf{u}^0)(\mathbf{x}) + O(\varepsilon_0). \tag{186}$$

In practice, the last formula often converges as $O(\varepsilon_0^2)$ because $\langle \mathbf{u}^1 \rangle$ is small.

Remark: In the periodic case, the major symmetry of the effective elastic tensor can be demonstrated (e.g., Sanchez-Palencia, 1980 or section 2.4 of Pavliotis & Campus, 2004). However, this is not currently the case in the nonperiodic case: (181) does not warranty the symmetry of $\mathbf{c}^{*,\varepsilon_0}$. In most cases, $\mathbf{c}^{*,\varepsilon_0}$ is very close to being symmetric, but some cases can lead to a significant nonsymmetry, which contradicts the energy conservation of the wave equation. There is currently no solution to this problem and, in practice, we just use the symmetric part of $\mathbf{c}^{*,\varepsilon_0}$ without tampering the convergence and the accuracy of the method.

6.2 The 2-D and 3-D acoustic case

The acoustic case is interesting because it leads to a nonintuitive concept: the density anisotropy. The full description of this case can be found in Cance and Capdeville (2015).

The primary variable of the acoustic wave equation often is the pressure p. Here, we use the velocity potential q, defined by $p = \dot{q}$, so that the acoustic wave equation in a domain Ω is

$$\frac{1}{\kappa} \partial_{tt} q - \boldsymbol{\nabla} \cdot \dot{\mathbf{u}} = \dot{g}, \tag{187}$$

$$\dot{\mathbf{u}} = \frac{1}{\rho} \boldsymbol{\nabla} q. \tag{188}$$

The natural boundary condition on $\partial\Omega$ is $q = 0$, that is zero pressure.

In the following, it is useful to define the inverse density tensor,

$$\mathbf{L} = \frac{1}{\rho} \mathbf{I}, \tag{189}$$

where $\mathbf{I}$ is the identity tensor. (188) can then be rewritten as

$$\dot{\mathbf{u}} = \mathbf{L} \cdot \boldsymbol{\nabla} q. \tag{190}$$

Following the elastic case, we can use an asymptotic expansion for q and $\mathbf{u}$:

$$q(\mathbf{x}, t) = \sum_{i=0}^{\infty} \varepsilon_0^i q^i(\mathbf{x}, \mathbf{y}, t), \tag{191}$$

$$\mathbf{u}(\mathbf{x}, t) = \sum_{i=0}^{\infty} \varepsilon_0^i \mathbf{u}^i(\mathbf{x}, \mathbf{y}, t), \tag{192}$$

still with $\mathbf{y} = \mathbf{x}/\varepsilon_0$. The series of equations to be solved can be obtained in a very similar way to the elastic case. Once solved, we obtain that, to the leading order, the effective velocity potential q^* and the effective displace $\mathbf{u}^*$ are solution of the following effective acoustic wave equations:

$$\frac{1}{\kappa^*} \partial_{tt} q^* - \boldsymbol{\nabla} \cdot \dot{\mathbf{u}}^* = \dot{g}^* \tag{193}$$

$$\dot{\mathbf{u}}^* = \mathbf{L}^{*,\varepsilon_0} \cdot \boldsymbol{\nabla} q^* \tag{194}$$

where

$$q^* = q^0, \tag{195}$$

$$\mathbf{u}^* = \langle \mathbf{u}^0 \rangle. \tag{196}$$

The effective bulk modulus is obtained similar to effective density of the elastic case:

$$\frac{1}{\kappa^*} = \mathcal{F}^{\varepsilon_0} \left(\frac{1}{\kappa} \right). \tag{197}$$

The procedure for effective inverse density tensor is the same as the elastic tensor

$$\mathbf{L}^{*,\varepsilon_0} = \mathcal{F}^{\varepsilon_0}(\mathbf{P}_s) \cdot \mathcal{F}^{\varepsilon_0}(\mathbf{Q}_s), \tag{198}$$

where $\mathbf{P}_s$ and $\mathbf{Q}_s$ are obtained by first solving the cell problem to obtain the starting corrector χ_s:

$$\nabla \cdot \mathbf{u}^i = -\nabla \cdot (\mathbf{L} \cdot \hat{\mathbf{i}}) \tag{199}$$

$$\mathbf{u}^i = \mathbf{L} \cdot \nabla \chi_s^i, \tag{200}$$

with periodic boundary conditions. Then, $\mathbf{P}_s$ and $\mathbf{Q}_s$ are built as:

$$\mathbf{Q}_s = \mathbf{I} + \nabla \chi_s, \tag{201}$$

$$\mathbf{P}_s = \mathbf{L} \cdot \mathbf{Q}_s. \tag{202}$$

As it can be seen, the principle and process is very similar to the elastic case. Nevertheless, it leads to interesting differences:

- the anisotropy is carried by the inverse density tensor and not the elastic tensor. It implies that the effective density is anisotropic even though the fine-scale density is isotropic. As a result, the apparent density depends on the direction of the wave propagation, which is unusual. For a finely layered horizontal medium, we have

$$\mathbf{L}^{*,\varepsilon_0} = \begin{pmatrix} \mathcal{F}^{\varepsilon_0}\left(\dfrac{1}{\rho}\right) & 0 \\ 0 & \dfrac{1}{\mathcal{F}^{\varepsilon_0}(\rho)} \end{pmatrix} \tag{203}$$

which implies a simple but real anisotropy;

- the physical acoustic anisotropy (in contrast to the nonphysical anisotropy introduced in exploration geophysics to mimic elastic anisotropy in acoustic media (Alkhalifah, 2000)) is different from the elastic anisotropy. It is always purely elliptic (the wavefront is elliptic), whereas the simplest elastic anisotropy (vertically transverse anisotropy, VTI) is rather complex (the wavefront has a diamond shape). See Fig. 16 and Cance and Capdeville (2015) for more details.

- The order 0 displacement behaves similarly to the elastic stress and depends on the fine-scale $\mathbf{y}$:

$$\mathbf{u}^0(\mathbf{x}, \mathbf{y}) = (\mathbf{I} + \nabla \chi(\mathbf{x}, \mathbf{y})) \cdot \nabla q^*(\mathbf{x}). \tag{204}$$

It can be seen that, to the leading order, the acoustic displacement is sensitive to small structures unlike in the elastic case. It implies that the acoustic displacement is strongly affected by small scale heterogeneities similar to strain in the elastic case.

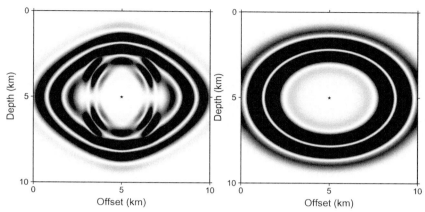

Fig. 16 Energy snapshot of a 2-D wave propagation for an isotropic source (an explosion) at the center of each plots in an elastic VTI medium (*left*) and in an anisotropic acoustic media with similar main axis velocities (*right*).

From the point of view of small scales, the acoustic and the elastic cases are physically different despite being mathematically very similar. These differences can be important: for example, many full waveform inversion tests are performed in the acoustic case, using a spatially constant density and inverting for the acoustic velocity (or κ). This case is a poor proxy of the elastic case as it is similar to only invert for the density in the elastic case. Based on homogenization, we see that it is important to allow density variations so that such tests make sense.

7. What we skipped

Some important aspects of the homogenization theory have not been addressed in the present article.

7.1 Spatial filters

In this paper, we have used a single type of spatial filter (Fig. 1). Many other types of filters could be used, depending on applications, and we have not discussed this aspect so far.

The choice of the filter is always a compromise to address as best as possible the following points:
1. the filter spectrum must be as close as possible to 1 in $[0, k_{max}]$;
2. the filter spectrum must go down to zero as quickly as possible after k_{max};
3. in the time domain, the filter must be as compact as possible;
4. the effective medium must be defined.

Only a strictly positive filter wavelet can make sure point 4 is achieved for all cases. This is the choice of Backus (1962) using a Gaussian filter. Even if for some very strongly contracted media a Gaussian filter might be the only option available, we try to avoid this possibility because it makes point 1 and 2 more difficult to achieve. Most often, we use the filter presented in Fig. 1 because of its flat spectrum norm between the 0 spatial frequency and $1/\lambda_0$, which is optimal to leave the medium intact in the desired spatial frequency band. Nevertheless, for strongly contrasted media, its negative lobes in the space domain may break point 4. In such cases, it can be necessary to use a positive filter, such as a Gaussian filter.

7.2 Boundary conditions

Domain boundaries are always a difficulty for the homogenization theory because the periodicity of the cell problem makes it impossible to introduce them in a simple way. A classical solution to tackle this difficulty is the matched asymptotic approach, mainly developed for the static periodic case (David, Marigo, & Pideri, 2012; Dumontet, 1990; Marigo & Pideri, 2011; Nevard & Keller, 1997; Sanchez-Palencia, 1987). The idea is to use two asymptotic expansions: one in the volume (let us say ($\mathbf{u}^i$, $\boldsymbol{\sigma}^i$)) and one specifically for the boundary (let us say ($\mathbf{v}^i$, $\boldsymbol{\tau}^i$)), and to match them. The area near the boundary where ($\mathbf{v}^i$, $\boldsymbol{\tau}^i$) is valid is the boundary layer. Fig. 17 displays a sketch of the situation with a fine-scale topography. The solution in the

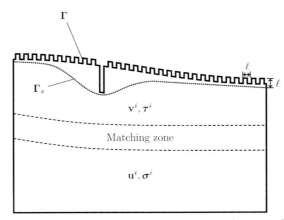

Fig. 17 Sketch of the matched asymptotic approach. Γ is the fine-scale free surface; Γ_s is the effective free surface; ($\mathbf{v}^i$, $\boldsymbol{\tau}^i$) is the asymptotic expansion valid close to the free surface; ($\mathbf{u}^i$, $\boldsymbol{\sigma}^i$) is the asymptotic expansion valid away from the free surface. In this sketch, the topography presents some local periodicity, but it is not necessarily the case as shown by Capdeville and Marigo (2013).

boundary layer and the matching conditions provide the boundary condition for the effective volumetric solution, which, to the order 1, is in general a Dirichlet-to-Neumann condition:

$$\boldsymbol{\sigma}^0(\mathbf{x}) \cdot \mathbf{n} = \varepsilon_0\big(h(\mathbf{x})\partial_{tt}\mathbf{u}^0(\mathbf{x}) + \alpha(\mathbf{x})\boldsymbol{\nabla}_\mathbf{p} \cdot \boldsymbol{\sigma}^0 \cdot \mathbf{p}\big) \text{ for } \mathbf{x} \in \boldsymbol{\Gamma}_s, \tag{205}$$

where $\boldsymbol{\Gamma}_s$ is the effective topography, $\mathbf{n}$ the outward unit vector normal to $\boldsymbol{\Gamma}_s$, $\mathbf{p}$ the unit vector tangential to $\boldsymbol{\Gamma}_s$, and h and α two smooth coefficients which depend on the fine-scale topography and on the fine-scale heterogeneities near the free surface. These two coefficients are obtained thanks to the matched conditions. If there is no fine-scale topography and just fine-scale heterogeneities below a smooth free surface, then $\boldsymbol{\Gamma}_s = \boldsymbol{\Gamma}$.

This method has been applied to the dynamic nonperiodic case for layered media (Capdeville & Marigo, 2008) and for fine-scale topography (Capdeville & Marigo, 2013). For fine-scale topography, it appears that the effective topography is not the average topography but rather a smooth lower envelop of the fine-scale topography, as sketched in Fig. 17. A casual explanation for this is that surface waves propagate below the fine-scale topography, and the only effect of the fine-scale topography is to weigh on the effective topography (through the coefficient h in (205)). Note that a solution exists to make the boundary condition valid up to the order 1 without changing it if lateral variations below the free surface are smooth (Capdeville & Marigo, 2007). Finally, the case of fine-scale heterogeneity below a smooth or a fine-scale topography remains to be studied, but it is expected to be similar to the fine-scale topography case alone.

7.3 Other aspects to address

Many other aspects have not been mentioned or studied. Among them, the fact that λ_{min} can strongly vary with location in $\boldsymbol{\Omega}$, would need to be studied. For example, in most of the kilometer scale geological models, the S wave speed can vary by a factor 10 from the top to the bottom, implying a factor 10 change of λ_{min} through the domain. Knowing that λ_0 is fixed, the ε_0, and therefore the quality of the solution, changes with depth. This is a problem, and an "adaptive" homogenization, making ε_0 constant through the domain, is needed. A solution valid for layered media exists and is based on a normal mode solution decomposition, a kind of adaptive filtering (see Capdeville et al., 2013, appendix B). It needs to be extended to general media and this aspect will be addressed in the near future.

Another aspect that would need a careful study is the solid–fluid coupling when both the solid and fluid domains are larger than λ_{min}.

8. Examples of applications

In this section, we show a series of applications of the two-scale non-periodic homogenization, one in the context of inversion and the rest in the context of forward modeling. Most of them are 2-D examples and one of them is in 3-D (Section 8.4). The latter is a typical example of what can be expected from the method to simplify the forward modeling and make it much faster. More examples can be found in the two-scale nonperiodic homogenization bibliography already cited.

8.1 2-D refracted waves

One of the concerns often expressed about homogenization is its capacity to model back-scattered or refracted waves. It is not really clear why such a concern exists because there is nothing specific about those waves: the convergence of homogenization is true for the wavefield as a whole and it should work equally well for all waves within the bounds of the method hypothesis. Perhaps the concern about back-scattered waves is linked to the fact that it is more difficult to obtain reflected waves than transmitted waves in smooth media. Similarly, the concern about refracted waves may be linked to the belief that a sharp contrast is important because such waves propagate along the interface for a long time. In any case, we show in the following example that both back-scattered and refracted waves can be accurately modeled in the homogenized models.

To verify that the reflected and refracted waves are correctly modeled in the homogenized model, we present two tests in a simple 2-D two layer elastic setting with a free surface on the top, as presented in Fig. 18A. The elastic properties are given in Table 1. The vertical transition between the two layers is not a simple step, but something a bit more complex. In model a, the elastic properties alternate vertically between the upper and the lower layer properties for 65 m (Fig. 18B). In model b, at a height of 65 m, the elastic properties alternate horizontally between the upper and the lower layer (Fig. 18C). We compute the reference solutions using SEM, meshing

Table 1 Elastic properties for the refracted wave tests (see Fig. 18).

	V_P (km s^{-1})	V_S (km s^{-1})	ρ (kg m^{-3})
Upper layer	2.4	1.2	1500
Lower layer	5.6	2.8	2800

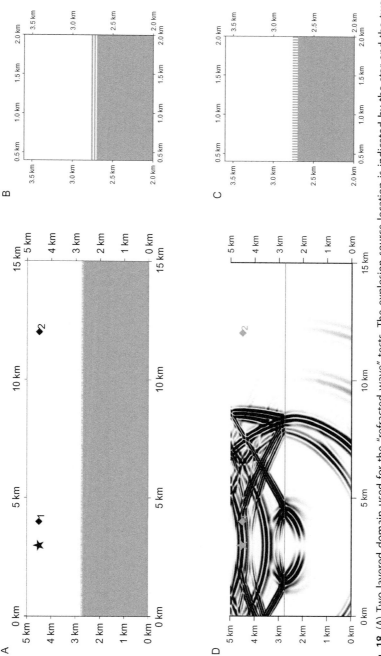

Fig. 18 (A) Two layered domain used for the "refracted wave" tests. The explosion source location is indicated by the star and the two receivers locations by the diamonds. The gray background corresponds to the elastic properties in the "lower layer" in Table 1 and the white background to properties in the "upper layer." (B) zoom on the transition area between the lower and upper layer for the "model a." (C) same as B but for "model b." (D) Energy snapshot in model a at $t = 2.6s$.

all the interfaces and using perfectly matched layers (PML, Festa & Vilotte, 2005) as absorbing boundaries around the domain. The wavefield is generated by an explosion located 500 m below the free surface using a Ricker source time wavelet with a central frequency of 5 Hz and a maximum frequency of 15 Hz. An energy snapshot of the wavefield at $t = 2.6$ s is displayed in Fig. 18D for model a. Because of the very fine elements imposed by the heterogeneities, the numerical cost of the simulations is about 100 times more expensive than a similar simulation without the fine structures.

Once the reference solutions are obtained, we compute the two homogenized models with $\varepsilon_0 = 0.6$ where we use the smallest V_S velocity to evaluate $\lambda_{\min}$ (note that we obtain $\varepsilon_0 = 0.25$ if we use the lower layer V_S value instead). Vertical cross section in the homogenized models are shown in Fig. 19.

The "model a" is purely layered and the Backus homogenization is enough for such a model. The "model b" is not layered and we rely on the two-scale homogenization method presented in the work, using the finite-element version of our homogenization tool. The effective model cross sections show smooth elastic models with a significant anisotropy localized around the interface for both cases.

Vertical velocity traces for the two receivers and for both models are plotted in Fig. 20. The traces computed in a model obtained by low-pass filtering

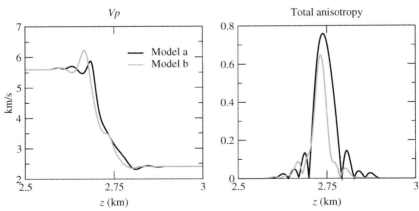

Fig. 19 Vertical cross section in the homogenized effective from "model a" (*black*) and "model b" (*gray*). The isotropic projection, as defined in Section 6.1, of the effective elastic tensor corresponding to P wave velocity (*left panel*) and the total anisotropy (*right panel*) are shown.

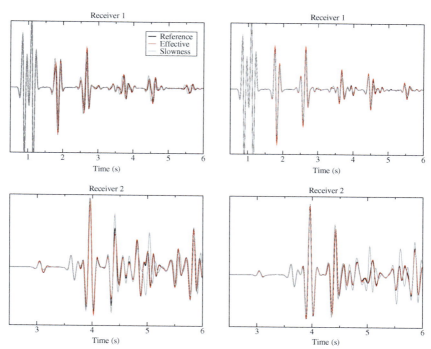

Fig. 20 Vertical velocity traces computed in "model a" (*top panels*) and "model b" (*bottom panels*) for receivers 1 (*left panels*) and receivers 2 (*right panels*). The reference traces are displayed (*black*) as well as the homogenized effective solution (*red*) and the slowness averaged solution (*gray*).

the slowness and the density using the same ε_0 as for the homogenization are also displayed. These traces are plotted to show what happens if one decides to approximate the true model by its low-pass filtered version. These naive up-scaled models are smooth but purely isotropic. Receiver 1 is close to the source and mainly records reflected waves. Receiver 2 has a larger offset and records also many refracted waves. While the "slowness average" traces show large time shifts with respect to the reference solution with increasing time, the homogenized effective traces show a better accuracy for the whole time window. If a better accuracy is needed, it is enough to lower ε_0 and the error will decrease as ε_0^2 in the case of homogenization but will converge poorly in the slowness average case.

To conclude this example, we can say that homogenized effective models can model back-scattered and refracted waves with no difficulty whatsoever.

8.2 A solid with a fluid inclusion

Media with both solid and fluid parts are common in geophysics. At very large scales, the outer core or the oceans are fluid, and at local scales, fluid regions can be, for example, gas reservoirs. The classical solution to compute the wavefield in such cases is to separate the fluid and solid into two distinct domains, to use the appropriate set of equations in each of them and to couple them. As usual, it creates meshing difficulty and it would be very interesting to be able to homogenize both domains at once. Unfortunately, this is not possible because one of the mathematical conditions to solve the cell problem is that the elastic or acoustic tensor must be positive–definite. This implies that setting V_S to zero in a solid to make it fluid is, in general, not possible and so is the option to mix solid and fluid in a single domain.

Even if it is in principle not possible, we will show in this section that we can homogenize a solid with fluid inclusions if the inclusions are small compared to λ_{min}. As a test example, we use here single elliptic inclusion in an homogeneous media (see Fig. 21 for the geometry and Table 2 for the mechanical properties of the domain). The effective medium can be computed using finite element and domain decomposition, with the acoustic equations in the fluid domain and the elastic equation in the solid. Here we have used very small value for V_S in the fluid. Such a simplification is, in general, a bad idea but, as we will see it below, as long as the inclusion remains small, it gives a good result. An example of the obtained effective media is shown in Fig. 22. It is fully anisotropic and, to represent it, we follow the quantities defined in Section 6.1.

For the ellipse presented in Fig. 21, we computed three reference solutions, using three different source wavelet maximum frequencies, such that the minimum wavelengths in the background medium are $\lambda_{min} = 10$, 5 and 2.5 m. We computed the 3 effective media for each λ_{min}, with $\varepsilon_0 = 0.5$. Then, for each of these effective media, we computed the effective wave equation solutions and compared them with the respective reference solutions as shown in Fig. 23 for receiver 4 (which records the back–scattered wavefield from the ellipse inclusion). It can be seen that the effective solution is of good quality when the size of the inclusion is very small compared to λ_{min}(top panel in Fig. 23), but is of poorer quality when it is not that small (bottom panel in Fig. 23). Although it depends on the desired accuracy, we can say that the effective solution is of good quality if the inclusion is smaller than $\lambda_{min}/3$. Note that the effective solution is obtained using a very sparse mesh (a sample of the mesh is shown in Fig. 22, left plot) compared to the reference solution.

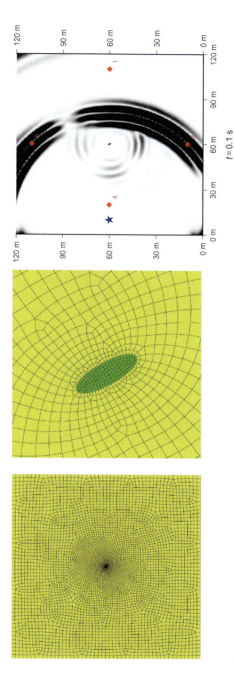

Fig. 21 *Left*: quadrangular mesh of the domain (120 × 120 m²). The *gray lines* represent the mesh used to model the wave propagation in the effective medium. *Middle*: zoom on the elliptic fluid inclusion mesh (the long axis is 1.5 mm and the short axis is 0.5 mm long). *Right*: energy snapshot for a moment tensor source (*blue star*) at $t = 0.1$ s.

Table 2 Elastic properties for the fluid ellipse inclusion test (see Fig. 21).

	V_P (km s^{-1})	V_S (km s^{-1})	ρ (kg m^{-3})
Background	1.8	1	2000
Ellipse	1	0	1000

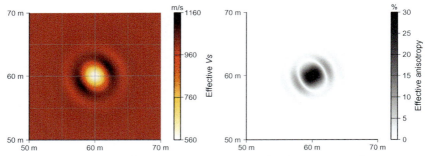

Fig. 22 Zoom in on the effective medium computed with $\lambda_0 = 5$m. *Left*: Isotropic V_S velocity. *Right*: total anisotropy (see Section 6.1 for the "total anisotropy" and isotropic V_P definitions).

Consequently, the effective solution is obtained more than 100 times faster than the reference solution. This example is interesting but it is clear that this option will fail if one tries to go beyond the small inclusion case, for example, if the small inclusions are connected. The other way (solid inclusions into a fluid) works similarly: it can be homogenized as a single fluid as long as S waves cannot develop in the solid inclusion, that is as long as they are small compared to $\lambda_{\min}$. Finally, note that void inclusions can be also treated the same way.

8.3 A failing case: Helmholtz resonators

Helmholtz resonators are open cavities that can resonate at a surprisingly low frequency compared to the fundamental eigenfrequency of a similarly closed cavity. In Fig. 24 a numerical example in 2-D is shown: it is a simple open ring made of aluminum immersed in air (see the SEM mesh in the left and middle panels in Fig. 24). If a pressure source is placed a few wavelengths away after the ballistic wave has passed, the Helmholtz resonator continues ringing for a long time as it can be seen on the energy snapshot in the right plot of Fig. 24 and in the pressure trace for receiver 3 in the left plot of

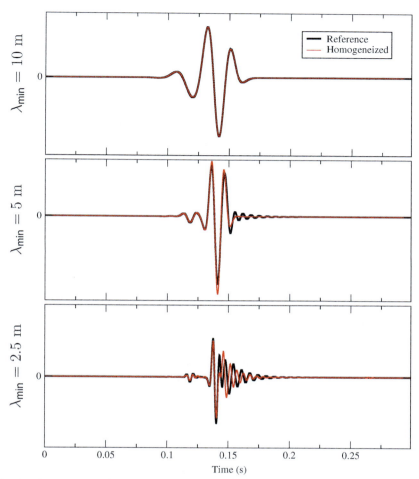

Fig. 23 Comparison of vertical velocity traces between the reference solution (*black*) and the effective solution (*red*) for receiver 4, which is at a back-scattering location. $\varepsilon_0 = 0.5$ and is fixed. *Top*: the source maximum frequency is such $\lambda_{min} = 10$ m; *middle*: $\lambda_{min} = 5$ m; *bottom*: $\lambda_{min} = 2.5$ m.

Fig. 25. This acoustic resonance is produced by a process analogous to the oscillation of a mass–spring oscillator. The resonator must contain a neck connected to the cavity filled with a large volume of fluid. The large volume of fluid acts as the spring, while the neck acts as the mass. For the specific shape of the Helmholtz resonator example used here (see Fig. 24), the natural frequency of a 2-D Helmholtz resonator as given by Mechel (2013) is approximately

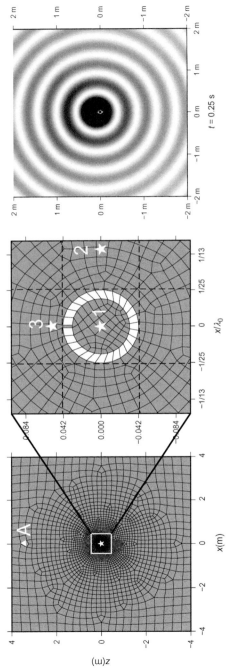

Fig. 24 Homogeneous acoustic domain with an Helmholtz ring inclusion at the center. *Left*: spectral-element mesh of the domain. *Middle*: zoom in on the Helmholtz resonator part. The λ_0 scaling factor for the horizontal axis corresponds to the wavefield dominant wavelength (about $3\lambda_{min}$). *Right*: energy snapshot for $t = 0.25$s. Even if the source is in A, the energy clearly rings from the Helmholtz resonator after some time.

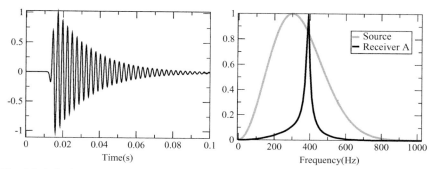

Fig. 25 *Left*: Pressure signal recorded for receiver 3 and source A (see Fig. 24). *Right*: Source wavelet and the amplitude spectra of receiver 3.

$$f_H \simeq \frac{V_P}{2\pi}\sqrt{\frac{d}{\pi r^2 l}} \qquad (206)$$

where l is the length of the neck, r is the diameter of the ring, d is the width of the neck. In the example used here, $f_H \simeq 383$ Hz, which is much lower than the fundamental resonance frequency of an equivalent closed cavity ($\simeq 4$ kHz).

This unusual resonance is a remarkable property for such a simple linear elastic system. It can be shown that, when reaching the resonance frequency, the medium dispersion relation becomes singular (Lemoult, Kaina, Fink, & Lerosey, 2013) and the wavefield no longer exhibits any minimum wavelength. This breaks the main hypothesis of homogenization ($\lambda_{\min}$ is undefined in that case) and makes any interpretation of the obtained effective medium beyond that frequency difficult and probably meaningless. Indeed, the effective $\mathbf{L}^{*,\varepsilon_0}$ becomes locally negative and cannot be used anymore. In Fig. 26, we show cross sections in the quantity V_h, the sound velocity in an isotropic fluid, $V_h = \sqrt{\kappa^{*,\varepsilon_0} L_{22}^{*,\varepsilon_0}}$ for different ε_0. To compute ε_0, we used the background air $\lambda_{\min}$. When ε_0 is such that the Helmholtz resonance wavelength is filtered out by the homogenization filter, i.e., for large ε_0, the effective medium can be computed. However, the resonance cannot be obtained in the effective medium in that case. Lowering ε_0, $\mathbf{L}^{*,\varepsilon_0}$ becomes locally negative and the effective medium cannot be obtained anymore.

8.4 A 3-D geological model

Numerical simulation of seismic wave propagation in 3-D geological media can be extremely challenging. First, meshing geological structures usually

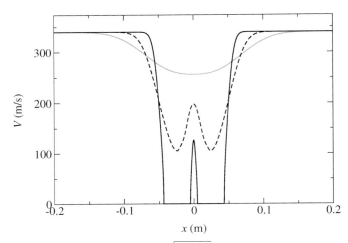

Fig. 26 Horizontal cross section in $V_h = \sqrt{\kappa^{\varepsilon_0 *} L_{22}^{*,\varepsilon_0}}$ for $y = 0.02$ m (at the Helmholtz resonator neck level) and for three values of ε_0 (computed with the background velocity). $\varepsilon_0 = 0.5$ (*solid gray*), $\varepsilon_0 = 0.25$ (*dashed*) and $\varepsilon_0 = 0.125$ (*solid*) are plotted. For, $\varepsilon_0 = 0.125$, the zero values correspond to places where L_{22}^{*,ε_0} is negative and V_h cannot be computed.

requires huge efforts for both seismologists and computers: seismologists have to make sure that the geological structures in study (e.g., horizons, faults, intrusions, cracks) verify all the geological rules (e.g., Caumon, Collon-Drouaillet, de Veslud, Viseur, & Sausse, 2009; Wellmann & Caumon, 2018) and computers then run meshing algorithms which can possibly requires considerable convergence time or even fail in providing a result, especially when dealing with hexahedra. Second, the obtained mesh can contain extremely small elements because of the complexity of the geological structures to be honored and/or the inefficiency of the meshing algorithm, yielding to gigantic, sometimes prohibitive, computational cost for just a single wavefield simulation. In this section, we illustrate these issues and show how the homogenization enables to overcome most of them. To do so, we use the SEG-EAGE overthrust model (Fig. 27A) as a case study.

The SEG-EAGE overthrust model is 20 km × 20 km × 4 km large. It is made of twelve faulted and folded layers. All of them are isotropic. The P-wave velocity ranges from 2 500 ms^{-1} to 6 000 ms^{-1}; the S-wave velocity ranges from 1 600 ms^{-1} to 3 500 ms^{-1} (Fig. 27A). To perform wave propagation simulations in such a medium, the finite-difference method can be considered at the price of using an extremely small space-step to capture the effect of the discontinuities. Finite-element methods are more suitable for

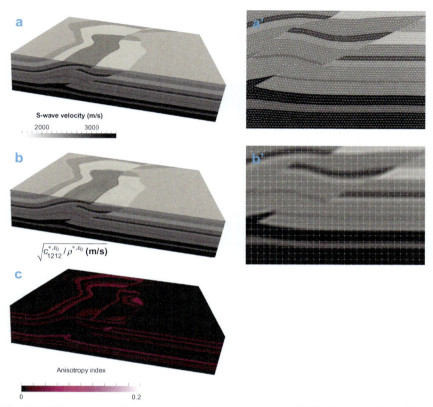

Fig. 27 (A) S-wave velocity structure of the overthrust model. (A'): Zoom in a lateral border of the model meshed with tetrahedra. (B) The homogenized version of the model; the plotted quantity is the SH-wave velocity. (B'): Zoom in a lateral border of the homogenized model meshed with hexahedra. (C) Total anisotropy of the homogenized overthurst.

taking discontinuities into account. This kind of methods relies on meshing the medium with simple shapes like tetrahedra, hexahedra, pyramids, prisms, or a mix of those. Because tetrahedra enable the maximum flexibility, we use them to mesh the overthrust (Fig. 27A') and we rely on a mass–lumped finite-element method (Geevers, Mulder, & van der Vegt, 2019) to compute seismic wave propagation. After an important work of building the best possible mesh, using various surface and volume meshing algorithms in a row (Kononov, Minisini, Zhebel, & Mulder, 2012; Pellerin, Lévy, Caumon, & Botella, 2014; Si, 2015) to benefit from the relevant features of each of them, we obtain 75.5 cm for the inner-sphere radius of the smallest element. Such a small value leads to a small time step within the wave propagation simulation, for numerical stability reasons. Computing a 12s long wavefield (hereafter denoted by $\mathbf{u}^{\mathrm{ref}}$) for a single $f_0 = 3.125$ Hz Ricker source therefore requires 12.6 days on 40 Skylake cores.

Let us now compute the effective overthrust medium. To do so, we use a tetrahedral finite-element method to solve the cell problem (178) and (179). This problem is static, so there is no time step involved and small elements in the mesh do not lead to computational limitations. The main difficulty when solving the cell problem is to handle the large memory requirement: degree-3 polynomials in 16,391,195 tetrahedra (10,904,725 in the model itself along with 5,486,470 mirrored elements to get a solution everywhere, see Section 7.2) indeed means a very large linear system to solve. We therefore cut the model into 200 overlapping subdomains, each of them being treated independently from the others (Capdeville et al., 2015; Cupillard & Capdeville, 2018). $\lambda_{\min}$ and ε_0 are set to 200m and 0.75, respectively, so $\lambda_0 = 150m$. In this configuration, the whole computation (i.e., the resolution of the cell problem followed by the filtering of the stress and strain concentrators, see steps 1 and 2 in Section 6.1.5) requires 3 hours and 100 Gb on a single PowerEdge M610 for each subdomain. The result is shown in Fig. 27B and C.

By construction, the homogenized overthrust is smooth: it has no spatial variations smaller than λ_0. Computing waveforms in it therefore is very light: the mesh no longer needs to honor geological structures, and the size of the elements is constrained only by λ_0. Fig. 27B' shows a zoom in a regular hexahedral mesh of the homogenized overthrust. All the elements are 200 m^3 large. They are used in a degree-6 SEM (Cupillard et al., 2012) to compute the zeroth-order displacement $\mathbf{u}^0$ corresponding to the wavefield $\mathbf{u}^{\mathrm{ref}}$ generated in the original overthrust model. The computation cost of the spectral-element simulation is 4 163s on two Xeon Gold 6130 processors, which is 260 times less than the computation cost required for $\mathbf{u}^{\mathrm{ref}}$. A comparison between $\mathbf{u}^0$ and $\mathbf{u}^{\mathrm{ref}}$ at a given point in space is shown in Fig. 28. The error $\sqrt{\frac{\int (\mathbf{u}^{\mathrm{ref}} - \mathbf{u}^0)^2 dt}{\int (\mathbf{u}^{\mathrm{ref}})^2 dt}}$ at this point is 8.73 %. Such an error averaged over 200 randomly-positioned points is 7.57 %.

Using the same regular hexahedral mesh, we perform a spectral-element simulation in the original overthrust model. In this case, the geological structures are not honored by the mesh. They are smoothed by the numerical method itself, which is not a physical smoothing. As a consequence, the obtained wavefield $\mathbf{u}^{\mathrm{brutal}}$ does not match $\mathbf{u}^{\mathrm{ref}}$. The error between the

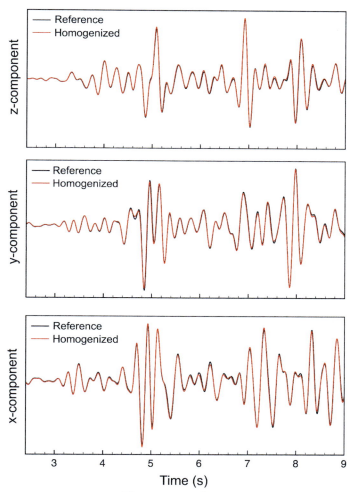

Fig. 28 Comparison between $\mathbf{u}^{\text{ref}}$ (*black*) and $\mathbf{u}^0$ (*red*) at a randomly chosen point in space. Many wiggles are observed because there is no absorbing boundaries in the two simulations.

two wavefields reaches 21.8 %. Refining the regular mesh (using 150 m^3, 120 m^3, 100 m^3, 80 m^3, and 60 m^3 large elements), the error decreases but the computation cost increases drastically, as shown in Fig. 29. This figure also shows that the refinement has no impact on the wavefield computed in the homogenized medium. This is because all the heterogeneities of the medium are properly captured by the coarsest mesh (i.e., 200 m^3 large elements).

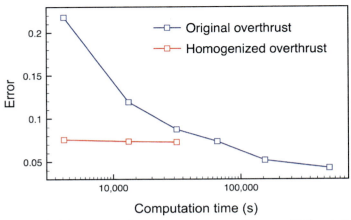

Fig. 29 Error of $\mathbf{u}^{brutal}$ (*blue*) and $\mathbf{u}^0$ (*red*) with respect to $\mathbf{u}^{ref}$. Both $\mathbf{u}^{brutal}$ and $\mathbf{u}^0$ are computed using a regular hexahedral mesh within a spectral-element method. The error of these two wavefields is plotted as a function of the computation time associated with the size of the hexahedra (200 m^3, 150 m^3, 120 m^3, 100 m^3, 80 m^3, and 60 m^3).

8.5 Rotational receivers corrector

From Section 4.1.5 and Figs. 4 and 9, we observed that to the leading order, the wavefield, i.e., displacement (and their time derivatives, such as velocities and accelerations) is smooth while wavefield gradients like strain and rotation (and their time derivatives, such as strain rate and rotation rate) are not smooth. A direct consequence of this observation is the relation between data collected in the field and synthetics derived from numerical methods. Here we present an example from the G Ring laser in Wettzell, Germany which records the vertical component of rotation rates. A collocated seismometer also measures the respective displacements.

The Earth's crust is highly heterogeneous and it is impossible to have a complete information on this heterogeneity. Despite this, we can model very long period surface waves using PREM with good accuracy because per the theory of homogenization, displacements, velocities, and accelerations are not sensitive to small-scale structures. For the wavefield gradient measurements, however, small scales are as important as the large scales: the effect of the structure, which we call the corrector, appears at the order 0 for the wavefield gradients (see (104) and (185)) whereas it only appears at the order 1 for the wavefield (see (95) and (186)). This implies that, in order to model rotation rates, a complete information on the heterogeneity of the crust is needed. Since we do not have such information, the agreement

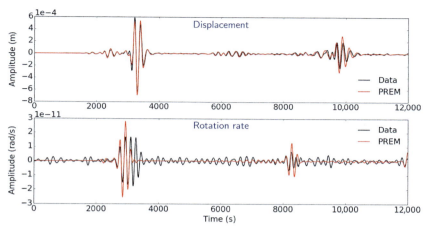

Fig. 30 Data vs synthetics derived in PREM for the 201210280304A Queen Charlotte Islands. *Top*: displacements. *Bottom*: rotation rates.

between rotation rates derived in PREM, i.e., without the corrector, and rotation rate data will not be as good as the agreement between displacements. This is the case in Fig. 30, which shows one of several events recorded at Wettzell where the agreement between displacements is better compared to the agreement between rotation rates.

One way to resolve this issue is of course, to study the behavior of small-scale structures at the ring laser and understand how they react to wavefield. The problem with this is that for any given scale on the Earth, we can always find something smaller. The simpler solution comes from (104), repeated below for rotation rates.

$$\dot{\omega}_0^\varepsilon(\mathbf{x}_r, t; \mathbf{x}_s) = \dot{\omega}^0(\mathbf{x}_r, t; \mathbf{x}_s) + \mathbf{J}(\mathbf{y}_r) : \dot{\epsilon}^0(\mathbf{x}_r, t; \mathbf{x}_s), \qquad (207)$$

where $\dot{\omega}_0^\varepsilon$ is the rotation rate data while $\dot{\omega}^0$ and $\dot{\epsilon}^0$ are the rotation rate and the strain rate computed in PREM. $\mathbf{x}_s$ denotes the location of the source and $\mathbf{y}_r = \frac{\mathbf{x}_r}{\varepsilon_0}$ denotes the location of the receiver. $\mathbf{J}$ is the corrector and the unknown of the following inverse problem. Since this term depends only on the small scales located at the receiver, it captures the effect of small-scale structures at that location. Since it does not depend on time or source location, we can invert for $\mathbf{J}$ by solving the over-determined linear problem in the least-squares sense (Tarantola & Valette, 1982) by minimizing the L^2 misfit between the synthetics and the data through

$$\mathbf{J} \sim ({}^t\mathbf{F}\,\mathbf{F})^{-1}\mathbf{F}\,\delta d, \qquad (208)$$

where $\delta d = \dot{\omega}_0^\varepsilon - \dot{\omega}^0$ and $\mathbf{F}$ is the matrix with strain rates derived in PREM. Accordingly, $\mathbf{J}$ can be used as a measure of the effect of small-scale structures at the receiver location.

For the ring laser, we invert for $\mathbf{J}$ using 32 events and use it to correct the synthetic rotation rates, i.e., we add the second term on the right-hand side of (207) to the rotation rates derived in PREM. To truly assess the quality of our results, we use the $\mathbf{J}$ to also correct rotation rates not used for the inversion.

From Fig. 31, it can be seen that the agreement between the corrected rotation rates are a better match to data compared to the rotation rates in PREM, also for events not previously used in the inversion. Disagreements are still present but they stem from the fact that the method is asymptotic and implies some unavoidable errors.

Note that correction is needed not just for point-measured wavefield gradient measurements but also for array-derived wavefield gradient measurements. The full study on this topic can be found in Singh, Capdeville, and Igel (2020).

8.6 Full waveform inversion (FWI)

One of the most interesting and promising applications of the homogenization is for the inverse problem. In the inverse problem, the objective is to retrieve information about the medium from seismic data. One of the critical issues, in that case, is the nonuniqueness of the solution. Indeed, we can distinguish two main approaches to solve seismic inverse problems: a global approach and a local approach. In a global approach, the objective is to sample the possible model space and to give statistics about all possible solutions. This is the Bayesian global search method. It is a good solution to handle ill-posed problems such as seismic inverse problems. Unfortunately, the associated numerical cost is out of reach for most applications. The second approach, which is the local optimization, tries to find the best solution based on a least-square minimization scheme. This approach is more commonly used in seismology because it is accessible based on our current numerical power (but is still very challenging). Unfortunately, such method (a) can easily be trapped in a local minimum of the misfit function (which measures the difference between synthetic data and real data) and (b) does not provide any information about other possible solutions.

In that context, homogenization raises an interesting point: for limited frequency band data, we already know that two models can fit the observed data: the true model and the homogenized version of the true model.

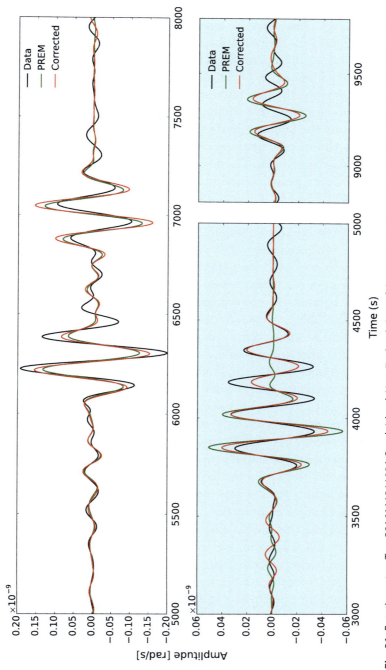

Fig. 31 Rotation rates. *Top*: 201611131102A South Island, New Zealand. One of the 32 events that are used to invert for the coupling tensor **J**. *Bottom*: 201210280304A Queen Charlotte Islands. An event from the validation data set. It is shown in *blue* to indicate that this event has not been used for the inversion. Data, synthetic, and corrected rotation rates are given in *black*, *green*, and *red*, respectively.

Two solutions are already too many for local inversion methods and only one can be found. This aspect is discussed in Capdeville and Métivier (2018) and it appears that the only solution that can be found is the homogenized model: indeed, the true model belongs to an infinite dimension space (we would need an infinite set of parameters, or at least an unknown set of parameters, to describe the true microscale model), whereas the homogenized model lives in a finite (known) dimension space. The only possibility is therefore to find the homogenized model and not the true model.

Setting up a basis for the homogenized model space is not trivial and is only possible for layered media (Capdeville et al., 2013). For higher dimensions, we need to set a parameterization such that the model space can contain the homogenized space. The inverted model then needs to be projected into the homogenized model space using a homogenization operator.

In the following, we show a simple 2-D elastic full waveform inversion (FWI) example. The objective is to show that inversion cannot retrieve the true model but can only find its homogenized version. We first generate synthetic data with a good coverage setting of the area we want to image. The geometry of the sources, the receivers and the heterogeneity is shown in Fig. 32A and C and in Table 3. The sources are single forces point sources, using a Ricker source time function with a maximum frequency such that the background λ_{min} is the one used to measure distances as shown in Fig. 32A (this example is dimensionless). An example of the energy wavefield propagating in the domain to generate the data as well as the SEM mesh used is displayed in Fig. 32B. This elastic model (the target model) voluntarily contains heterogeneities of scales much smaller than the minimum wavelength. For example, the slow heterogeneity layers "d" around circle "A" has a thickness of $\lambda_{min}/8$. Knowing that a FWI has, at best, a resolution of $\lambda_{min}/2$, the inversion has no chance to recover such a thin layer based on those data. This simple example could mimic a damage zone around an iron bar in concrete. It is a representative example of most real applications: there are always scales much smaller than the minimum wavelength of the data.

To invert the full waveform, we rely on the Gauss–Newton iterative least-square optimization scheme (Pratt, Shin, & Hicks, 1998). To describe the inverted model, we rely on a piece-wise polynomial basis for the spatial parameterization. More precisely, a square inversion subdomain $I \subset \Omega$ is chosen and divided into $n \times n$ non overlapping elements:

$$I = \cup_{e=1}^{n^2} I_e^n.$$

An example of inversion the domain I and the associated inversion mesh for $n = 10$ is shown in Fig. 32A. For each element I_e^n, similar to what is

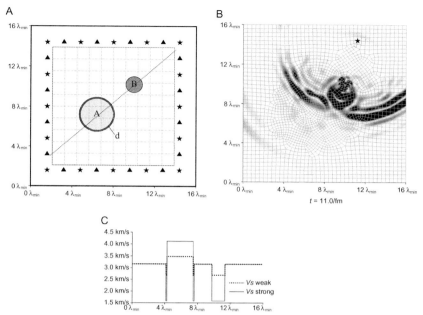

Fig. 32 (A) Setting of the numerical experiment. The location of sources (*stars*) and receivers (*triangles*) are shown. The *gray-dashed lines* represent the elements of an inversion mesh used for that example. The *dotted line* is the *cross section line* used for panel C. (B) kinetic energy snapshot for one of the sources shown by the *star*. (C) V_S cross section along the *dotted line* in the panel A for weak and strong heterogeneity models. Only the "weak" heterogeneity case is used here.

Table 3 Material properties used in the inversion test.

	V_P (km/s)	V_S (km/s)	ρ (10³kg/m³)
Background	5.6	3.17	2.61
Circular inclusions, weak heterogeneities case:			
A	6.27	3.48	2.73
B & d	4.85	2.69	2.47

done in SEM, elastic parameters and density are represented using a 2-D tensorial product polynomial approximation of degree N in each direction. This defines the parameterization $\boldsymbol{P}_n^N(\boldsymbol{I})$ ($n \times n$ elements of degree $N \times N$). We do not impose the continuity of the fields between elements, which implies that $\boldsymbol{P}_n^N(\boldsymbol{I})$ has $n^2 \times (N+1)^2$ degrees of freedom for each scalar. We perform three different inversions, each of them using a different parameterization:

- $M_{30}^{1,\text{iso}}$: $P_{30}^{1}(I)$ with ρ, V_P and V_S;
- $M_{10}^{3,\text{ani}}$: $P_{10}^{3}(I)$ with ρ and $\mathbf{c}$;
- $M_{20}^{1,\text{ani}}$: $P_{20}^{1}(I)$ with ρ and $\mathbf{c}$;

These three parameterizations have roughly the same number of free parameters (unknowns), $30^2 \times 2^2 \times 3 = 10800$ for $M_{30}^{1,\text{iso}}$, $10^2 \times 4^2 \times 7 = 11200$ for $M_{10}^{3,\text{ani}}$ and $20^2 \times 2^2 \times 7 = 11200$ for $M_{20}^{1,\text{ani}}$.

The raw results of the inversions are shown in Figs. 33 and 34. From the figures it can be seen that the results make sense, but they are

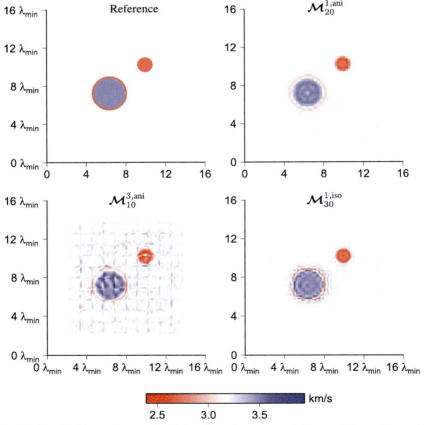

Fig. 33 Raw FWI inversion results. V_S for the reference model (*upper left panel*), raw V_S inversion results for the $M_{20}^{1,\text{ani}}$ parameterization (*upper right panel*), $M_{10}^{3,\text{ani}}$ (*lower left panel*), and $M_{30}^{1,\text{iso}}$ (*lower right panel*) parameterizations are plotted. For $M_{20}^{1,\text{ani}}$ and $M_{10}^{3,\text{ani}}$, V_S is the based on the isotropic projection of the anisotropic elastic tensor.

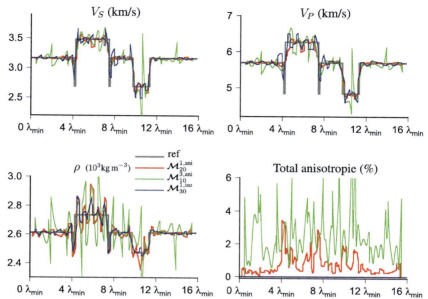

Fig. 34 FWI raw inversion results. Cross sections along the *dotted line* shown in Fig. 32A, for the reference model ("ref") and for the raw inversion results for the $\mathcal{M}_{30}^{1,iso}$, $\mathcal{M}_{20}^{1,ani}$, and $\mathcal{M}_{10}^{3,ani}$ parameterizations. V_S, V_P, ρ and the total anisotropy are presented. For anisotropic parameterizations, V_P, V_S and the total anisotropy are defined in Section 6.1.

different and with a significant noise (it is striking on the cross section in Fig. 34). The impact of the parameterization choice on the results is obvious. Nevertheless, the three inversion results explain the data equally well. Moreover, their ability to model data that have not been used in the inversion is equally good. In other words, the three results are equally good.

We then homogenize the three different results as well as the target model and compare them in Figs. 35 and 36. It appears that all results, once homogenized, are the same and fit the homogenized target model very well. Interestingly, the isotropic inversion, once homogenized, manages to reproduce the necessary effective anisotropy. This is a manifestation of the nonuniqueness of small scales: the inversion was able to find small scales in the parameterization (mainly due to discontinuities between the

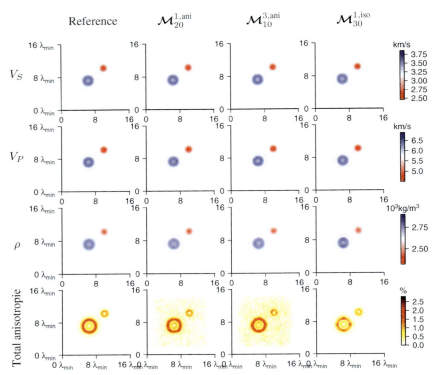

Fig. 35 FWI test reference model and three different inversion results, all homogenized with $\varepsilon_0 = 1$. V_S, V_P, ρ and the "total anisotropy" are presented (*lines* of panels from *top* to *bottom*, respectively). Reference, $\mathcal{M}_{20}^{1,\text{ani}}$, $\mathcal{M}_{10}^{3,\text{ani}}$, and $\mathcal{M}_{30}^{1,\text{iso}}$ inverted models are plotted (*left* to *right* columns of panels, respectively). V_S, V_P, and the "total anisotropy" are defined in Section 6.1

elements in the parameterization) that have nothing to do with the fine scales in the true model but produce the same anisotropy once homogenized. Through this example, we numerically confirm that results for FWI with a limited frequency band data is at best, the homogenized version of the true model.

This result has important consequences for the FWI design but also the interpretation of the results. It opens the door to a new inverse problem for image interpretation: the downscaling (Hedjazian, Capdeville, & Bodin, 2020). A more complete study and discussion about this topic can be found in Capdeville and Métivier (2018).

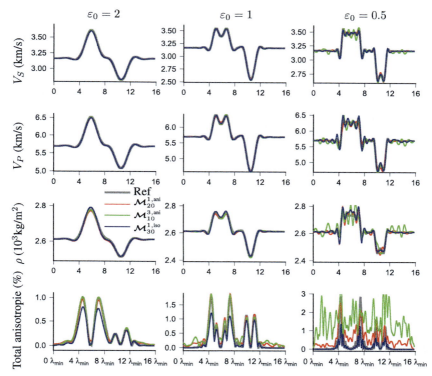

Fig. 36 Weak contrast circular inclusion test reference and homogenized inversion results. Cross sections along the *dotted line* in Fig. 32A for the reference model ("ref") and for the inversion results for the $\mathcal{M}_{30}^{1,\text{iso}}$, $\mathcal{M}_{20}^{1,\text{ani}}$, and $\mathcal{M}_{10}^{3,\text{ani}}$ parameterizations are represented. Three different values of ε_0 are used ($\varepsilon_0 = 2$, 1 and 0.5). V_P, V_S, and the "total anisotropy" are defined in Section 6.1.

9. Discussion and conclusions

We have presented an introduction to the two-scale nonperiodic homogenization method for the elastic and acoustic wave equations. It is based on the classical asymptotic two-scale homogenization method for periodic or stochastic media. Geological media are neither periodic nor stochastic, but still multiscale. The objective of the method introduced here is to fill the gap, for two-scale homogenization methods, between the periodic or stochastic media cases and multiscale deterministic media with no scale separation case. In general, homogenization methods require a microscale (e.g., the periodic cell) and a macro-scale (e.g., the whole domain). For geological media, there is no scale separation and there is no simple way to tell what is a small scale and what is a large scale. Nevertheless, in the dynamic case (for the wave equations), another scale is present which makes it possible to separate the scales: the minimum wavelength of the propagating wavefield λ_{min}. Using this quantity as a scale separator here, we have developed and presented an extension of the two-scale homogenization to the nonperiodic case, valid for geological media.

Compared to the classical periodic case, one of the weaknesses of the nonperiodic homogenization is that it does not have its mathematical rigor, i.e., no convergence theorem exists. There is, therefore, no warranty that the method works for every case, and indeed, the Helmholtz resonator case shows that the method can fail. Nevertheless, the method has been working for the numerous cases treated for the last 10 years and sometimes, even beyond the hypothesis of the underlying method as it is the case for the small fluid inclusion case example. The failure of the Helmholtz resonator case can be physically expected: one of the main hypotheses of the nonperiodic homogenization is the presence of a minimum wavelength and, amazingly enough, the Helmholtz resonator has none at its eigenfrequency. It is, therefore, not that surprising that the nonperiodic homogenization fails in that specific case. We can conclude that the nonperiodic homogenization method is robust enough for all media faced in geosciences and nondestructive testing sciences.

Many aspects of the nonperiodic homogenization remain to be studied. One of them is that the effective tensor symmetry is not mathematically ensured, even though it is numerically symmetric for most of the cases we have been working on. This may be a sign that the method is not fully general and in the Helmholtz resonator case, for example, the effective medium becomes strongly nonsymmetric just before becoming negative. A deeper study of this aspect would be useful. Another aspect that would deserve attention is that of cases for which the minimum wavelength varies strongly from one place to another. This situation is very common in geophysics where a factor 10 between the top and the bottom model wave speeds is not exceptional. Using the minimum wave speed in such cases is not optimal and a variable filtering solution is needed. Finally, the two-scale homogenization of the seismic source is also needed. The objective, in that case, is to obtain an effective source (that would not be a point source anymore) and the associated correctors. Development in that direction is currently in preparation and will be submitted soon.

Despite the remaining points to be studied, useful nonperiodic homogenization tools are already available. Their first obvious application is forward modeling. By removing interfaces in the media, they significantly ease the meshing and often make the modeling possible (when the meshing is impossible). The gain in computing time is often very significant (several orders of magnitude). Nevertheless, the two-scale nonperiodic homogenization does not remove all the meshing difficulties: we still have to adapt the mesh size when the dominant wavelength changes significantly (or at least it is best to) and we still have to mesh the free topography (even if it is smoother once homogenized and therefore easier to mesh) and solid–fluid interfaces.

Another important application of the nonperiodic homogenization is the inverse problem. We can speculate and check numerically (Capdeville & Métivier, 2018) that the solution of the full waveform inverse problem is, at best, the homogenized version of the true model. This has important consequences for the inverse problem design. For example, trying to explicitly invert for material discontinuities from limited frequency band data is known to be a difficult ill-posed problem. Still, attempts are made to retrieve sharp discontinuities using sophisticated regularizations (e.g., total variation regularization in Lin and Huang (2014)). Nevertheless, based on the homogenization theory, we know that discontinuities map into an effective smooth anisotropic transition. Inverting for such a smooth media is stable and simple: compared to inverting sharp discontinuities, it drastically reduces the nonlinearity of the problem. Homogenization can, therefore, drive our way to design inverse problems and help to stabilize them as shown in Capdeville and Métivier (2018). It also raises the question of the interpretation of the FWI images. If the images do not represent the true model but only its effective version, they must be interpreted with care as the homogenization can have counter-intuitive effects. Their interpretation can be seen as another inverse problem: the downscaling inverse problem. It corresponds to trying to find possible small scale models compatible with the effective image obtained from the FWI and some external a-priori information. This inverse problem is highly nonunique can only be tackled with Monte-Carlo type methods (Hedjazian et al., 2020).

To conclude, we hope to have convinced the reader that, even if the nonperiodic homogenization method is not trivial, it is worth the effort and will be useful for many research topics in seismology and geophysics.

Acknowledgments

The authors would like to thank Cédric Schmelzbach and Mike Afanasiev for their comments, helping to improve the manuscript. This work has benefited from the contributions of many co-authors since 2007: Jean-Jacques Marigo, Laurent Guillot, Philippe Cance, Gaël Burgos, Ming Zhao and Ludovic Métivier. This work was funded by the three ANR (Agence Nationale pour la Recherche) projects (NR-06-BLAN-0283 "MUSE", ANR-10-BLAN-613 "mémé" and ANR-16-CE31-0022-01 "HIWAI"). Most of the computations were performed using the "Centre de Calcul Intensif des Pays de la Loire" (CCIPL) resources. We thank the European COST action TIDES (ES1401) for fruitful discussions.

References

Abdelmoula, R., & Marigo, J. J. (2000). The effective behavior of a fiber bridged crack. *Journal of the Mechanics and Physics of Solids*, *48*(11), 2419–2444.

Afanasiev, M., Boehm, C., May, D., & Fichtner, A. (2016). Using effective medium theory to better constrain full waveform inversion. In *78th EAGE conference and exhibition 2016*.

Alder, C., Bodin, T., Ricard, Y., Capdeville, Y., Debayle, E., & Montagner, J. P. (2017). Quantifying seismic anisotropy induced by small-scale chemical heterogeneities. *Geophysical Journal International*, *211*(3), 1585–1600.

Alkhalifah, T. (2000). An acoustic wave equation for anisotropic media. *Geophysics*, *65*, 1239–1250.

Allaire, G. (1992). Homogenization and two-scale convergence. *SIAM Journal on Mathematical Analysis*, *23*, 1482–1518.

Auriault, J.-L., Borne, L., & Chambon, R. (1985). Dynamics of porous saturated media, checking of the generalized law of Darcy. *The Journal of the Acoustical Society of America*, *77*(5), 1641–1650.

Auriault, J. L., & Sanchez-Palencia, E. (1977). Étude du comportement macroscopique d'un milieu poreux saturé déformable. *Journal de Mécanique*, *16*(4), 575–603.

Babuška, I. (1976). Homogenization and its application. Mathematical and computational problems. In *Numerical solution of partial differential equations—III*. Elsevier, pp. 89–116.

Backus, G. E. (1962). Long-wave elastic anisotropy produced by horizontal layering. *Journal of Geophysical Research*, *67*(11), 4427–4440.

Bensoussan, A., Lions, J. L., & Papanicolaou, G. (1978). *Asymptotic analysis of periodic structures*. North Holland.

Bodin, T., Capdeville, Y., Romanowicz, B., & Montagner, J.-P. (2015). Interpreting radial anisotropy in global and regional tomographic models. In *The Earth's heterogeneous mantle*. Springer, pp. 105–144.

Boutin, C., & Auriault, J. L. (1990). Dynamic behaviour of porous media saturated by a viscoelastic fluid. application to bituminous concretes. *International Journal of Engineering Science*, *28*(11), 1157–1181.

Boutin, C., & Auriault, J. L. (1993). Rayleigh scattering in elastic composite materials. *International Journal of Engineering Science*, *31*(12), 1669–1689.

Bowen, R. M. (1976). *Continuum physics, chap. theory of mixtures*. New York: Academic Press.

Browaeys, J. T., & Chevrot, S. (2004). Decomposition of the elastic tensor and geophysical applications. *Geophysical Journal International*, *159*, 667–678.

Burgos, G., Capdeville, Y., & Guillot, L. (2016). Homogenized moment tensor and the effect of near-field heterogeneities on nonisotropic radiation in nuclear explosion. *Journal of Geophysical Research*, *121*(6), 4366–4389.

Burridge, R., & Keller, J. B. (1981). Poroelasticity equations derived from microstructure. *The Journal of the Acoustical Society of America*, *70*(4), 1140–1146.

Cance, P., & Capdeville, Y. (2015). Validity of the acoustic approximation for elastic waves in heterogeneous media. *Geophysics*, *80*(4), T161–T173.

Capdeville, Y. (2000). Méthode couplée éléments spectraux—solution modale pour la propagation d'ondes dans la terre à l'échelle globale (Unpublished doctoral dissertation). Université Paris.

Capdeville, Y., Guillot, L., & Marigo, J. J. (2010a). 1-D non periodic homogenization for the wave equation. *Geophysical Journal International*, *181*, 897–910.

Capdeville, Y., Guillot, L., & Marigo, J. J. (2010b). 2D nonperiodic homogenization to upscale elastic media for P-SV waves. *Geophysical Journal International*, *182*, 903–922.

Capdeville, Y., Gung, Y., & Romanowicz, B. (2005). Towards Global Earth tomography using the spectral element method: A technique based on source stacking. *Geophysical Journal International*, *162*, 541–554.

Capdeville, Y., & Marigo, J. J. (2007). Second order homogenization of the elastic wave equation for non-periodic layered media. *Geophysical Journal International*, *170*, 823–838.

Capdeville, Y., & Marigo, J. J. (2008). Shallow layer correction for spectral element like methods. *Geophysical Journal International*, *172*, 1135–1150.

Capdeville, Y., & Marigo, J.-J. (2013). A non-periodic two scale asymptotic method to take account of rough topographies for 2-D elastic wave propagation. *Geophysical Journal International*, *192*(1), 163–189.

Capdeville, Y., & Métivier, L. (2018). Elastic full waveform inversion based on the homogenization method: Theoretical framework and 2-D numerical illustrations. *Geophysical Journal International*, *213*(2), 1093–1112.

Capdeville, Y., Stutzmann, E., Wang, N., & Montagner, J.-P. (2013). Residual homogenization for seismic forward and inverse problems in layered media. *Geophysical Journal International*, *194*(1), 470–487.

Capdeville, Y., Zhao, M., & Cupillard, P. (2015). Fast Fourier homogenization for elastic wave propagation in complex media. *Wave Motion*, *54*, 170–186.

Caumon, G., Collon-Drouaillet, P., de Veslud, C. L. C., Viseur, S., & Sausse, J. (2009). Surface-based 3D modeling of geological structures. *Mathematical Geosciences*, *41*, 927–945.

Chaljub, E., Komatitsch, D., Capdeville, Y., Vilotte, J. P., Valette, B., & Festa, G. (2007). Spectral element analysis in seismology. In R.-S. Wu & V. Maupin (Eds.), *Vol. 48. Advances in wave propagation in heterogeneous media* (pp. 365–419). Elsevier. (Eds.).

Cupillard, P., & Capdeville, Y. (2018). Non-periodic homogenization of 3-D elastic media for the seismic wave equation. *Geophysical Journal International*, *213*(2), 983–1001.

Cupillard, P., Delavaud, E., Burgos, G., Festa, G., Vilotte, J.-P., Capdeville, Y., & Montagner, J.-P. (2012). RegSEM: A versatile code based on the spectral element method to compute seismic wave propagation at the regional scale. *Geophysical Journal International*, *188*(3), 1203–1220.

David, M., Marigo, J. J., & Pideri, C. (2012). Homogenized interface model describing inhomogeneities located on a surface. *Journal of Elasticity*, *109*, 153–187.

Davit, Y., Bell, C. G., Byrne, H. M., Chapman, L. A. C., Kimpton, L. S.,Lang, G. E. … (2013). Homogenization via formal multiscale asymptotics and volume averaging: How do the two techniques compare? *Advances in Water Resources*, *62*, 178–206.

De Giorgi, E. (1984). G-operators and γ-convergence. In *Proceedings of the international congress of mathematicians (Warsazwa, August 1983)* (pp. 1175–1191). PWN Polish Scientific Publishers and North Holland.

De Giorgi, E., & Spagnolo, S. (1973). Sulla convergenza degli integrali dell energia per operatori ellittici del secondo ordine. *Bollettino della Unione Matematica Italiana Series IV*, *8*, 191–411.

Dumontet, H. (1990). Homogénéisation et effets de bords dans les materiaux composites (Unpublished doctoral dissertation). Univertité Paris.

Dziewonski, A. M., & Anderson, D. L. (1981). Preliminary reference Earth model. *Physics of the Earth and Planetary Interiors, 25*, 297–356.

Engquist, B., & Souganidis, P. E. (2008). Asymptotic and numerical homogenization. *Acta Numerica, 17*, 147–190.

Faccenda, M., Ferreira, A. M. G., Tisato, N., Lithgow-Bertelloni, C., Stixrude, L., & Pennacchioni, G. (2019). Extrinsic elastic anisotropy in a compositionally heterogeneous earth's mantle. *Journal of Geophysical Research, 124*(2), 1671–1687.

Festa, G., & Vilotte, J. P. (2005). The Newmark scheme as velocity-stress time-staggering: An efficient implementation for spectral element simulations of elastodynamics. *Geophysical Journal International, 161*, 789–812.

Fichtner, A., & Hanasoge, S. M. (2017). Discrete wave equation upscaling. *Geophysical Journal International, 209*(1), 353–357.

Fichtner, A., Kennett, B. L. N., & Trampert, J. (2013). Separating intrinsic and apparent anisotropy. *Physics of the Earth and Planetary Interiors, 219*, 11–20.

Fish, J., Chen, W., & Nagai, G. (2002). Nonlocal dispersive model for wave propagation in heterogeneous media. Part 1: One-dimensional case. *International Journal for Numerical Methods in Engineering, 54*, 331–346.

Francfort, G. A., & Murat, F. (1986). Homogenization and optimal bounds in linear elasticity. *Archive for Rational Mechanics and Analysis, 94*(4), 307–334.

Geevers, S., Mulder, W. A., & van der Vegt, J. J. W. (2019). Efficient quadrature rules for computing the stiffness matrices of mass-lumped tetrahedral elements for linear wave problems. *SIAM Journal on Scientific Computing, 41*(2), A1041–A1065.

Gold, N., Shapiro, S. A., Bojinski, S., & Müller, T. M. (2000). An approach to upscaling for seismic waves in statistically isotropic heterogeneous elastic media. *Geophysics, 65*(2), 1837–1850.

Grechka, V. (2003). Effective media: A forward modeling view. *Geophysics, 68*(6), 2055–2062.

Guillot, L., Capdeville, Y., & Marigo, J. J. (2010). 2-D non periodic homogenization for the SH wave equation. *Geophysical Journal International, 182*, 1438–1454.

Hashin, Z. (1972). *Theory of fiber reinforced materials*. No. NASA Report No. CR1974.

Hashin, Z., & Shtrikman, S. (1962). On some variational principles in anisotropic and nonhomogeneous elasticity. *Journal of the Mechanics and Physics of Solids, 10*(4), 335–342.

Hedjazian, N., Capdeville, Y., & Bodin, T. (2020). *Geophysical Journal International* (submitted for publication).

Hill, R. (1965). A self–consistent mechanics of composite materials. *Journal of the Mechanics and Physics of Solids, 13*, 213–222.

Igel, H. (2017). *Computational seismology: A practical introduction*. Oxford University Press.

Jordan, T. H. (2015). An effective medium theory for three-dimensional elastic heterogeneities. *Geophysical Journal International, 203*(2), 1343–1354.

Komatitsch, D., & Vilotte, J. P. (1998). The spectral element method: An effective tool to simulate the seismic response of 2D and 3D geological structures. *Bulletin of the Seismological Society of America, 88*, 368–392.

Kononov, A., Minisini, S., Zhebel, E., & Mulder, W. A. (2012). A 3D tetrahedral mesh generator for seismic problems. In *74th EAGE conference and exhibition incorporating EUROPEC 2012, cp–293*. European Association of Geoscientists & Engineers.

Lemoult, F., Kaina, N., Fink, M., & Lerosey, G. (2013). Wave propagation control at the deep subwavelength scale in metamaterials. *Nature Physics, 9*(1), 55–60.

Lin, C., Saleh, R., Milkereit, B., & Liu, Q. (2017). Effective media for transversely isotropic models based on homogenization theory: With applications to borehole sonic logs. *Pure and Applied Geophysics, 174*(7), 2631–2647.

Lin, Y., & Huang, L. (2014). Acoustic-and elastic-waveform inversion using a modified total-variation regularization scheme. *Geophysical Journal International, 200*(1), 489–502.

Marigo, J. J., & Pideri, C. (2011). The effective behavior of elastic bodies containing microcracks or microholes localized on a surface. *International Journal of Damage Mechanics, 20*(8), 1151–1177. https://doi.org/10.1177/1056789511406914.

Mechel, F. P. (2013). *Formulas of acoustics*. Springer Science & Business Media.

Meng, S., & Guzina, B. B. (2018). On the dynamic homogenization of periodic media: Willis' approach versus two-scale paradigm. *Proceedings of the Royal Society A: Mathematical, Physical and Engineering Sciences, 474*(2213), 20170638.

Michel, J. C., Moulinec, H., & Suquet, P. (1999). Effective properties of composite materials with periodic microstructure: A computational approach. *Computer Methods in Applied Mechanics and Engineering, 172*(1), 109–143.

Moulinec, H., & Suquet, P. (1998). A numerical method for computing the overall response of nonlinear composites with complex microstructure. *Computer Methods in Applied Mechanics and Engineering, 157*(1), 69–94.

Murat, F., & Tartar, L. (1985). Calcul des variations et homogénéisation. In *Homogenization methods: Theory and applications in physics (Bréau-sans-Nappe, 1983): Vol. 57.* (pp. 319–369). Paris: Eyrolles.

Tartar, L., & Murat, F. (1997). *Homogénéisation et compacité par compensation (cours peccot).* College de France. Partially written in F. Murat, H-convergence, Séminaire d'Analyse Fonctionnelle et Numérique de l'Université d'Alger, 1977-78. English translation in Mathematical modeling of composite materials, edited by R.V. Kohn; *Progress in nonlinear differential equations and their applications*, Birkhiiuser, Boston (1994).

Nassar, H., He, Q. C., & Auffray, N. (2016). On asymptotic elastodynamic homogenization approaches for periodic media. *Journal of the Mechanics and Physics of Solids, 88*, 274–290.

Nevard, J., & Keller, J. B. (1997). Homogenization of rough boundaries and interfaces. *Journal of Applied Mathematics, 67*(6), 1660–1686.

Nguetseng, G. (1989). A general convergence result for a functional related to the theory of homogenization. *SIAM Journal on Mathematical Analysis, 20*(3), 608–623.

Nissen-Meyer, T., Fournier, A., & Dahlen, F. A. (2007). A two-dimensional spectral-element method for computing spherical-earth seismograms–I. Moment-tensor source. *Geophysical Journal International, 168*(3), 1067–1092.

Owhadi, H., & Zhang, L. (2008). Numerical homogenization of the acoustic wave equations with a continuum of scales. *Computer Methods in Applied Mechanics and Engineering, 198*(3–4), 397–406.

Papanicolaou, G. C., & Varadhan, S. R. S. (1979). Boundary value problems with rapidly oscillating random coefficients. In *Proceedings of conference on random fields, Esztergom, Hungary, 27.* North Holland, 1981, pp. 835–873.

Pavliotis, G. A., & Campus, S. K. (2004). *Homogenization theory for partial differential equations.* London: Department of Mathematics, South Kensington Campus, Imperial College.

Pellerin, J., Lévy, B., Caumon, G., & Botella, A. (2014). Automatic surface remeshing of 3D structural models at specified resolution: A method based on Voronoi diagrams. *Computers and Geosciences, 62*, 103–116. https://doi.org/10.1016/j.cageo.2013.09.008.

Pratt, R. G., Shin, C., & Hicks, G. J. (1998). Gauss-Newton and full Newton methods in frequency domain seismic waveform inversion. *Geophysical Journal International, 133*, 341–362.

Pride, S. R., Gangi, A. F., & Morgan, F. D. (1992). Deriving the equations of motion for porous isotropic media. *The Journal of the Acoustical Society of America, 92*(6), 3278–3290.

Sanchez-Palencia, E. (1980). *Non homogeneous media and vibration theory*. Berlin: Springer. No. 127.

Sanchez-Palencia, E. (1987). Elastic body with defects distributed near a surface. In *Homogenization techniques for composite media* . Berlin, Heidelberg: Springer, pp. 183–192.

Shapiro, S. A., Schwarz, R., & Gold, N. (1996). The effect of random isotropic in homogeneities on the phase velocity of seismic waves. *Geophysical Journal International, 123*, 783–794.

Si, H. (2015). TetGen, a Delaunay-based quality tetrahedral mesh generator. *ACM Transactions on Mathematical Software, 41*(2), 1–36. https://doi.org/10.1145/2629697.

Singh, S., Capdeville, Y., & Igel, H. (2020). Correcting wavefield gradients for the effects of local small-scale heterogeneities. *Geophysical Journal International, 220*(2), 996–1011.

Tarantola, A., & Valette, B. (1982). Generalized nonlinear inverse problems solved using the least squares criterion. *Reviews of Geophysics, 20*, 219–232.

Tartar, L. (1978). Quelques remarques sur l'homogénéisation Proc. of the Japan-France Seminar 1976. In *Functional analysis and numerical analysis* (pp. 469–482). Japan Society for the Promotion of Sciences.

Thomsen, L. (1972). Elasticity of polycrystals and rocks. *Journal of Geophysical Research, 77*(2), 315–327.

Tiwary, D. K., Bayuk, I. O., Vikhorev, A. A., & Chesnokov, E. M. (2009). Comparison of seismic upscaling methods: From sonic to seismic. *Geophysics, 74*(2), wa3–wa14.

Wang, N., Montagner, J.-P., Fichtner, A., & Capdeville, Y. (2013). Intrinsic versus extrinsic seismic anisotropy: The radial anisotropy in reference Earth models. *Geophysical Research Letters, 40*(16), 4284–4288.

Watt, J. P. (1988). Elastic properties of polycrystalline minerals: Comparison of theory and experiment. *Physics and Chemistry of Minerals, 15*(6), 579–587.

Wellmann, F., & Caumon, G. (2018). 3-D Structural geological models: Concepts, methods, and uncertainties. *Advances in Geophysics, 59*, 1–121.

Willis, J. R. (1981). Variational principles for dynamic problems for inhomogeneous elastic media. *Wave Motion, 3*, 1–11.

Willis, J. R. (1983). The overall elastic response of composite materials. *Journal of Applied Mechanics, 50(4b)*, 1202–1209.

Willis, J. R. (1985). The nonlocal influence of density variations in a composite. *International Journal of Solids and Structures, 21*(7), 805–817.

Willis, J. R. (1997). Dynamics of composites. In P. Suquet (Ed.), *Continuum Micromechanics International Centre for Mechanical Sciences (Courses and Lectures): 377*. (pp. 265–290). Springer.

Willis, J. R. (2009). Exact effective relations for dynamics of a laminated body. *Mechanics of Materials, 41*(4), 385–393.